Engineering Materials and Processes

Series Editor

Professor Brian Derby, Professor of Materials Science
Manchester Materials Science Centre, Grosvenor Street, Manchester, M1 7HS, UK

Other titles published in this series

Fusion Bonding of Polymer Composites
C. Ageorges and L. Ye

Composite Materials
D.D.L. Chung

Titanium
G. Lütjering and J.C. Williams

Corrosion of Metals
H. Kaesche

Corrosion and Protection
E. Bardal

Intelligent Macromolecules for Smart Devices
L. Dai

Microstructure of Steels and Cast Irons
M. Durand-Charre

Phase Diagrams and Heterogeneous Equilibria
B. Predel, M. Hoch and M. Pool

Computational Mechanics of Composite Materials
M. Kamiński

Gallium Nitride Processing for Electronics, Sensors and Spintronics
S.J. Pearton, C.R. Abernathy and F. Ren

Materials for Information Technology
E. Zschech, C. Whelan and T. Mikolajick

Fuel Cell Technology
N. Sammes

Casting: An Analytical Approach
A. Reikher and M.R. Barkhudarov

Computational Quantum Mechanics for Materials Engineers
L. Vitos

Modelling of Powder Die Compaction
P.R. Brewin, O. Coube, P. Doremus and J.H. Tweed

Silver Metallization
D. Adams, T.L. Alford and J.W. Mayer

Microbiologically Influenced Corrosion
R. Javaherdashti

Modeling of Metal Forming and Machining Processes
P.M. Dixit and U.S. Dixit

Electromechanical Properties in Composites Based on Ferroelectrics
V.Yu. Topolov and C.R. Bowen

Charged Semiconductor Defects
E.G. Seebauer and M.C. Kratzer

Modelling Stochastic Fibrous Materials with Mathematica®
W.W. Sampson

Ferroelectrics in Microwave Devices, Circuits and Systems
S. Gevorgian

Porous Semiconductors
F. Kochergin and H. Föll

Chemical Vapour Deposition
C.-T. Yan and Y. Xu

Composite Materials
D.D.L. Chung

Jaroslav Pokluda · Pavel Šandera

Micromechanisms of Fracture and Fatigue

In a Multiscale Context

Springer

Prof. Jaroslav Pokluda
Prof. Pavel Šandera

Brno University of Technology
Faculty of Mechanical Engineering
Technicka 2
61669 Brno
Czech Republic

pokluda@fme.vutbr.cz
sandera@fme.vutbr.cz

ISSN 1619-0181
ISBN 978-1-4471-2562-4 ISBN 978-1-84996-266-7(eBook)
DOI 10.1007/978-1-84996-266-7

British Library Cataloguing in Publication Data
A catalogue record for this book is available from the British Library

Softcover reprint of the hardcover 1st edition 2010

Cover design: eStudioCalamar, Figueres/Berlin

Printed on acid-free paper

Springer is part of Springer Science+Business Media (www.springer.com)

There are only a few things that may improve when getting older. One of these is a grasp on the rules of Nature.

Preface

Classical concepts and numerical methods of mechanical engineering such as, for example, fracture mechanics and finite element analysis focus on macroscale problems where the microstructure is incorporated by using averaging constitutive laws. On the other hand, the domain of materials science and solid state physics focuses on investigation of atomic bonds, lattice defects, microstructural elements and their interactions at nano, micro and meso scales. However, a recent development in fracture and fatigue research clearly indicates that the most promising and effective concepts are based on coupling engineering mechanics with materials science within so-called multiscale fracture models. The objective of these approaches is to bridge the enormous gap between time and space scales and, therefore, they constitute a great challenge in the sense of scientific knowledge. Moreover, they still drive at psychological barriers of conservative mechanical engineers and/or material scientists. Therefore, an overwhelming majority of books about fracture and fatigue were written from the point of view of either mechanical engineers or material scientists. To our knowledge, a pioneering attempt to produce a successful integrated concept of fracture was made by Kelly and Macmillan [1]. Since that time, however, many new methods and concepts have been developed which should be incorporated into advanced multiscale models of fracture and fatigue.

This book was written as an overview of scientific results achieved by the authors during about 40 years of their research. However, another strong motivation was to support advanced trends in fracture and fatigue which lead to the development of multiscale concepts for securing the integrity of engineering components and structures. This second aim has always prevailed over the first. Therefore, the book is composed in a compact manner and provides a rather comprehensive survey of fracture micromechanisms and related multiscale models. Although these models were predominantly proposed by the authors of this book, many passages devoted to models that were published by other authors are included in order to ensure a consistent presentation of the subject. A prevailing part of the book reflects the joint work of authors at the Brno University of Technology. However, several results and models originate from the research of the first author, performed at the Military Institute of Material Science and Technology from 1973 to 1985. The opportunity to present these results is much appreciated since, for obvious reasons, they were not allowed to be published in international scientific journals at that time.

The book addresses students at graduate and postgraduate levels, lecturers, materials scientists and mechanical engineers, as well as materials physicists and chemists. Any kind of criticism or advice that can help to improve the text will be very welcome.

Many results presented in this book were achieved either in the frame of international scientific collaboration or appeared as a consequence of stimulating discussions with colleagues from foreign universities and research institutes. Our very special thanks go to Prof. R. Pippan from the Institute of Materials Science, Austrian Academy of Sciences, in Leoben, Austria, Prof. V. Vitek from the University of Pennsylvania in Philadelphia, Pennsylvania, USA and Prof. Y. Murakami from the Kyushu University in Fukuoka, Japan, for our stimulating discussions during our long-term collaboration. Our warm thanks go to Dr. A. Doig from the Military Academy in Shrivenham, England, and Dr. R. Gröger from the Los Alamos National Laboratory, USA, for their fruitful comments on the scientific content and English language of this book. We are also deeply indebted to Prof. O. Kolednik, Prof. J. Janovec, Prof. M. Jenko, Prof. Y. Kondo, Dr. G. Chai, Prof. Y. Kitamura, Prof. C. Laird, Prof. S. Stanzel-Tschegg, Dr. M. Sauzay, Prof. M. Zehetbauer, Prof. A. Krasowski, Prof. E. Macha, Prof. A. Shaniavsky and Prof. L. Tóth for helpful and friendly discussions associated with joint publications and/or scientific meetings and visits.

There are also a number of Czech colleagues who directly or indirectly contributed to this book. Let us first mention a long-term collaboration, fruitful discussions and extended joint work with Prof. M. Šob from the Masaryk University in Brno and Prof. P. Lejček from the Institute of Physics, Academy of Sciences of the Czech Republic in Prague. Furthermore, Dr. P. Lukáš from the Institute of Physics of Materials, Academy of Sciences of the Czech Republic in Brno and Dr. F. Kroupa (†) from the Institute of Plasma Physics, Academy of Sciences of the Czech Republic in Prague have helped us very much particularly during the first periods of our research activities. We are also grateful to Prof. J. Švejcar, Dr. I. Saxl (†), Prof. I. Dlouhý, Prof. B. Vlach, Prof. M. Kotoul, Dr. P. Ponížil, Prof. K. Stránský, Prof. I. Dvořák, Dr. J. Siegl, Dr. L. Obdržálek, Prof. J. Zeman, Prof. L. Kunz, Prof. Z. Knésl, Prof. P. Lukáč, Dr. A. Machová, Dr. V. Paidar, Prof. V. Navrátil, Prof. J. Polák, Dr. M. Holzmann, Prof. J. Kohout, Prof. R. Foret, Dr. P. Staněk, Dr. A. Buchal, Dr. K. Obrtlík, Prof. M. Šlesár (†), Prof. M. Bílý, Dr. V. Oliva and Dr. H. Lauschmann for helpful discussions associated with joint publications and/or scientific meetings.

We are deeply grateful to our friends and colleagues from the Department of Materials Micromechanics and Applied Acoustics, especially to Dr. M. Černý, who was our partner and consultant in most topics of the first chapter. We have also very much appreciated the help of Dr. J. Horníková and Dr. K. Slámečka with experimental and theoretical investigations in mixed-mode fracture and fatigue as well as with the design and preparation of many figures. Finally, we would like to thank Anthony Doyle from Springer UK, who has offered us the opportunity to write this book.

Brno,
December 2009

Jaroslav Pokluda
Pavel Šandera

Contents

Chapter 0
Introduction

A systematic investigation of fracture and fatigue started during the nineteenth century mainly as a consequence of repeating failures of dynamically loaded engineering components and structures. In the first half of the last century, pure empirical and mechanical views on fracture prevailed and this explains the rather empirical understanding of the failure of engineering components and structures under applied static or dynamic loads. Results of experiments were evaluated by using linear–elastic fracture mechanics and the resistance of structural materials to fracture was believed to be predominantly determined by their chemical composition. However, fundamental experience revealed that the fracture resistance of materials is to a large extent influenced by their microstructure. Since about 1950, therefore, the materials science approaches based on physics, chemistry and mechanical engineering became very important in fracture research. Advanced experimental methods such as electron microscopy, atomic-force microscopy, Auger spectroscopy, X-ray methods, electron-beam microstructure analyses and servohydraulic close-loop testing devices started to be utilized in fracture research. At the same time, the employment of numerical methods in materials science and fracture was promoted by the rapidly increasing efficiency of computer systems as well as by growing demands for quantitative predictions in industrial research. Thus, the theoretical models of fracture and fatigue processes started to be developed and experimentally supported on all spatial scales from atomistic to macroscopic approaches.

Phenomenological fracture models beyond the atomic scale are based on constitutive relationships that usually involve many *variables*. These variables can be categorized into independent variables, state variables and parameters [2]. As a rule, time and space coordinates serve as independent variables while state variables represent physical quantities as functions of the independent variables. Quantities such as stress, strain, dislocation density, vacancy concentration, particle size or interphase spacing are typical examples of state variables. A further distinction between *explicit* and *implicit* state variables can often be encountered, especially in microstructure mechanics. The explicit state variables define *local* microstructural features such as grain size or particle volume. On the other hand, the implicit variables are defined as mesoscopic or even macroscopic *averages.* In kinematic equations describing the evolution of microstructural damage during the fracture process, the state variables are usually weighted, i.e., based on parameters. The parameters should be measurable quantities of well–defined physical meaning.

As a consequence of both advanced experimental and theoretical studies, the fracture of engineering components is, nowadays, commonly grasped as a process of damage accumulation in the materials microstructure. This process consists of four fundamental stages: Localization of plastic deformation, nucleation of cracks, propagation of cracks and final fracture of the component. However, the duration of these stages can substantially differ depending on microstructure, chemical composition, loading, temperature and environment. Some of these stages might be absent as well. The materials microstructure should be understood in terms of a spatial hierarchy ranging from the *nanoscale* (electrons, atoms, individual lattice defects), through the *microscale* (crystallography, lattice defects ensembles and secondary phases below the grain scale) up to the *mesoscale* (defects and phase ensembles at the grain scale). The *macroscopic* level embraces the geometry of samples and components. The state variables that quantify microstructural elements are often called *microstructural characteristics.* In order to illustrate the role of microstructural elements and characteristics in individual stages of the fracture process, the physical background of uniaxial tension tests of metallic materials of the same chemical composition but possessing a different microstructure will be briefly discussed hereafter. The influence of microstructure on fracture characteristics will also be mentioned.

Let us first consider a perfect single crystal as a highly ordered solid possessing the lowest possible internal energy per atom. In this case, the first three stages of the fracture process are absent. Indeed, such a crystal would experience a nonlinear–elastic deformation until it fails either by homogeneous unstable decohesion along the atomic planes perpendicular to the applied load (when reaching the ideal tensile strength) or, much more probably, by a shear instability (when reaching the ideal shear strength). In both cases, however, this stress would significantly depend on the direction of the applied load, i.e., on the crystallography (see Chapter 1 for details). The related uniaxial fracture stress σ_u would lie close to an upper limit of the material strength of the order of 10 GPa, whereas the fracture strain would approach the lower limit of the order of 0.1. The cohesive energy per atom U is another important mechanical characteristic related to this fracture process. This energy is proportional to the fracture (surface) energy γ that refers to the work necessary to create a unit area of an internal free surface (crack), and also depends on the crystallography.

Real single crystals exhibit much lower fracture stresses. This is, however, not a consequence of a lower cohesive (or fracture) energy due to the presence of dislocations. Indeed, the cohesive energy associated with a dislocation core is only slightly lower than that of the perfect crystal. On the other hand, real crystals exhibit much higher fracture strains due to a long mean free path of dislocation. The reason is that the fracture process involves all four fracture stages, thereby being completely different from that of the perfect crystal. In the first stage, the mobile primary dislocations on selected glide planes move rather long distances thus producing large macroscopic strains under very small applied stresses of the order of 10 MPa. In the second stage, the interaction of dislocations at intersections of primary and secondary slip planes builds long pile-ups associated with very high local internal stresses. When a superposition of internal stresses with the external stress locally approaches the ideal strength, microcracks nucleate in the bulk. Growth and coalescence of these cracks lead to a final fracture at a rather high fracture strain ($\varepsilon_f \gg 1$). In body centred cubic (bcc) metals the interaction of pile-ups creates sessile dislocations that initiate cleavage fracture along (100) crystallographic planes (see Chapter 2 for more details).

Polycrystalline single-phase materials contain grain boundaries that restrict the dislocation motion. Therefore, lower fracture strains ($\varepsilon_f \geq 1$) are usually observed. On the other hand, higher external stresses ($\sigma \approx 100\,\text{MPa}$) are needed to build dislocation pile-ups inside the grains. The high-angle grain boundaries possess a rather low cohesive energy. This energy and the related ideal strength can be further reduced by a diffusion of interstitial atoms like hydrogen or oxygen to the grain boundary sites where high triaxial tensile stresses are built by the superposition of internal and external stresses. This can lead to grain boundary cracking (see Chapter 3 for more details).

Engineering polycrystalline materials contain secondary phase particles (precipitates and inclusions) both inside the grains and at the grain boundaries. This restricts dislocation motion to an even greater extent so that the fracture strain is further reduced ($\varepsilon_f \leq 1$). On the other hand, higher external stresses are needed to build dislocation pile-ups which leads to a higher fracture stress. Incoherent matrix/particle interfaces exhibit a very low cohesive energy and a reduced ideal strength. These characteristics can be further worsened by interstitial diffusion and/or segregation of impurity atoms (see Chapter 2 for more details). Therefore, the microcracks can be initiated at both the particle/matrix interfaces and the grain boundaries. Consequently, the fracture process can be accomplished by diverse micromechanisms: Transgranular cleavage, intergranular decohesion, coalescence of voids or their mixture (see Chapter 2 for more details). In comparison to previously discussed microstructures, the stress-strain diagram of a typical high-strength engineering material ($\sigma_u \approx 1\,\text{GPa}$, $\varepsilon_f \approx 0.3$) is closer to that of the perfect crystal.

The radical engagement of dislocations in the microcrack nucleation might be understood in terms of their negative role in fracture processes. However, this statement is generally not true. When the microcrack is already formed, dislocations start to serve in a very positive way particularly in the case of brittle or quasi-brittle fracture. Indeed, the emission of dislocations from the crack tip (or their absorption at the crack tip) increases the fracture energy and blunts the crack tip, thus increasing the material resistance to unstable fracture. This favourable effect is appreciated in the case of engineering components containing microdefects that have been produced during their production or nucleated by corrosion and/or fatigue. On the other hand, a repeated emission and absorption of dislocations enables stable crack propagation under fatigue loading. These examples demonstrate a miscellaneous role of lattice defects in fatigue and fracture. Simultaneously, they highlight the importance of recognizing the fracture micromechanism in the research and development of advanced materials with high resistance to fracture and fatigue.

Table 0.1 Space scales and simulation methods

Scale range [m]												
10^{-12}	10^{-11}	10^{-10}	10^{-9}	10^{-8}	10^{-7}	10^{-6}	10^{-5}	10^{-4}	10^{-3}	10^{-2}	10^{-1}	10^{0}
Subnano		Nano			Micro			Meso		Macro		
ab initio, phonon spectra												
		Monte Carlo, cluster variation, molecular dynamics										
			dislocation dynamics, **discrete dislocations**									
				tessellation, **topology**, grain boundary dynamics								
			percolation									
				finite element, **boundary element**, finite difference, rheology, cellular automata								

There are many simulation methods used in modelling deformation and fracture processes on different scales. In Table 0.1, selected methods are collected according to their applicability in subnano-nano-micro-meso-macro regimes. Many of these methods (emphasized in Table 0.1) were also employed by us. Advanced microstructure models of fracture and fatigue processes based on simulation methods should enable us to use, or even derive, path-independent relationships between microstructural parameters and fracture characteristics. They should provide an insight into physical principles that govern the damage accumulation in materials microstructure as well as allow a proper interpretation of experimental results. To fulfil these requirements, the advanced models are increasingly complemented by the concept of integrated modelling and simulation. This means that computer codes and methods are coupled with the aim of bridging the scale differences between adjacent simulation levels. Such models belong to the *multiscale* category. To our knowledge, there is no generalized rule for how to design the multiscale approach. However, there are two principally different integration methods widely employed in such simulations: *simultaneous* integration and *sequential* integration [2]. The first method means that various interacting simulation codes are utilized simultaneously in one computer experiment. The second principle is based on an adequate transfer of parameters between simulation that are used sequentially. Multiscale models of fracture processes, presented in the works of the authors of this book, are preferentially built on the second approach.

It should be emphasized that partial submodels of the sequential multiscale model should not involve too many state variables serving as *fitting* parameters. Indeed, too complicated mathematical descriptions can lead to a physical nontransparency of the model. Similarly, a deficiency of experimental verification of partial theoretical concepts usually results in a misleading phenomenology of the multiscale model. Nevertheless, without analyzing the fracture micromechanisms in the frame of relevant multiscale approaches, many rather surprising experimental results (or even those standardly accepted) could not be reasonably elucidated. Since these models embrace a wide range of microstructure states, they are able to predict materials behaviour and fracture properties on different scales. Let us briefly mention a few examples of such unique capabilities.

In 1970–1974, numerical simulations of an arrangement of atoms in screw dislocations in bcc metals confirmed that their core is spatially extended into three slip planes that lie in the zone of a common slip direction [3, 4]. Much later, the coupling of codes based on atomistic, crystallographic and finite element approaches revealed that such dislocations exhibit asymmetry in their mobility under tensile and compressive loading [5]. Moreover, they do not follow the Schmid's law at low temperatures. These unique properties could explain an asymmetric mechanical hysteresis observed in bcc single crystals or the anomalous temperature dependence of the yield stress in some intermetallic materials [6]. This multiscale analysis led to the development of a physically based yield criterion with generally asymmetric yield surface that is directly applicable in macroscopic analyses of yielding [7].

In fatigue, the existence of the threshold ΔK_{th} of long crack growth has been well known and widely accepted since nearly 1970. Unfortunately, the theoretical methods based on the macroscopic theory of plasticity did not predict such a threshold behaviour at all. Starting from 1990, however, discrete dislocation concepts could be brushed up and applied to fracture research owing to a massive involvement of new numerical methods (e.g., [8]). One of the most important results that followed from such approaches was the exact theoretical confirmation of the threshold behaviour [9]. Afterwards, the discrete dislocation concept proved to be a very useful tool for elucidation and quantification of various phenomena near the fatigue threshold, e.g., crack closure effects under plane strain conditions [10, 11]. These concepts were also able to provide a physically clear background of the size effect observed in micro and nanocomponents [12].

During 1975–1985, many experimental results verified an anomalous fracture behaviour of ultra-high-strength low-alloy (UHSLA) steels. The fracture toughness of these steels significantly increased with coarsening of the mean prior austenite grain size (raising austenitizing temperature). On the other hand, the absorbed energy in impact tests revealed the opposite trend. These contradictions could be elucidated by an extreme geometrical shielding of the crack tip that was produced by decohesion of coarse grain boundaries during the fracture toughness tests. A multiscale model that coupled methods based on topology, stereology and metallography with finite element analysis quantitatively predicted fracture behaviour in very good agreement with experimental data. A similar model could quantitatively reproduce a steep increase in fracture toughness of borosilicate glasses with increasing concentration of reinforced particles (see also Chapter 2).

Besides a survey of our work, this book aims to present a rather consistent overview of fracture micromechanisms. Indeed, all the fundamental kinds of fracture processes and related micromechanisms are considered except for high-temperature creep damage. The description of fracture processes ranges from atomistic up to macroscopic levels. Therefore, the multiscale context can be easily found in many models of fracture processes that couple such approaches by means of sequential integration. Although emphasis is given to metallic materials, the fracture behaviour of ceramics and composites is also discussed.

The topic of the first chapter is the deformation and fracture of perfect crystals. Their mechanical behaviour under various kinds of monotonously increasing (static) loading is preferentially investigated by means of *ab initio* (first principles) methods based on electronic structure calculations. It is our strong belief that engineers can also learn a lot from the results of these apparently academic studies. They provide a clear distinction between intrinsic lattice properties and those induced by defects and secondary phases in engineering materials. They also constitute physical benchmarks for engineering multiscale models such as upper and lower limits of fracture and fatigue characteristics (ideal strength and fracture toughness). Three-scale models coupling electronic structure with atomic arrangements and crystallography are applied to calculate the ideal strength of crystals and nanocomposites under various loading conditions. Multiscale models predicting intrinsic ductile/brittle behaviour of crystals are also presented. These models couple atomistic, crystallographic and fracture-mechanics approaches. Finally, a multiscale model of nanoindentation is presented in order to quantify pop-in effects observed in load-penetration diagrams. This model links numerical methods and results embracing all spatial scales from nano to macro.

The second chapter is divided into sections devoted to micromechanisms of brittle, quasi-brittle and ductile fracture under static (low strain rate) and dynamic (high strain rate) loading. The leading theme of the first topic is the geometrically-induced crack tip shielding effect and the related fracture micromechanism. A two-scale analytical model based on quantitative fractography and mixed-mode fracture mechanics is applied to explain the fracture behaviour of borosilicate glasses. In order to quantify the quasi-brittle fracture, the geometrically-induced shielding effect is related to the ratio of a characteristic microstructure dimension and the crack-tip plastic zone size (the size ratio) in a two-scale analytical model based on quantitative fractography and fracture mechanics. This model is employed to describe the anomalous fracture behaviour of UHSLA steels as well as to assess the value of fracture energy of grain boundaries with segregated phosphorus in ferritic alloys. Finally, analytical three-scale models of ductile fracture processes based on dislocation dynamics, microvoids formation kinetics and fracture mechanics are presented. These models are able to predict the fracture strain in the tensile testing of ductile metallic materials and to assess the values of fracture toughness for steels exhibiting the ductile fracture micromechanism.

The last chapter is dedicated to fatigue fracture of metallic materials. The micromechanisms of mechanical hysteresis, crack initiation and crack propagation under all modes of crack-tip loading are described in the multiscale concept from nanoscale to mesoscale levels. An analytical multiscale model of crack closure was developed in order to evaluate individual crack-tip shielding components and to assess the intrinsic fatigue threshold value under the opening loading mode. Micromechanisms of crack propagation under shear and mixed-mode loading are also discussed from the point of view of both theoretical and experimental approaches. For example, a two-scale model connecting three-dimensional topology with linear–elastic fracture mechanics is able to explain a formation of factory-roof morphological patterns on the fracture surfaces generated by cyclic torsion. The initiation and propagation of fish-eye cracks is studied under combined bending-torsion loading in specimens made of nitrided steel. A method enabling a quantitative reconstitution of the fatigue process from the fracture morphology is described in the final part of the chapter. This method of engineering failure analysis was successfully applied in many practical cases.

The book is complemented by three appendices devoted to *ab initio* methods utilized in atomistic models, criteria employed in mixed-mode fracture and derivation of void-induced dislocation dynamics in the model of ductile fracture.

Chapter 1
Deformation and Fracture of Perfect Crystals

Perfect crystals constitute an idealization of real single crystals that always contain some crystal defects. This holds even for almost perfect single crystals, so-called whiskers, where the presence of an equilibrium concentration of vacancies is inevitable. On the other hand, the mechanical behaviour of perfect crystals can be very successfully simulated and predicted by theoretical *ab initio* approaches based on electronic structure calculations. Because many of the crystallographic, elastic, electric, magnetic and thermodynamic characteristics of crystals do not depend on crystal defects, the *ab initio* results can still be experimentally verified. The practical importance of such studies lies in several general aspects.

First, knowledge of the behaviour of perfect crystals clearly identifies the role of crystal structure and chemical composition. This enables us to separate the role of defects due to the difference in the mechanical behaviour of perfect and real crystals. Second, the ideal (theoretical) strength of perfect crystals of a particular chemical composition represents an upper bound of the strength of solids. Consequently, engineers can see the gap remaining between the strength of contemporary high-strength materials and that of the theoretical limit. Third, the models of processes of dislocation creation near stress concentrators such as cracks in perfect crystals provide us with lower bounds of macroscopic characteristics used in fracture mechanics. Thus, the *ab initio* results yield both upper and lower benchmarks defining physically possible ranges of mechanical characteristics of engineering materials. Fourth, the characteristics of ideal crystals can be utilized in multiscale models of deformation and fracture processes of engineering materials. Finally, these results can be used as fitting data for the construction of sophisticated semi-empirical interatomic potentials that are utilized for prediction of the behaviour of extended defects in real crystals and polycrystals.

All of the above-mentioned aspects are demonstrated in the three sections of this chapter devoted to the mechanical behaviour of perfect crystals. In the first section, the application of *ab initio* methods to calculations of elastic stress-strain response and, particularly, to the determination of ideal strength under various loading conditions is demonstrated. One should note that these concepts are, in fact, two scale models based on both electronic and atomic structures. Section 1.2 deals with another important problem concerning the physics of brittle/ductile behaviour of perfect crystals. The models are three-scale approaches dealing with electronic, atomic and crystallographic structures of crystals. The third section illustrates the modelling of nanoindentation tests with regard to the physical interpretation of pop-ins observed at the end of the elastic part in the force-indentation depth diagram. These multiscale models couple electronic, atomic, crystallographic and continuum approaches to provide a unique tool for experimental measurements of the ideal shear strength.

1.1 Ideal Strength of Solids

The strength of any solid has an upper limit called the ideal (theoretical) strength (IS). This value corresponds to the failure of an infinite perfect single crystal loaded in a defined mode. The strength of engineering materials is usually controlled by nucleation and propagation of dislocations and/or microcracks. If such defects were not present, the material would only fail when the IS is reached. Until recently loads of this magnitude were only approached in studies of the mechanical behaviour of whiskers of very pure metals and semiconductors [1,13,14]. Starting from the beginning of the last century, there has been a more or less continuous effort expended in order to obtain theoretical and experimental data concerning IS of various solids. The IS values set an upper limit to the envelope of attainable stresses and its knowledge enables us to assess the gap remaining to upper strength values of advanced engineering materials in each period of time. However, this is not the only reason for IS investigation. From the theoretical point of view, the IS plays a decisive role in the fundamental theory of fracture. For example, the stress necessary for nucleation of a dislocation loop can be identified with the shear IS value. This has been proved most eloquently by nanoindentation experiments (see e.g., [15–18]) which suggest that the onset of yielding at the nanoscale is controlled by a homogeneous nucleation of dislocations in a small, dislocation free, volume under the nanoindenter, where the stresses approach the shear IS. Similarly, the local stress needed for unstable propagation of a cleavage crack should overcome the value of the tensile IS [19–22]. The ratio of shear and tensile IS expresses a tendency of the crystal matrix to become brittle or ductile [23–25] (see Section 1.2). The values of IS may also be used in the construction or checking of semi-empirical interatomic potentials. From the practical point of view, the shear IS appears to control both the onset of fracture and the dislocation nucleation in defect-free thin films and, in particular, in nano-structured materials that are currently being developed. Perfect single crystal wires (whiskers) are used as reinforcements in advanced composite materials and large metallic and ceramic single crystals start to be important in special engineering components, e.g., in turbine blades [26].

Section 1.1.1 provides a picture of the historical development including remarks on sophisticated calculation methods. The principles of *ab initio* methods and stability procedures, utilized in recent computations, are presented in Section 1.1.2. Some results of *ab initio* calculations of IS under simple loading modes reported in our works are outlined in Section 1.1.3. Section 1.1.4 is dedicated to IS of crystals under multiaxial loading. An analysis of mechanical properties of ideal nanocomposites is outlined in Section 1.1.5. A brief discussion concerning the influence of lattice defects and temperature on the crystal strength is presented in the last subsection.

Section 1.1 is complemented by two appendices focused on *ab initio* methods and mixed-mode criteria of crack stability.

1.1.1 From Classics to Recent Concepts

For every particular loading, the stress state is characterized by six stress tensor components and, consequently, an infinite number of ideal strengths exists for a given crystal. For practical reasons, therefore, the IS was usually evaluated only for several special cases of loading, each defined by a single value of the stress tensor component, specifically, for the uniaxial tensions and compressions along various crystallographic directions, the isotropic (hydrostatic) tension and compression and for the pure shear in certain planes and directions. The respective IS values denoted here as $\sigma_{iut}, \sigma_{iuc}, \sigma_{iht}, \sigma_{ihc}$ and τ_{is} cover, to a reasonable degree, the most important cases occurring in the engineering practice.

1.1.1.1 Classical Theories

Historically the first calculations of τ_{is} were performed in 1926 by Frenkel [27]. The model of an ideal crystal subjected to block-like shear is, along with the related behaviour of the shear stress τ under applied shear deformation, expressed by the plane shift s in Figure 1.1. The values of ideal shear strength for the block-like model will be denoted $\tau_{is,b}$. The $\tau(s)$ dependence was assumed to be of a sinusoidal shape.

According to specification of variables in Figure 1.1, the stress behaviour could be described by the relation

$$\tau = \tau_{max} \sin \frac{2\pi}{a} s.$$

For a small shift ($\sin \frac{2\pi}{a} s \approx \frac{2\pi}{a} s$), the shear modulus should be $G = \mathrm{d}\tau/\mathrm{d}\xi$, where $\xi = s/b$. Since $\tau_{is,b} = \tau_{max}$ this leads to

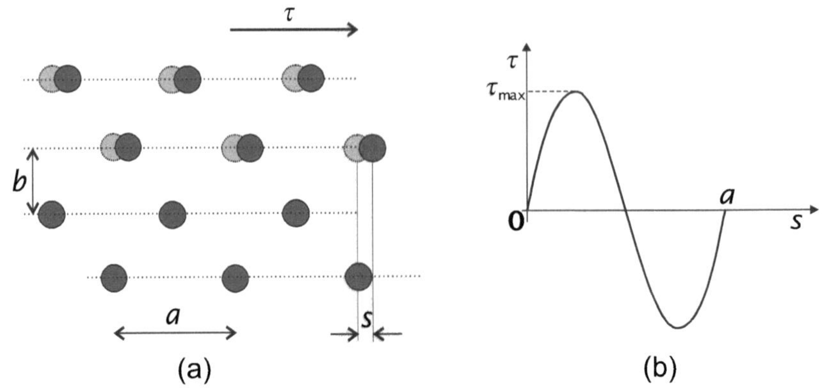

Figure 1.1 (a) Model of a block-like shear deformation (*dark spheres* represent atomic positions within the upper block, that is, as a whole, shifted by s towards the lower block along the shear plane, *light spheres* show their original positions), and (b) shear stress τ as a function of the shift s of two adjacent planes

$$\tau_{is,b} = \frac{Ga}{2\pi b}. \tag{1.1}$$

Equation 1.1 gave values $\tau_{is,b} \approx \frac{1}{9}G$ for the $\{111\}\langle 11\bar{2}\rangle$ shear of face centred cubic (fcc) metals, and $\tau_{is,b} \approx \frac{1}{5}G$ for $\{110\}\langle 1\bar{1}1\rangle$ shear of bcc metals as well as for the $\{111\}\langle 1\bar{1}0\rangle$ shear of fcc metals. Because the yield stress of real crystals was found to be about three orders lower, the only plausible explanation of this discrepancy was the presence of line defects (dislocations). Thus, the Frenkel's result created a milestone for a development of the dislocation theory.

First attempts to compute the ideal strength in uniaxial tension σ_{iut} were performed by Polanyi [28] and Orowan [29]. They were based on an assumption of tearing fracture of a stretched crystal along a crystallographic plane. Forces between two adjacent atomic planes of a perfect solid vary with the interplanar distance as in Figure 1.2. This dependence was approximated by a sinusoidal function

$$\sigma = \sigma_{\max} \sin \pi \frac{x - a_0}{d}$$

and the expected deviation from this trend for high strain values was neglected. The function was parametrized according to the following assumptions: 1) the work of deformation per unit area corresponds to energy 2γ of the two new surfaces

$$\int_{a_0}^{a_0+d} \sigma \mathrm{d}x = 2\gamma;$$

and 2) in the vicinity of the equilibrium state ($x = a_0$), the stress is proportional to Young's modulus and the relation $E = \dfrac{d\sigma}{d\varepsilon}$ must be valid for the strain $\varepsilon = \dfrac{x - a_0}{d}$. Then, the maximum value of the tensile stress can be simply evaluated as

$$\sigma_{iut} = \sqrt{\frac{E\gamma}{a_0}}. \tag{1.2}$$

The corresponding ideal tensile (tear) strengths of metals are mostly very high (several tens of GPa).

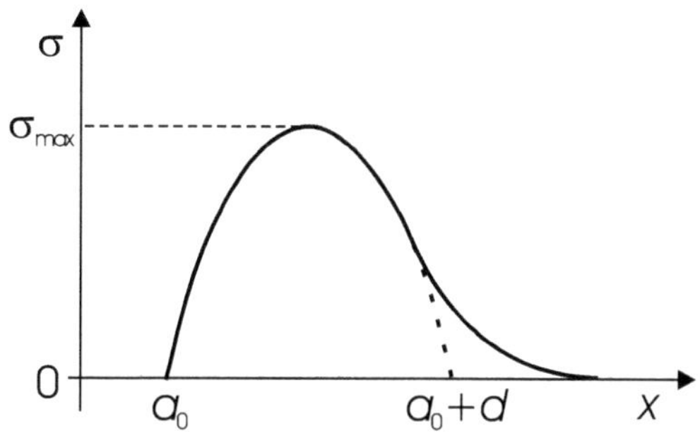

Figure 1.2 Stress as a function of the distance between atomic planes

Mackenzie presented in 1949 a more extended study of the shear IS based on a variation of potential energy U per unit area of a shear plane with the plane shift s [30]. The shear stress can be calculated from the energy U as

$$\tau = \frac{dU}{ds}. \tag{1.3}$$

From this point of view, the Frenkel's approach described above refers only to the first two terms in the Fourier series for $U(s)$. Therefore, Mackenzie took further terms into consideration. This theory gave a very low value $\tau_{is,b} \approx \frac{1}{30}G$ for $\{111\}\langle 11\bar{2}\rangle$ shear in fcc lattice [30]. As found by Šandera and Pokluda [31], however, some assumptions used in that theory were not physically legitimate. Indeed, more recent calculations [31, 32] based on more sophisticated atomistic approaches confirmed a much better validity of Frenkel's estimation.

Further IS calculations were performed by means of so-called empirical interatomic potentials [33]. Most of them used an analogy to Equation 1.3. The potential energy U was calculated as a sum of pair-potentials of various kinds such as the Morse potential, the Lennard–Jones potential, etc. (e.g., [23, 24, 34–36]). As an example, some results of calculations of σ_{iht} are introduced in Section 1.1.3.

1.1.1.2 *Ab initio* Methods

Nowadays, so-called *ab initio* (or first principle) approaches enable us to compute the crystal energy in a very accurate manner. This particularly holds for single crystals of pure elements or compounds, crystals with periodical arrangements of atoms of various kinds and also for local crystal defects. These methods utilize the density functional theory [37, 38] in which the problem of many interacting electrons is transformed into a study of single electron motion in an effective potential. A brief description of principles of such methods is presented in Appendix A.

Most probably, the first *ab initio* calculation of the uniaxial IS σ_{iut} was that of Esposito *et al.* [39] for the copper crystal. However, these authors did not perform relaxations of atomic positions inside the loaded crystal in directions perpendicular to the loading axis (Poisson's type of expansion or contraction). Probably the first *ab initio* simulation of a tensile test that included the relaxation in perpendicular directions to the loading axis was performed by Price *et al.* [40] for TiC along the [001] axis. Later, σ_{iut} was calculated for [001] and [111] loading axes for a variety of cubic crystals by Šob *et al.* [41–43]. Kitagawa and Ogata [44,45] studied the tensile IS of Al and AlN but also did not include the Poisson's contraction. Further calculations of σ_{iut}, performed for α-SiC, diamond, Si, Ge, Mo, Nb and Si_3N_4, have already taken that effect into account by allowing a transversal relaxation of atoms [46–51].

The values $\tau_{is,b}$ of the shear IS were first calculated by Paxton *et al.* [32] for V, Cr, Nb, Mo, W, Al, Cu and Ir. The values τ_{is}^* calculated according to the model of a uniform shear (see Figure 1.3) were later reported by Moriarty *et al.* [52,53] for Mo and Ta. These calculations did not include any relaxations. Recently, the relaxed values of τ_{is} were calculated for many crystals such as TiC, TiN, HfC, Mo, Nb, Si, Al, Cu and W by groups of Morris *et al.*, Kitamura *et al.* and Pokluda *et al.* [54–58]. In these models, the interplanar distance was relaxed towards the minimum energy during deformation. More advanced models also enabled relaxations of the arrangement of atomic positions within the planes [48, 59] – see also Sections 1.1.2 and 1.1.3.

Since 1997, *ab initio* calculations of σ_{iht} have been performed by Pokluda *et al.* [60–62] and other authors (e.g., [63]). Since spherical symmetry was maintained during deformation, the relaxation procedures were not necessarily applied in these models.

In the majority of older studies on IS, the deformation process was assumed to proceed in a stable manner until the applied stress reached its maximum value. This means that the crystal failed in the same mode in which it was originally deformed from the very beginning. However, this assumption was disputed in many works [64–66]. Under both tensile and compressive loadings, the shear stresses in some slip systems can exceed their critical values (corresponding to the related τ_{is}) well before the stress reaches its maximum. This was also observed in tensile tests on whiskers [30, 33]. Indeed, some of the whiskers evidently failed by shear across some favourable crystallographic

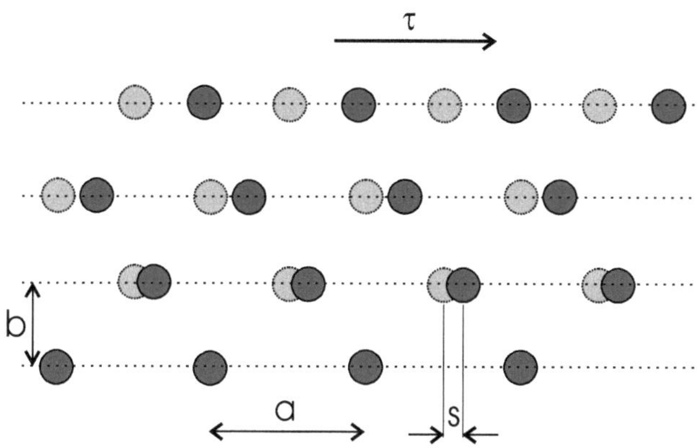

Figure 1.3 Scheme of the uniform shear applied to a perfect crystal

plane resulting in an atomically smooth fracture surface. Thus, the definition of IS as a maximum attainable stress along the deformation path was assigned to a stress related to the first onset of instability. Many stability studies were based on an analysis of the elastic response of crystals subjected to small homogeneous deformations. Such analyses of the mechanical stability, based on calculations of independent elastic moduli, led to a significant decrease in computed values of IS. This was the case for [001] uniaxial loading in Al, Nb and Cu [49, 67, 68]. More attention to this issue will be paid in Sections 1.1.2 and 1.1.3.

1.1.1.3 Other Advanced Methods

Besides *ab initio* approaches several other methods were also utilized for IS computations. Although these methods lie beyond the scope of this book, it is worthwhile making a brief comment on the topic.

In the last 20 years, very sophisticated semi-empirical methods such as many-body potentials of Finnis–Sinclair type [69], embedded atom method [69] or bond-order potentials [70] were used for analyses of extended crystal defects [71–73]. These concepts represent a hybrid between empirical and *ab initio* approaches. At present, the *ab initio* methods are only capable of giving a sufficiently precise prediction of the mechanical behaviour in very simple cases (but still far enough from the unstressed equilibrium state). This is the reason why, starting from the early 1990s, they are used for the calibration of semi-empirical potentials [42].

A further step towards identification of the initial onset of an instability was made by studies on the phonon spectra of crystal states along investigated deformation paths. This approach has further reduced the calculated values of IS [74]. The phonons are quasi particles used to express a particle aspect of lattice vibrations. They play a major role, e.g., in the theory of both thermal and electric conductivity. However, they can also serve as an indicator of the lattice instability related to so-called soft phonon modes, at which the phonon frequency becomes an imaginary number. Such instabilities are responsible for various structural transitions [75,76] and, in general, they can be understood as an irreversible non-uniform (heterogeneous) distortion of a crystal. In this aspect, phonon analysis represents a generalization of elastic stability analysis since any observed elastic instability corresponds to a soft phonon mode with an infinite wavelength.

Once the crystal becomes unstable, it will follow a trajectory in a configurational space that will eventually violate the harmonic approximation inherent in the phonon calculation. In order to find such trajectories, molecular dynamic (MD) methods can be utilized. These approaches account for variations of the unit cell shape as well as the positions of constituent atoms. MD methods are probably the most promising tools for an investigation of the system stability and eventual structure evolutions during spontaneous structural transitions. They can give a sufficient number of degrees of freedom to studied systems and, furthermore, they can also incorporate finite temperatures and pressures thus bringing the simulations closer to reality. However, certain limitations related to computational capacity hinder them from wider applications. Present MD methods are mostly based on empirical or semi-empirical interatomic potentials. Some results obtained by means of these methods are mentioned in Section 1.2.

Let us finally note that the most sophisticated methods, including so-called correlated electron-ion dynamics, are even more computationally demanding. On the other hand, they may represent a reliable tool for atomistic simulations in the near future [77].

1.1.2 Calculation Principles

When applying atomistic approaches to a particular crystalline system, the dependence of its total energy on the deformation state constitutes the main output.

In order to describe the deformation of a crystalline system, it is necessary to define appropriate deformation parameters (strain variables). However, there is no unique way of defining a set of parameters which would provide a measure of a pure finite strain related to the crystal reference state. Therefore, one can find several different definitions used in the technical literature. A homogeneous strain of a crystal can be specified, e.g., by any six parameters that define a primitive cell. Some authors use lengths of cell edges (a_i) and their included angles (α_i) [66,78–80]. Although the deformation is described by changes of these parameters rather than themselves, they are also widely used in stability analyses. Such a natural set of variables is sometimes called Milstein's variables [79].

A simple description of any deformation that is used throughout this work employs a general rule on how to change position of an arbitrary point (e.g., atomic position) within the chosen coordinate system. This rule is expressed by a transformation matrix (deformation gradient) which is also called the Jacobian matrix J [81]. The practical advantage of such a description is apparent when one simulates a homogeneous deformation of a crystal through the deformation of its primitive cell. The primitive cell comprises both the motif and the set of vectors that determine the translational symmetry of a crystal. The transformation of vectors is then performed by multiplying by the Jacobian matrix.

Let us consider a crystalline system in a reference state with the corresponding set of primitive translational vectors $\mathbf{a_r}, \mathbf{b_r}, \mathbf{c_r}$. Applying a deformation described by J to the system, the set is transformed to a new set of vectors corresponding to the deformed state

$$\mathbf{a} = \mathsf{J}\mathbf{a_r}; \quad \mathbf{b} = \mathsf{J}\mathbf{b_r}; \quad \mathbf{c} = \mathsf{J}\mathbf{c_r}.$$

The tensor of a finite deformation (also called Lagrangian strain tensor) is defined according to the relation

$$\hat{\eta} = \frac{1}{2}(\mathsf{J}^T\mathsf{J} - \mathsf{I}),$$

where I means the identity matrix.

An equivalent definition of the finite strain tensor uses a rule that describes changes of lattice points positions via the displacement vector $\mathbf{u} = \mathbf{a} - \mathbf{a}_r$ and the matrix $e_{ij} = \partial u_i/\partial a_j$ [82]. When using the Einstein summation rule, the components of the finite strain tensor can then be written as

$$\eta_{ij} = \frac{1}{2}\left(\frac{\partial u_i}{\partial a_j} + \frac{\partial u_j}{\partial a_i} + \frac{\partial u_k}{\partial a_i}\frac{\partial u_k}{\partial a_j}\right). \tag{1.4}$$

The components η_{ii} refer to stretches and η_{ij} stands for shear strains if $i \neq j$.

A small deformation can also be depicted by a small strain tensor (known also as Euler strain, Green tensor or Cauchy infinitesimal strain):

$$\varepsilon_{ij} = \frac{1}{2}\left(\frac{\partial u_i}{\partial a_j} + \frac{\partial u_j}{\partial a_i}\right). \tag{1.5}$$

Hence, the finite strain differs from the small strain by the cross-term $\frac{\partial u_k}{\partial a_i}\frac{\partial u_k}{\partial a_j}$. Consequences of this difference with respect to the IS analysis are discussed in [83]. As the tensor represents a symmetric part of the matrix e_{ij}, any infinitesimal transformation can be expressed by a linear combination of the tensor ε_{ij} (describing pure deformation) and the antisymmetric part of the matrix e_{ij}

$$\omega_{ij} = \frac{1}{2}\left(\frac{\partial u_i}{\partial a_j} - \frac{\partial u_j}{\partial a_i}\right)$$

that represents the rotation as can be seen in Figure 1.4.

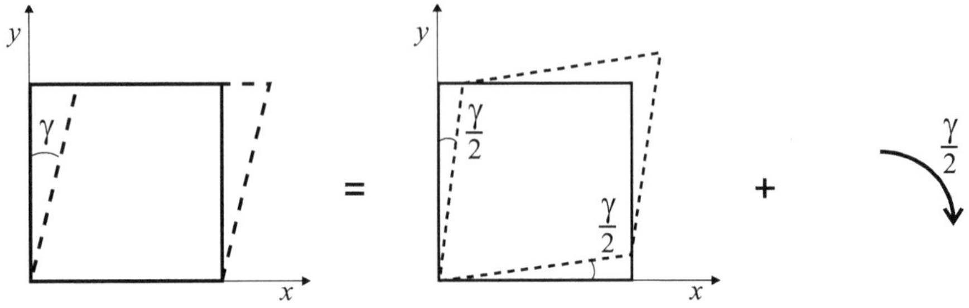

Figure 1.4 Two-dimensional illustration of lattice distortion (simple shear) as a pure shear plus a rotation $\hat{e} = \hat{\varepsilon} + \hat{\omega}$

As follows from Equations 1.4 and 1.5, both the strain tensors are symmetric (even if J is generally asymmetric). Therefore, it is useful to use Voigt notation that reduces the number of indices of symmetric tensors according to the prescription

$$\begin{array}{ll} 11 \to 1 & 23 \to 4 \\ 22 \to 2 & 13 \to 5 \\ 33 \to 3 & 12 \to 6. \end{array}$$

Thus, the second-rank tensor is reduced to a six-dimensional vector. Regrettably, there are two different standards used in the literature. The first one exactly follows the above-mentioned substitution $\eta_{ij} = \eta_\alpha$ (where indices i, j run from 1 to 3 and the index α runs from 1 to 6) [84,85] whereas the other one uses substitution $\eta_{ij} = \frac{1}{2}\eta_\alpha(1 + \delta_{ij})$ that leads to the following difference for the shear components: $\eta_4 = 2\eta_{23}$, $\eta_5 = 2\eta_{13}$ and $\eta_6 = 2\eta_{12}$ [82, 86–89]. The latter case is usually called standard Voigt notation. This simplifies the expression for energy expansion at Equation 1.7.

When a crystal system is subjected to a deformation, its energy changes (in the case of a stable state it increases, of course). The crystal energy can be expanded into Taylor series with respect to the finite strain η_{ij} as

$$U = U_0(V) + V\sigma_{ij}\eta_{ij} + \frac{1}{2}VC_{ijkl}\eta_{ij}\eta_{kl} + O(\eta^3), \tag{1.6}$$

where V represents a system volume, C_{ijkl} are elastic moduli and σ_{ij} are the related stress tensor components. In classical continuum mechanics, C_{ijkl} are usually called elastic constants [82]. However, the analysis of crystal stability (see Section 1.1.2.2) requires calculation of C_{ijkl} also at states far from equilibrium, where their values depend on the applied strain (or stress). In this book, therefore, the term *elastic moduli* is used for C_{ijkl}.

The symmetry of C_{ijkl} and σ_{ij} with respect to interchange of indices $(i \leftrightarrow j)$ and $(k \leftrightarrow l)$ enables us to use the Voigt notation for both of them.

The standard Voigt notation for strain $\eta_{ij} = \frac{1}{2}\eta_\alpha(1+\delta_{ij})$ is used here. The corresponding energy expansion is then

$$U = U_0(V) + V\sigma_\alpha\eta_\alpha + \frac{1}{2}VC_{\alpha\beta}\eta_\alpha\eta_\beta + O(\eta^3). \tag{1.7}$$

Note that the energy expansion at Equation 1.7 contains a double sum instead of a quadruple sum in Equation 1.6.

1.1.2.1 Elastic Moduli

The 6×6 matrix of elastic moduli generally contains 21 independent elements that do not transform like the second-rank tensor components. According to the number of point group symmetry operations, the amount of independent elastic moduli can be lower. With respect to the energy expansion, the elastic moduli can be defined as

$$C_{\alpha\beta} = \frac{1}{V}\left(\frac{\partial^2 U}{\partial\eta_\alpha\partial\eta_\beta}\right)$$

or

$$c_{\alpha\beta} = \frac{1}{V}\left(\frac{\partial^2 U}{\partial\varepsilon_\alpha\partial\varepsilon_\beta}\right).$$

From now on, the C_{ij} will denote the elastic moduli defined on the basis of the finite strain and c_{ij} will stand for the elastic moduli based on the small strain.

When, for example, the crystal is subjected to small isotropic deformation, the lattice parameter a is related to the reference parameter a_r as $a = a_r(1+e)$. Here e is a small stretch that represents diagonal components of the small strain tensor. Then, the finite strain is related to e according to $\eta_1 = \eta_2 = \eta_3 = \eta = e + \frac{e^2}{2}$, and the deformation gradient can be expressed as

$$\mathsf{J}_{iso} = \begin{pmatrix} 1+e & 0 & 0 \\ 0 & 1+e & 0 \\ 0 & 0 & 1+e \end{pmatrix} \quad \text{or} \quad \mathsf{J}_{iso} = \begin{pmatrix} \sqrt{1+2\eta} & 0 & 0 \\ 0 & \sqrt{1+2\eta} & 0 \\ 0 & 0 & \sqrt{1+2\eta} \end{pmatrix}.$$

The corresponding energy expansion at Equation 1.7 gives

$$U = U_0(V_r) + 3V_r\sigma\eta + \frac{1}{2}V_r(3C_{11}\eta^2 + 6C_{12}\eta^2) + \ldots$$

from which the following combination of elastic moduli:

$$C_{11} + 2C_{12} = \frac{1}{3V_r}\frac{\mathrm{d}^2 U}{\mathrm{d}\eta^2}$$

can be derived. Thus, this combination defines the bulk modulus

$$B = \frac{1}{3}(C_{11} + 2C_{12})$$

which expresses the elastic response of a crystal to isotropic (hydrostatic) loading.

Several other elastic moduli can be related to simple types of loading. For example, the modulus C_{11} can be determined by a simple lattice stretch in the [100] direction (see Figure 1.5). The Young's modulus expresses the crystal response to uniaxial loading and, therefore, it depends on the crystal orientation with respect to the loading direction. For particular orientations one can derive

$$E_{\langle 100\rangle} = \frac{(C_{11} - C_{12})(C_{11} + 2C_{12})}{C_{11} + C_{12}}, \tag{1.8}$$

$$E_{\langle 110\rangle} = 4\frac{C_{44}}{C_{11}}\left[1 + \frac{2C_{44}}{(C_{11} - C_{12})(C_{11} + 2C_{12})}\right]^{-1},$$

$$E_{\langle 111\rangle} = \frac{3C_{44}(C_{11} + 2C_{12})}{C_{44} + C_{11} + 2C_{12}}.$$

The shear modulus G can be expressed as

$$G = \frac{3C_{44}(C_{11} - C_{12})}{4C_{44} + C_{11} - C_{12}}$$

for $\{111\}\langle\bar{2}11\rangle$, $\{111\}\langle\bar{1}10\rangle$ and $\{110\}\langle\bar{1}11\rangle$ slip systems,

$$G = \frac{1}{2}(C_{11} - C_{12})$$

for $\{110\}\langle\bar{1}10\rangle$ slip system and $G = C_{44}$ for $\{110\}\langle 001\rangle$ slip system.

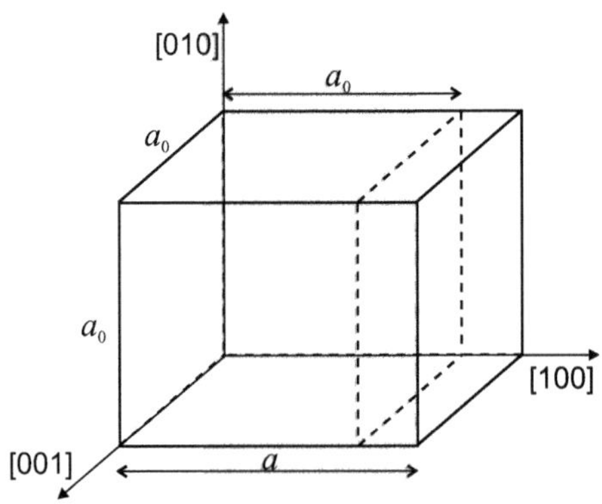

Figure 1.5 Illustration of a lattice distortion for calculation of the elastic modulus C_{11}

1.1.2.2 Mechanical Stability

Hereafter, the crystal potential energy U is determined from the electronic structure as E_{tot}. Thus, by using the internal energy per unit volume $E_u = E_{tot}/V$, one can also write

$$C_{ij} = \frac{\partial^2 E_u}{\partial \eta_i \partial \eta_j}$$

and

$$\sigma_i = \frac{\partial E_u}{\partial \eta_i}. \tag{1.9}$$

When no other instability (unstable phonon modes, phase transformations, elastic shear instabilities, etc.) occurs in the crystal, the relation at Equation 1.9 shows that IS corresponds to the first point of inflection on the energy *vs* strain curve ($C_{ij} = 0, \sigma_i = \sigma_{imax}$).

Probably the first attempt to formulate general criteria of crystal stability based on its elastic moduli was made in 1940 in the work of Born [64] and Born and Fürth [65]. They showed that by expanding the internal energy of a crystal in a power series in the strain and requiring positivity of the energy, one obtains a set of conditions on the elastic constants appropriate to the crystal that must be satisfied to maintain structural stability. It can be briefly said that, in the Born criterion, the system is stable if the matrix of elastic moduli C_{ij} is positive definite, i.e., all its eigenvalues are positive. Their results are valid only when the lattice is not under external stress.

In general, two basic cases of elastic (homogeneous) instability behaviour related to IS can be distinguished when analysing the crystal deformation:

1. instability occurs along the original deformation path,
2. instability changes the loading mode or the type of the deformation path.

The instability of the first kind (so-called volumetric instability) means that the process of unstable crystal collapse starts at the above-mentioned point of inflection on the original deformation path. Assuming the constant stress ensembles (i.e., the stress-controlled loading), the crystal starts to disintegrate spontaneously after reaching this point. During this process, however, strain induced phase transformations (so-called displacive transformations) may appear along the deformation path [90, 91]. These transformations proceed by means of cooperative displacements of atoms away from their lattice sites and alter the crystal symmetry without changing the atomic ordering or composition. They are of the first order and, therefore, accompanied by a symmetry-dictated extrema on the stress–strain curve. For example, the tetragonal Bain's path also induces typical displacive transformations (see Section 1.1.3). Moreover, additional extrema that are not dictated by the symmetry may occur, and reflect properties of the specific material. Consequently, more "IS values" can be found related to different points of inflection on the energy–strain curve. However, the IS is determined by the stress associated with the first point of inflection on the original energy–strain curve which corresponds to the maximal energy gradient. Note that atomic configurations related to energy minima behind the first point of inflection may mimic stable or metastable atomic arrangements that could be encountered when investigating thin films or extended defects such as interfaces or dislocations. Similar configurations can also be reached during the strain-controlled deformation path of a crystal (the constant strain ensembles), provided that they are not preceded by any instabilities of the second kind.

The instabilities of the second kind (so-called shear instabilities) can be derived by considering a requirement that the free energy (and at $T = 0$ also the total internal energy) be minimum in subsequent constant stress ensembles in accordance with the second law of thermodynamics [66, 78–80, 87, 92–94]. The main point of such an analysis was the assumption that the crystal is subjected both to the applied load and to an infinite variety of small perturbing forces. Any of the forces can make the crystal fail in a different mode than that related to the main loading force. The proposed stability assessment requires information about an elastic response of the system to small deviations from the current state (let us call it the reference state). Therefore, in the case of a quasi-static loading, the stability assessment is in a sense independent of the deformation path which led the system to this state, because the same atomic arrangement can be obtained via many various transformations of an arbitrary original state. For that reason, the further deformations used to investigate the stability have nothing to do with the original deformation path. If the solid is strained infinitesimally from the reference state associated with the stress σ_{ij} (in the standard notation) by a strain tensor ε_{ij}, the related Cauchy (true) stress τ_{ij} can be written as

$$\tau_{ij} = \sigma_{ij} + B_{ijkl}\varepsilon_{kl},$$

where

$$B_{ijkl} = C_{ijkl} + \frac{1}{2}\left(\delta_{ik}\sigma_{jl} + \delta_{jk}\sigma_{il} + \delta_{il}\sigma_{jk} + \delta_{jl}\sigma_{ik} - 2\delta_{kl}\sigma_{ij}\right) \tag{1.10}$$

is the elastic stiffness matrix ($i, j, k, l = 1, 2, 3$) introduced by Wallace [95] that is generally asymmetric with respect to the interchange of indices. Construction of this matrix is crucial for the stability assessment. As was shown in [92, 94, 96], the system becomes unstable once its symmetrized counterpart

$$\mathsf{A} = \frac{1}{2}\left(\mathsf{B}^T + \mathsf{B}\right)$$

attains a zero determinant, i.e.,

$$|\mathsf{A}| = 0 \tag{1.11}$$

during the loading. It should be emphasized that the elastic moduli in Equation 1.10 are the local ones, i.e., corresponding to different points of the deformation path. Thus, in order to assess the stability, their values should be determined by introducing a sufficient number of independent small deviations (strain increments) away from the original deformation path at each point, in accordance with the symmetry of the particular crystal lattice. The solution of Equation 1.11 gives a different number of possible stability conditions for different crystal lattice symmetries as well as different loading modes. The higher the symmetry (the more point group symmetry operations), the fewer the stability conditions are to be tested. The smallest possible number of necessary stability conditions (only two) corresponds to the isotropic solid.

The stability conditions for cubic crystals loaded in uniaxial tension or compression along the [001] direction, the so-called Bain's path, can serve as a suitable example. The tetragonal symmetry induced by the uniaxial loading means $C_{11} = C_{22} \neq 0, C_{33} \neq 0, C_{12} \neq 0, C_{13} = C_{23} \neq 0, C_{44} = C_{55} \neq 0, C_{66} \neq 0$ and $C_{ij} = 0$, other, and a simple relationship $\sigma_{ij} = \sigma\delta_{i3}\delta_{j3}$ stands for the stress tensor. By introducing these relations into Equation 1.11, one obtains the following stability conditions:

$$(C_{33} + \sigma)(C_{12} + C_{11}) - 2(C_{13} - \sigma/2)^2 > 0, \tag{1.12}$$

$$C_{11} - C_{12} > 0, \tag{1.13}$$

$$2C_{44} + \sigma > 0, \tag{1.14}$$

$$C_{66} > 0. \tag{1.15}$$

The left-hand side of Equation 1.12 differs from the tetragonal $E_{[001]}$ modulus for the stress-free state $\sigma = 0$ only by a multiplication constant. Therefore, the violation of that so-called volumetric condition is closely related to the first inflection point on the energy *vs* strain curve along the [001] deformation path (the instability of the first kind). In any case, the maximum value of the stress determining the IS along the [001] path is associated with the point of inflection. This is the main reason why the testing of the volumetric instability can actually be omitted. All the other conditions prevent the crystal from shear (second-kind) instabilities. Breaking the condition at Equation 1.13 causes a shear bifurcation from the tetragonal deformation path to the orthorhombic one [46, 85]. In the case of the fcc crystal, this instability induces branching to the tetragonal face centred orthorhombic path – it is the so-called Born's instability. The instabilities at Equations 1.14 and 1.15 are related to C_{44} and C_{66} shear moduli, respectively.

In order to test the shear-related criteria, special local Lagrangian deformations (determined by corresponding Jacobi matrices) are to be applied to the crystal at each point of the deformation path. For the tetragonal path [68], the Jacobi matrix

$$\mathsf{J}_{C'} = \begin{pmatrix} \sqrt{1+2e} & 0 & 0 \\ 0 & \sqrt{1-2e} & 0 \\ 0 & 0 & 1 \end{pmatrix}$$

and the corresponding deformation matrix

$$\hat{\eta}_{C'} = \begin{pmatrix} e & 0 & 0 \\ 0 & -e & 0 \\ 0 & 0 & 0 \end{pmatrix}$$

lead to the following change of the system energy (per volume unit) according to Equation 1.7:

$$\Delta E_u = (C_{11} - C_{12})e^2 + \ldots$$

This deformation changes the tetragonal symmetry to the orthorhombic one but the calculated $\Delta E_u(e)$ curve is symmetric ($\Delta E_u(e) = \Delta E_u(-e)$). With regard to the energy expansion, the tetragonal shear modulus $C' = 0.5(C_{11} - C_{12})$ can be expressed as

$$C' = \frac{1}{4}\frac{\partial^2 \Delta E_u}{\partial e^2}.$$

The condition at Equation 1.14 corresponds to the shear instability related to the C_{44} modulus. Using the Jacobi matrix and strain as

$$\mathsf{J}_{C_{44}} = \begin{pmatrix} 1 & 0 & 0 \\ 0 & \dfrac{\sqrt{1+2e}+\sqrt{1-2e}}{2} & \dfrac{2e}{\sqrt{1+2e}+\sqrt{1-2e}} \\ 0 & \dfrac{2e}{\sqrt{1+2e}+\sqrt{1-2e}} & \dfrac{\sqrt{1+2e}+\sqrt{1-2e}}{2} \end{pmatrix}, \quad \hat{\eta}_{C_{44}} = \begin{pmatrix} 0 & 0 & 0 \\ 0 & 0 & e \\ 0 & e & 0 \end{pmatrix},$$

one obtains the dependence $\Delta E_u(e)$ leading to

$$\Delta E_u = 2C_{44}e^2 + \ldots$$

and

$$C_{44} = \frac{1}{4}\frac{\partial^2 \Delta E_u}{\partial e^2}.$$

The stability condition at Equation 1.15 can be tested by using

$$\mathsf{J}_{C_{66}} = \begin{pmatrix} \dfrac{\sqrt{1+2e}+\sqrt{1-2e}}{2} & \dfrac{2e}{\sqrt{1+2e}+\sqrt{1-2e}} & 0 \\ \dfrac{2e}{\sqrt{1+2e}+\sqrt{1-2e}} & \dfrac{\sqrt{1+2e}+\sqrt{1-2e}}{2} & 0 \\ 0 & 0 & 1 \end{pmatrix}, \quad \hat{\eta}_{C_{66}} = \begin{pmatrix} 0 & e & 0 \\ e & 0 & 0 \\ 0 & 0 & 0 \end{pmatrix}.$$

The corresponding energy change reads

$$\Delta E_u = 2C_{66}e^2 + \ldots$$

and the modulus

$$C_{66} = \frac{1}{4}\frac{\partial^2 \Delta E_u}{\partial e^2}.$$

Let us also show the stability conditions applied for a cubic structure (with 48 symmetry operations) under isotropic loading ($\sigma_1 = \sigma_2 = \sigma_3 = \sigma$). Here the form of the Wallace matrix (Equation 1.10) becomes symmetric:

$$\mathsf{B} = \begin{pmatrix} C_{11}+\sigma & C_{12}-\sigma & C_{12}-\sigma & 0 & 0 & 0 \\ C_{12}-\sigma & C_{11}+\sigma & C_{12}-\sigma & 0 & 0 & 0 \\ C_{12}-\sigma & C_{12}-\sigma & C_{11}+\sigma & 0 & 0 & 0 \\ 0 & 0 & 0 & C_{44}+\sigma & 0 & 0 \\ 0 & 0 & 0 & 0 & C_{44}+\sigma & 0 \\ 0 & 0 & 0 & 0 & 0 & C_{44}+\sigma \end{pmatrix}$$

and its determinant

$$\det|\mathsf{B}| = (C_{44}+\sigma)^3(C_{11}-C_{12}+2\sigma)^2(C_{11}+2C_{12}-\sigma)$$

can be broken down into to a set of stability conditions:

$$\begin{aligned} &1. \quad C_{11}+2C_{12}-\sigma > 0 \\ &2. \quad C_{11}-C_{12}+2\sigma > 0 \\ &3. \quad C_{44}+\sigma > 0. \end{aligned} \tag{1.16}$$

Stability criteria for a tetragonal system under biaxial (epitaxial) loading as well as those for the simplest case of an isotropic solid under hydrostatic loading can be found elsewhere [83].

Let us recall that, in addition to the violation of the mechanical stability conditions, some phonon (heterogeneous) instabilities may occur along the deformation path. A more detailed description of this problem lies beyond the scope of this book.

The currently used methodology for the IS calculation can be, finally, briefly summarized in the following points:

1. construction of a suitable empirical interatomic potential or calculation of the electronic structure;
2. calculation of the energy–strain curve and the related stress *vs* strain dependence for a specific deformation path;
3. establishment of elastic and phonon instability ranges on the strain *vs* energy and/or stress *vs* strain curves;
4. determination of IS value as a stress related to the first instability point on the energy *vs* strain curve.

1.1.3 Simple Loading Modes

1.1.3.1 Ideal Isotropic Strength

A nearly isotropic, triaxial tensile stress state ($\sigma_1 = \sigma_2 \approx 1.6\sigma_3$) builds up at the tip of cracks in solids stressed by uniaxial tension (e.g., [8]). The value of σ_{iht} expresses a resistance to a brittle fracture (cleavage or tearing at a sharp crack tip), while the value of τ_{is} reflects a defiance to a ductile response (blunting of the crack tip). Consequently, the ratio σ_{iht}/τ_{is} can be used as a measure of the brittle/ductile behaviour of cracked perfect crystals [23–25] (see also Section 1.2). This is why the search for isotropic IS values also becomes worthwhile from the engineering point of view. To our best knowledge, however, there are no available experimental data on σ_{iht}. The reason lies in difficulties in experimental realization of an isotropic tensile loading. Thus, a theoretical assessment remains to be the only applicable tool to gain such information. When the isotropic deformation is applied to a system, its volume changes but the symmetry remains unchanged (unless a phase transition takes place). Under such conditions, a rather simple LMTO-ASA method is particularly suitable for *ab initio* calculations (see Appendix A). Indeed, the error of ASA approximation is nearly independent of the volume.

Let us consider a cubic crystal under applied isotropic stress σ. A natural parameter for a description of deformation is the crystal volume V. The deviation from the equilibrium (unstressed) volume V_0 can be expressed by the relative (normalized) volume $v = V/V_0$. The isotropic stress can be simply derived from the dependence of the crystal energy U (see Figure 1.6) on the relative volume v as

$$\sigma_{iht} = \frac{1}{V_0}\frac{\mathrm{d}U}{\mathrm{d}v}.$$

The stress σ_{iht} reaches its maximum value when

$$\frac{\mathrm{d}^2U}{\mathrm{d}v^2} = 0, \tag{1.17}$$

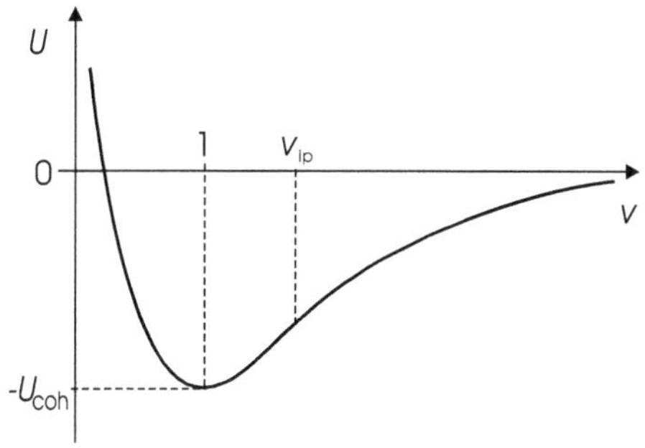

Figure 1.6 A schematic dependence of the crystal potential energy U on the relative volume v. The region with $v < 1$ corresponds to compression, $v > 1$ to tension and v_{ip} is assigned to the point of inflection. U_{coh} represents the cohesive energy of a crystal when a zero energy is related to a system of isolated atoms ($\lim_{v \to \infty} U = 0$)

Table 1.1 Theoretical isotropic strength σ_{iht} from *ab initio* and semi-empirical approximations

Element	Theoretical strength σ_{iht} (GPa)							
	ab initio results				Semi-empirical results			
	LMTO		VASP		Polynomial	Morse	Sinus	DVC
	[97]	[97]	[98]	[99]	[36]	[36]	[36]	[100]
Li bcc	3.53[a]	3.13[b]			5.06	2.49	4.91	1.92
C dia	66.1[a]	53.2[b]	88.5[a]	88.5[a]	84.7	69.7	138	
Na bcc	1.97[a]	1.55[b]			1.87	1.20	1.86	1.77
Al fcc	13.8[a]	12.0[b]		11.2[b]	22.2	11.9	23.0	
Si dia	15.0[a]	10.4[b]	15.5[a]	15.4[a]	15.1	13.7	28.2	
K bcc	0.955[a]	0.701[b]			0.99	0.659	1.28	0.10
V bcc	39.2[a]	33.2[b]	32.7[b]		32.6	23.5	45.4	38.3
Cr bcc	37.2[a]	21.0[b]			35.2	25.9	50.2	47.4
Fe bcc	37.7[a]	26.7[b]	27.7[b]	28.5[b]	33.8	24.1	48.1	
Ni fcc	39.5[a]	27.4[b]	28.9[b]	29.2[b]	44.7	26.9	51.2	
Cu fcc	28.8[a]	20.9[b]	19.8[b]	20.4[b]	32.7	19.9	38.4	
Ge dia	11.0[a]	6.46[b]	11.1[a]		11.3	10.1	21.4	
Nb bcc	36.3[a]	31.6[b]	31.6[b]		35.5	25.5	49.4	34.1
Mo bcc	49.3[a]	42.7[b]	42.9[b]	43.2[b]	48.2	35.0	68.9	42.2
Ag fcc	19.0[a]	12.6[b]		17.6[a]	20.2	13.7	26.7	
Ba bcc	2.69[a]	1.93[b]			2.36	1.64	3.17	
Ta bcc	41.3[a]	36.4[b]			39.2	28.5	55.1	41.3
W bcc	57.0[a]	50.6[b]	50.7[b]	50.2[b]	56.1	42.2	80.1	53.1
Ir fcc			40.1[b]		61.2	45.6	85.6	
Pt fcc	42.7[a]	33.6[b]	39.6[a]		48.5	35.1	68.0	
Au fcc	25.5[a]	17.6[b]	23.2[a]	23.5[a]	28.4	20.9	39.9	
Pb fcc	8.7[a]	6.98[b]			7.91	5.47	10.6	

[a] LDA
[b] GGA

i.e., at the inflection point of the $U(v)$ dependence ($v \to v_{ip}$). Considering the bulk modulus as a parameter expressing the elastic response of the crystal to the volume change as

$$B = \frac{\mathrm{d}\sigma_{iht}}{\mathrm{d}v} = \frac{1}{V_0}\frac{\mathrm{d}^2 U}{\mathrm{d}v^2}, \tag{1.18}$$

the volume at which the equality at Equation 1.17 is valid corresponds to vanishing of the bulk modulus. In other words, the first (volumetric) condition of a stability set at Equation 1.16 related to the isotropic loading of cubic crystals is fulfilled. Indeed, the bulk modulus is defined by the combination $C_{11} + 2C_{12}$ of elastic constants. In the further analysis, the crystal potential energy U will be substituted by a total energy E_{tot} that is evaluated from the electronic structure of investigated crystals. Thus, the first stability condition is related to the inflection point of the dependence of the total energy on the volume. The violation of the second stability condition corresponds to a shear instability (vanishing of the tetragonal shear modulus) when we can expect a bifurcation from the primary deformation (isotropic) path to a secondary one, where the lattice acquires tetragonal or orthorhombic symmetry. The third condition corresponds to another shear instability related to the trigonal shear modulus.

The values of σ_{iht} for various cubic crystals, as determined by *ab initio* approaches, are reported in our papers [60–62] and displayed in Table 1.1. The stability analysis, performed for crystals of Cu, Al, Ag, Fe, Ni and Cr, revealed that all these crystals fail under the volumetric instability [62]. This means that the values of the stress related to the point of inflection really correspond to σ_{iht}. The associated relative change of volume is about 1.5, which means that the relative elongation of the lattice parameter is about 1.15. In the case of aluminium crystal, however, the *ab initio* approach indicated a break in the trigonal shear stability before reaching the inflection point. Because this is in disagreement with results achieved by several other authors [80,94], the problem of the aluminium crystal is open for further investigations. For a majority of crystals, nevertheless, the values of σ_{iht} are higher than τ_{is} and σ_{iut}. Indeed, the values of τ_{is} are generally much lower since, to reach this value, the atoms in the shear plane need not be separated by higher distances. During uniaxial tension, the shear instabilities appear well before reaching σ_{iut} at the inflection point (see next subsections).

Table 1.2 Theoretical and experimental ideal strength data

Crystal	Ideal strength [GPa]												
	Isotropic		Uniaxial						Shear					
	σ_{iht}	Ref.	σ_{iut}	Direct.	Ref.	σ_{exp}	Direct.	Ref.	τ_{is}	Direct.	Ref.	τ_{exp}	Direct.	Ref.
Al (fcc)	11.0	[62]	9.2	⟨100⟩	[101]	2.27	bending	[33]	2.84	⟨112⟩	[99]			
			9.0	⟨111⟩	[101]				2.94	⟨112⟩	[59]			
			4.9	⟨110⟩	[101]				1.85	⟨112⟩	[85]			
			9.0	⟨100⟩	[102]				3.07	⟨112⟩	[102]			
			8.8	⟨111⟩	[102]				3.77	⟨110⟩	[103]			
			4.5	⟨110⟩	[102]									
Cu (fcc)	20.2	[62]	9.4	⟨001⟩	[68]	1.74	⟨001⟩	[33]	2.38	⟨112⟩	[59]	0.61	⟨011⟩	[30]
			9.3	⟨001⟩	[102]	1.59	⟨110⟩	[33]	3.77[b]	⟨011⟩	[103]			
			4.5	⟨110⟩	[102]	1.25	⟨111⟩	[30]						
			7.5	⟨111⟩	[102]	2.94	⟨111⟩	[33]						
Ni (fcc)	27.4	[61]	18.3	⟨001⟩	[102]				5.02	⟨112⟩	[59]			
			9.1	⟨110⟩	[102]				5.05	⟨112⟩	[99]			
			15.4	⟨111⟩	[102]									
Ag (fcc)	11.4	[62]				3.80	⟨001⟩	[30]	1.65	⟨112⟩	[99]	0.71	⟨011⟩	[30]
						1.73	⟨001⟩	[33]						
Au (fcc)	17.6	[97]							0.85	⟨112⟩	[99]			
									1.06	⟨112⟩	[59]			

[a] Inflection point
[b] Unrelaxed
[c] Semi-empirical

Table 1.2 (continued)

Crystal	Ideal strength [GPa]													
	Isotropic		Uniaxial						Shear					
	σ_{iht}	Ref.	σ_{iut}	Direct.	Ref.	σ_{exp}	Direct.	Ref.	τ_{is}	Direct.	Ref.	τ_{exp}	Direct.	Ref.
Na (bcc)	1.77	[63]							0.20	⟨112⟩{111}	[85]			
	2.18	[104]												
Fe (bcc)	26.7	[61]	14.2	⟨001⟩	[101]	13.1	⟨111⟩	[30]	7.51	⟨111⟩{112}	[99]	3.56	⟨111⟩{110}	[30]
			27.3[a]	⟨111⟩	[43]				8.14	⟨111⟩{110}	[99]			
Mo (bcc)	42.7	[97]	28.8[a]	⟨001⟩	[49]				15.18	⟨111⟩{110}	[99]			
									14.84	⟨111⟩{211}	[99]			
W (bcc)	53.1	[63]	28.9[a]	⟨001⟩	[41]	24.7	⟨001⟩	[23]	17.52	⟨111⟩{110}	[99]			
	57.4	[60]	40.1[a]	⟨111⟩	[41]				17.37	⟨111⟩{211}	[99]			
	50.6	[97]	54.3[a]	⟨110⟩	[41]									
			29.5[a]	⟨001⟩	[57]									
Cr (bcc)	21.0	[61]							20.5	⟨111⟩{112}	[105]			
Zn (hcp)									2.12	⟨111⟩{112}	[99]			
C (diam)			90.0[a]	⟨111⟩	[47]	20.7	(hex)	[30]	113.32	⟨110⟩{111}	[99]			
			95.0[a]	⟨111⟩	[48]	19.6	(hex)	[33]	99.60	⟨112⟩{111}	[106]			
			130.0[a]	⟨100⟩	[47]				93.0	⟨112⟩{111}	[48]			

[a] Inflection point
[b] Unrelaxed
[c] Semi-empirical

Table 1.2 (continued)

Crystal	Ideal strength [GPa]													
	Isotropic		Uniaxial						Shear					
	σ_{iht}	Ref.	σ_{iut}	Direct.	Ref.	σ_{exp}	Direct.	Ref.	τ_{is}	Direct.	Ref.	τ_{exp}	Direct.	Ref.
Ge (diam)			14.0[a]	⟨111⟩	[48]	3.82	(bend)	[33]	5.4 4.3	⟨112⟩{111} ⟨112⟩{111}	[106] [48]			
Si (diam)	26.4	[60]	22.0[a]	⟨111⟩	[48]	7.60	(bend)	[33]	8.6 6.8	⟨112⟩{111} ⟨112⟩{111}	[106] [48]			
Ni_3Al ($L1_2$)	28.1	[107]	17.5[a] 28.2[a]	⟨100⟩ ⟨111⟩	[108] [108]									
NiAl (B2)	25.6	[107]	46.0[a] 25.0[a]	⟨001⟩ ⟨111⟩	[109] [109]									
SiC (B3)			101.0[a] 50.8[a]	⟨001⟩ ⟨111⟩	[46] [46]				31.74	⟨110⟩{111}	[99]			
MgO (B1)	38.5[c]	[36]	37.0[c]	⟨100⟩	[36]	23.7	(bend)	[33]	17.09	⟨110⟩{110}	[99]			
Si_3N_4			55.0[a]	⟨100⟩	[83]	13.5		[33]	19	⟨0001⟩{1010}	[99]			

[a]Inflection point
[b]Unrelaxed
[c]Semi-empirical

It is to be expected that the values of σ_{ihc} could even be much higher than those of σ_{iht}. This was confirmed by a stability analysis of Ni, Cr and Fe crystals [58]. No instability was found in the compressive region up to 50% of unstressed volume V_0 in the cases of Ni and Cr which means that the values of σ_{ihc} must be, at least, several hundred GPa. The stress corresponding to the break of the third stability condition in Fe is 200 GPa. This is in agreement with previously predicted instability of bcc iron under pressure [110], at which the tetragonal meta-stable phase develops. As has been experimentally observed, however, that the bcc → hcp phase transition already starts at pressures of about 10–15 GPa [111]. This implies that this kind of instability cannot be revealed by mechanical stability analysis and, therefore, an identification of phonon instability ranges might be more helpful. However, we are not aware of any studies of this kind.

1.1.3.2 Ideal Tensile Strength

When a uniaxial tensile stress is applied to any cubic crystal, its symmetry becomes reduced. In the case of [001] loading, the crystal acquires a tetragonal symmetry with 16 symmetry operations. A model of two adjacent elementary fcc cells subjected to [001] loading is displayed in Figure 1.7.

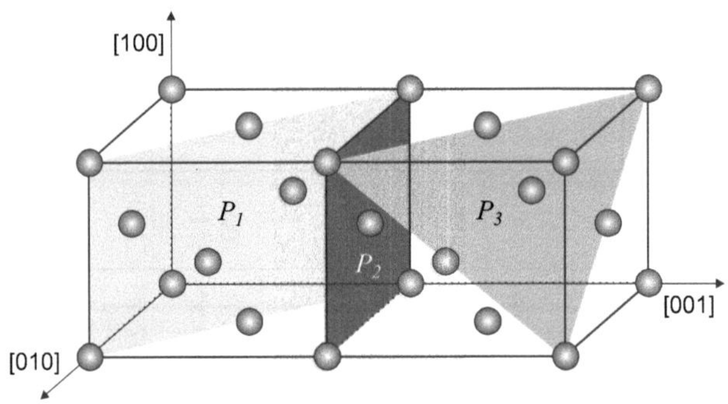

Figure 1.7 Illustration of fcc lattice extended in [001] direction with examples of one tear plane (P_2) and two possible shear (P_1 and P_3) planes

Experiments on whiskers [1,33] suggested that, under such particular loading, the lattice usually fails by the shear in the {111} plane (P_3 in Figure 1.7) in fcc systems or in the {110} plane (P_1 in Figure 1.7) in bcc systems rather than by the tear along the {001} plane (P_2 in Figure 1.7) – see also Section 1.1.4.3. The latter fracture process was assumed not only in the very first calculations of tensile strength [1] but even in a majority of later studies (e.g., [39, 41, 74, 112]). The elastic matrix of the tetragonal system contains six independent components:

$$\mathsf{C} = \begin{pmatrix} C_{11} & C_{12} & C_{13} & 0 & 0 & 0 \\ C_{12} & C_{11} & C_{13} & 0 & 0 & 0 \\ C_{13} & C_{13} & C_{33} & 0 & 0 & 0 \\ 0 & 0 & 0 & C_{44} & 0 & 0 \\ 0 & 0 & 0 & 0 & C_{44} & 0 \\ 0 & 0 & 0 & 0 & 0 & C_{66} \end{pmatrix}$$

and the corresponding stability analysis is presented in [68, 78, 113], particularly in Section 1.1.2. For example, the copper crystal was found to fail in shear at the tensile stress of 9.4 GPa, well before the volumetric instability at 24.3 GPa. This result is in good agreement with the calculations of Milstein and Chantasiriwan [78]. Unfortunately, practically all the previously calculated σ_{iut} values of different crystals are related to the volumetric instability (see Table 1.2) and, therefore, they substantially exceed the real ones. Moreover, the older studies did not include the atomic relaxation procedures enabling the transverse Poisson's contraction. Nevertheless, a mutual comparison of those values can serve as an orientation ordering of various crystals according to their uniaxial IS.

As can be seen in Table 1.2, there is a three-order of magnitude difference between the strongest and the weakest crystal. As expected, the highest IS value belongs to the diamond crystal with pure covalent bonds and the lowest values to the Van der Waals crystals of inert gases that are stable only in the low temperature range. High values of σ_{iut} of metals as W, Mo and Fe or intermetallics such as Ni_3Al and AlN are also understandable due to a high portion of covalent interatomic bonds in these crystals. It should be emphasized that this ranking of σ_{iut} is, more or less, compatible with that of ultimate strength σ_u for real crystals and engineering materials. This means that the intrinsic (matrix) strength properties predetermine the ultimate strength of the engineering materials in a rather significant manner.

1.1.3.3 Ideal Shear Strength

Unlike tensile and compressive deformation modes, the pure shear deformation process does not require any elastic stability assessments because it is expected to maintain, more or less, the original shear path. On the other hand, the symmetry of the crystal lattice is almost disturbed by the shear shifts and, therefore, the calculation procedures become rather cumbersome. Moreover, the full relaxation of atomic positions near the shear plane, allowing them to follow configurations of lowest total crystal energies along the whole shear trajectory, are essential for obtaining sufficiently precise values of τ_{is} [49, 99, 103].

A great majority of analyses published before the year 2000 did not utilize the relaxation procedure. From all these results let us mention only a rather general study made on fcc, bcc, diamond, B1 and B2 structures. In this study, the block-like shear was simulated by using both central and non-

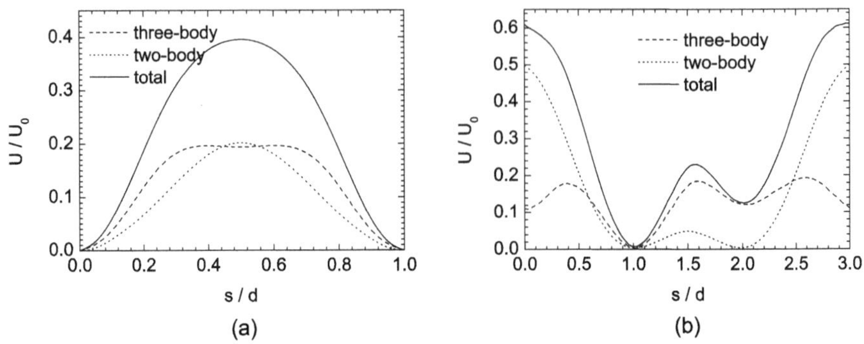

Figure 1.8 Shear potentials in the slip system (a) {111}⟨110⟩, (b) {111}⟨112⟩, non-central forces. U_0 is energy at the equilibrium (unstressed) position in utilized semi-empirical potentials

central empirical potentials [31]. Shapes of the potential energy along shear trajectories in fcc crystallographic systems $\langle\bar{1}10\rangle\{111\}$ and $\langle\bar{2}11\rangle\{111\}$, as computed using non-central potentials, are displayed in Figure 1.8. It can be seen that, in spite of the shallow local minimum created by the large non-central component (three-body), the total energy shape in Figure 1.8(a) is of a sinusoidal type, as originally presumed by Frenkel. Another investigated shear system in Figure 1.8(b) exhibits a secondary energy minimum that is related to the stable stacking fault. These results are in contradiction to the already mentioned Mackenzie theory that assumes an existence of a shallow minimum on the total energy curve in the range $s/d \in (1, 2)$. This is one of the main reasons why the calculated values of τ_{is} approach those already reported by Frenkel (see Table 1.3).

Table 1.3 Ideal strength in simple shear direction (a is the lattice parameter)

Lattice	Direction	Plane	d/a	b/a	τ_{id}/G
bcc	⟨111⟩	{110}	$\sqrt{3}/2$	$1/\sqrt{2}$	0.1949
fcc	⟨110⟩	{111}	$1/\sqrt{2}$	$1/\sqrt{3}$	0.1949
fcc	⟨112⟩	{111}	$1/\sqrt{6}$	$1/\sqrt{3}$	0.1025
B1	⟨110⟩	{110}	$1/\sqrt{2}$	$1/\sqrt{8}$	0.3183
B2	⟨001⟩	{110}	1	$1/\sqrt{2}$	0.2251
diam	⟨110⟩	{111}	$1/\sqrt{2}$	$\sqrt{3}/4$	0.2599

In most *ab initio* studies, a uniform shear of a perfect crystal was simulated, as is schematically depicted in Figure 1.3. Moreover, the first principles codes enable minimization of all stress tensor components, except for the resolved shear stress, to approximately zero values by appropriate shifting of atoms not only in the direction perpendicular to the shear plane, but also within this plane (so-called in-plane relaxation). This can be done by using the procedure based on the Hellman–Feynman forces, which is part of some *ab initio* program codes (VASP, Abinit). Consequently, the resulting (optimal) shear trajectory is usually slightly different from that prescribed by crystallography.

1.1.3.4 Comparison of Theoretical and Experimental Results

Because of the long history of IS calculations, a number of theoretical results concerning crystals of pure elements and compounds are available in the literature. On the other hand, the experimental data are rather limited which is related to problems associated with both specimen preparation and experimental arrangement. It is particularly difficult to measure the values of σ_{iht} and τ_{is}, although, for measuring the latter, nanoindentation appears as a good tool today (see Section 1.3).

Theoretical and experimental values of σ_{iht}, σ_{iut} and τ_{is} for various crystals are presented in Table 1.2. The lowest values of σ_{iut} for a particular crystal have always been found in calculations with full relaxations that also take the mechanical stability conditions into account. It can be seen that theoretical and experimental values can differ substantially according to the applied computational or experimental procedure. Nevertheless, the *ab initio* methods can be considered as the most reliable ones. It should be emphasized that a majority of σ_{iut} values correspond to the inflection point of the energy *vs* strain curve when omitting the stability analysis. Hence, most of such calculated values are substantially overestimated.

Rather rare experiments provided τ_{is} values of about 2–3 times lower than those calculated by means of the *ab initio* methods. Older experimental data on τ_{is} were obtained by recalculating the results of uniaxial tensile testing of whiskers by omitting the influence of the normal stress component (see Section 1.1.4 for more details). When taking this influence into account, the maximum resolved shear stresses at fracture are still substantially lower than the theoretical ones, but already of the same order of magnitude. Note that the most sophisticated methods based on nanoindentation tests render data almost in accordance with the recent theory (see Section 1.3 for more details). This is a very encouraging message showing that, at least in the case of τ_{is}, the advanced theoretical and experimental methods start to agree.

In the case of metals, a majority of experimental σ_{iut} data obtained on perfect large single crystals or whiskers are of an order of magnitude lower than those calculated even by means of *ab initio* methods considering the mechanical or phonon shear instabilities. This might be due to the dislocation-assisted shear instability controlling the final failure process. On the other hand, a very high Peierls–Nabarro stress in covalent and complex ionic ceramic crystals resists the nucleation and motion of dislocations. Indeed, for such crystals the difference between theory and experiment is relatively small. Probably the highest ever reported experimental value $\sigma_{iut} = 40\,\text{GPa}$ ($\approx E/20$) is for the ZnO whiskers [33]. However, experiments performed on tungsten whiskers by Mikhailovskii *et al.* [114] also approach the calculated results. Anyway, the discrepancy between theory and experiment for metallic crystals becomes much smaller when the mechanical and phonon stability conditions are taken into account (e.g., for Cu [68] and Al [101] crystals), though it still remains significant. It should be noted that the stability conditions imply that the loading mechanism should be able to maintain the Cauchy stress and the crystal symmetry along the whole deformation path. It particularly means that the shear instability $\langle\bar{2}11\rangle\{111\}$ in the case of [001] tension of the fcc crystal is not controlled by the stability conditions, although it is a usually observed failure mode in experiments. Indeed, this instability requires a finite shear in the $\{111\}$ plane changing the symmetry to triclinic or monoclinic. Additionally, a possible resonance of short-wavelength phonons should be considered. Such effects are clearly beyond the description supplied by the mechanical stability conditions that are based only on continuum mechanics. Nevertheless, the difference between theory and experiment can be understood particularly in terms of various imperfections of experimental procedure and the crystal defects.

Anyway, a lot of work is needed in both theoretical and experimental investigations of IS. Let us finally note that the ultimate strength of currently used ultra-high-strength steels still exhibits only a rather small fraction of the IS of the iron crystal. From the theoretical point of view, however, materials of an extreme dislocation density could, in principle, achieve more than half of the strength level of σ_{iut} [20]. This also holds for amorphous solids.

1.1.4 Multiaxial Loading

Although there is no doubt that crystals and whiskers used in industrial exploitation are usually subjected to multiaxial loading, little attention has been paid to the coupling of various stress tensor components. Another example of the practical importance is associated with reinforcing fibres used in engineering composite materials. Due to the matrix/reinforcement incompatibility strains, the single crystal fibres (or whiskers) are subjected to multiaxial loading even in the case of remote uniaxial tension of the composite. Consequently, *ab initio* modelling of the uniaxial tension of perfect crystals under superimposed tensile or compressive biaxial stresses is expected to be a reasonable theoretical simulation of the stress-strain behaviour of composite fibres (see Section 1.1.5). Several studies were also focused on the influence of superimposed hydrostatic or normal stress on the shear IS [103, 115, 116]. The results reveal, as a rule, an increase in the shear strength with increasing compressive normal stress. Such analyses allow us, for example, to predict the very onset of plastic deformation under the nanoindenter (see Section 1.3).

1.1.4.1 Coupling of Uniaxial and Transverse Biaxial Stresses

In order to describe the influence of superimposed transversal biaxial stress on the tensile IS of metallic fibres in composites, bulk systems of several metallic and covalent ideal crystals subjected to multiaxial loading were studied in [117]. The calculation procedure based on the VASP code was applied to ten cubic crystals (V, Fe, Ni, Cu, Nb, Mo, W, Ir, Pt and Au) and three crystals of a diamond structure (C, Si and Ge). All studied crystalline systems were subjected to the uniaxial tensile stress σ_{uni} along [100] direction combined with the transverse biaxial stress σ_{bi} in the (100) plane (see Figure 1.9). A relaxation procedure based on the Hellman–Feynman forces in (100) plane was applied in order to get the stress tensor in a simple form as

$$\hat{\sigma} = \begin{pmatrix} \sigma_{uni} & 0 & 0 \\ 0 & \sigma_{bi} & 0 \\ 0 & 0 & \sigma_{bi} \end{pmatrix}.$$

The relaxation process consisted of the following steps:

1. The crystal was subjected to σ_{bi} of a certain preset value.
2. The crystal was incrementally elongated in the [100] direction and the value of σ_{bi} was converged to the same preset value for any elongation.
3. The stress maximum $\sigma_{\max}$ and the related strain $\varepsilon_{\max}$ were found by a cubic spline interpolation of computed σ_{uni} values. If no other instability precedes, the $\sigma_{\max}$ value can be considered to be the theoretical tensile strength σ_{iut} under corresponding superimposed biaxial stress.
4. The relaxation procedure was repeated for several preset σ_{bi} values.

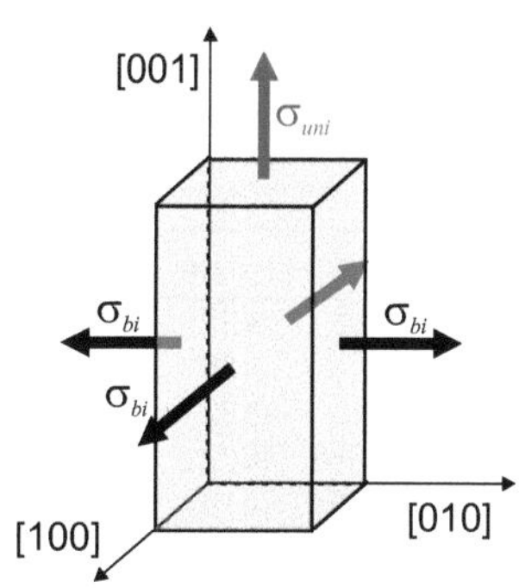

Figure 1.9 Illustration of a triaxial stress state that comprises a tensile stress in [001] direction σ_{uni} and superimposed transverse biaxial stresses σ_{bi}

The tensile strength of a majority of crystals increases almost linearly with the applied biaxial stresses (see Figure 1.10). The linear dependencies can be parametrized as

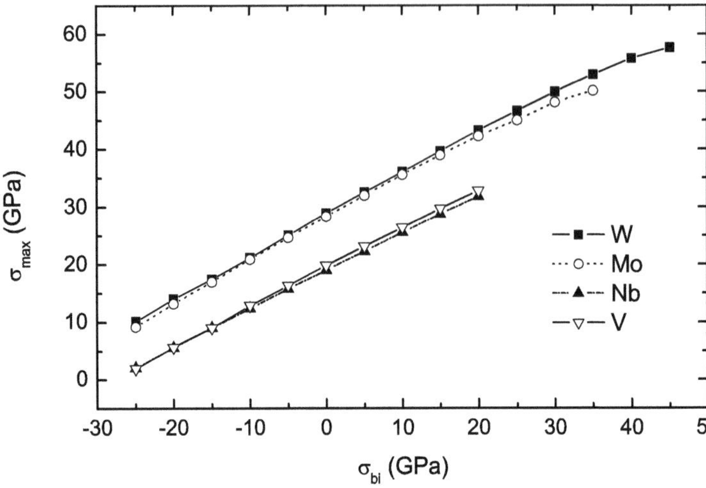

Figure 1.10 Dependence of the theoretical tensile strength $\sigma_{\max}$ on perpendicularly acting biaxial stresses σ_{bi} for W, Mo, Nb and V

$$\sigma_{\max} = \sigma_{\max,0} + k_{\max}\sigma_{bi}, \tag{1.19}$$

where $k_{\max}$ is the slope expressing the influence of the superimposed biaxial stress σ_{bi}. The parameters $k_{\max}$ are collected in Table 1.4. Values of the slope $k_{\max}$ seem to be higher for bcc crystals than for fcc ones.

Table 1.4 Slope of the linear regression lines $k = \dfrac{d\sigma_{\max}}{d\sigma_{bi}}$, maximum stress $\sigma_{\max,0}$ (GPa) and the ultimate strain $\epsilon_{\max,0}$ under pure uniaxial loading along with available literature data on σ_{iut} and computed values of theoretical isotropic strength σ_{iht} (GPa)

	$\sigma_{\max,0}$	$k_{\max}$	$\varepsilon_{\max,0}$	σ_{iut}	σ_{iht}
C	225	−1.08	0.37	225	88.5
Si	26.3		0.26		15.5
Ge	16.8		0.23		11.1
V	19.8	0.688	0.22		32.7
Fe	12.4	0.634	0.16	14.2	27.7
Ni	35.2		0.37	39.0	28.9
Cu	24.1		0.36	23.7	19.8
Nb	19.0	0.662	0.11	18.8	31.6
Mo	28.3	0.737	0.12	28.8	42.9
W	28.9	0.739	0.16	28.9	50.7
Ir	44.5	0.281	0.25		40.1
Pt	34.1	0.152	0.34		39.6
Au	18.6	0.163	0.33	22.5	23.2

A highly interesting exception of the studied elements is carbon with a diamond structure in which tensile strength decreases with increasing σ_{bi}. Thus, it exhibits a negative value of the slope $k_{\max} = -1.08$. An anomalous behaviour was also revealed in crystals of Si and Ge with a diamond structure and, to a certain extent, also in crystals of Ni and Cu. Both tensile and compressive applied biaxial stresses substantially reduce the σ_{iut}-values of Si and Ge (and less so those of Ni and Cu). Consequently, the $\sigma_{\max}$ *vs* σ_{bi} curves for Si and Ge exhibit sharp maxima near the zero applied σ_{bi} (see Figure 1.11) whereas the maximum of the diamond crystal is shifted by about 120 GPa towards the compressive region. This shift is a reason for the negative value of the slope $k_{\max}$ of the diamond crystal in the region around the zero value of σ_{bi}.

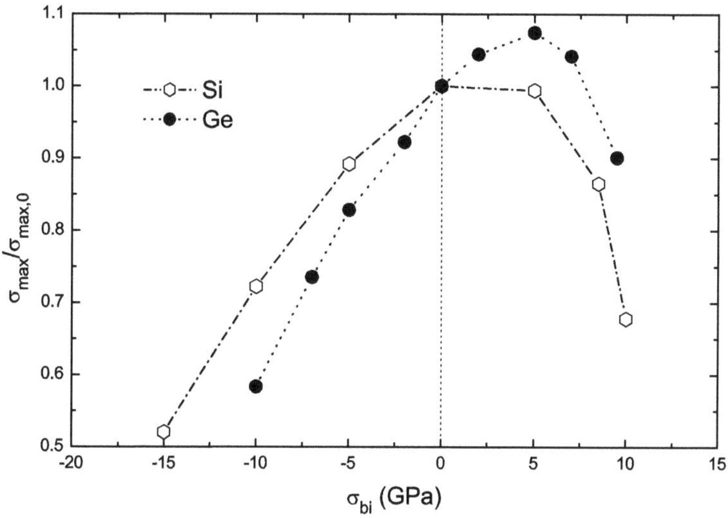

Figure 1.11 Normalized values of the theoretical tensile strength $\sigma_{\max}/\sigma_{\max,0}$ as functions of superimposed biaxial stress σ_{bi} for Si and Ge

One of practical consequences of the above-mentioned results is that the reinforcing fibres, subjected to tensile transverse stresses induced by incompatibility strains, can exhibit higher strength than those tested under uniaxial tension. In this way, a synergy effect can appear in composites (see Section 1.1.5).

Results concerning the dependence of the ultimate strain ε_u at the inflection point on the applied biaxial transverse stress were also obtained in the frame of the *ab initio* study. This dependence is depicted in Figure 1.12 for selected crystals. In general, the ultimate strain of fcc metals decreases with increasing biaxial stress (see the curves for Cu, Ni and Pt) whereas the opposite trend refers to bcc metals (Fe, Mo and V). The curves for diamond structures C, Ge and Si exhibit maxima near the zero biaxial stress. The average ductility of fcc metals is higher than that of bcc crystals but, rather surprisingly, it is comparable to that of diamond crystals. This result shows that the behaviour of perfect crystals in terms of the ultimate strain is qualitatively different from that of engineering materials. In the latter case, indeed, the increasing biaxial stress always leads to a decrease of the ductility. This effect can be attributed to a plasticity driven growth and coalescence of microvoids that initiate at secondary phase particles (see Section 2.3). Thus, the lattice imperfections play a decisive role in the ductility response of engineering materials.

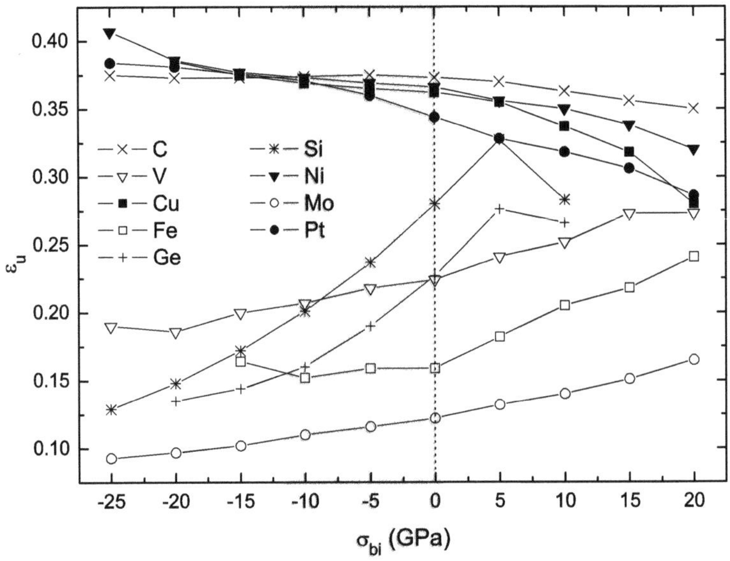

Figure 1.12 The ultimate uniaxial strain as a function of the applied transverse biaxial stress for selected perfect crystals

1.1.4.2 Coupling of Normal and Shear Stresses

Another important example of stress coupling is the influence of the stress component acting perpendicularly to slip planes during simple shear, as displayed in Figure 1.13 for a particular case of $\langle\bar{2}11\rangle\{111\}$ shear in fcc lattice. Results of such analyses are presented in [59, 103, 106, 115, 116] for selected metallic, intermetallic and ceramic crystals. It should be noted that the relaxation procedure used in [116] comprised just a relaxation of atomic positions in the direction perpendicular to the slip planes but did not include the relax-

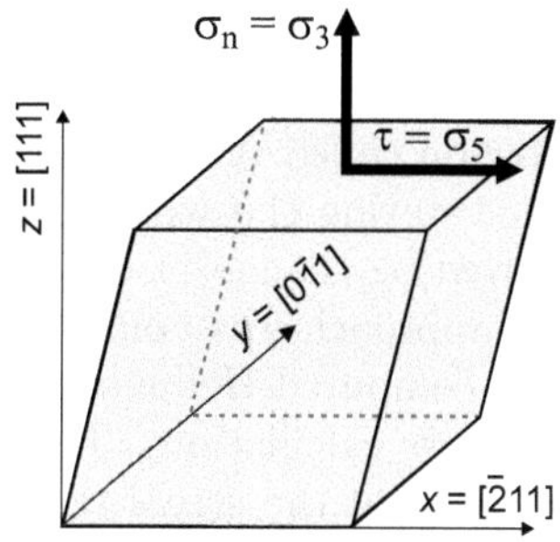

Figure 1.13 A crystal under superposition of shear and normal stresses

ation of ionic positions within the planes (in-plane relaxation). On the other hand, the in-plane relaxation was added to procedures used in [59, 103, 115].

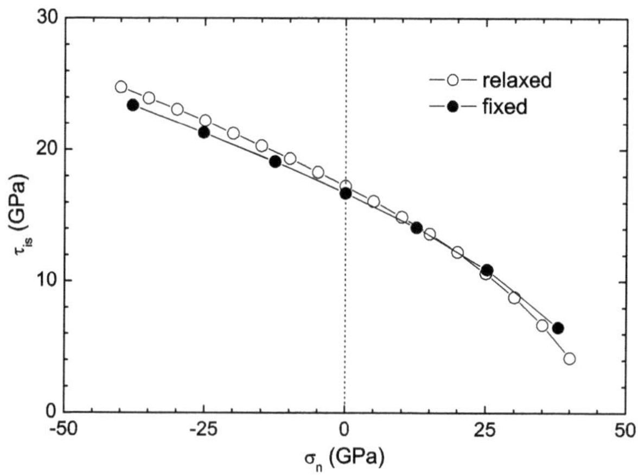

Figure 1.14 Influence of normal stress σ_n on the shear strength τ_{is} in Ir with and without in-plane relaxation

An example of results obtained for the crystal of iridium is shown in Figure 1.14. In general, the compressive normal force can substantially increase the shear strength. In the region of compressive normal stresses, the coupling effect can also be expressed by the linear function

$$\tau_{is} = \tau_{is,0} - b\sigma_n \tag{1.20}$$

in a formal analogy to the well known relation between the friction and the normal force exerted on sliding surfaces. In the region of tensile stresses, the coupling effect, at least for some crystals, changes to a parabolic shape (see Figure 1.14). This is in agreement with studies performed by Kelly *et al.* by means of empirical potentials [23].

In a more recent paper [118] six fcc crystals were subjected to homogeneous shear deformations in the {111} slip system in two distinct ways. In the first approach (from now on called the rigid-planes approach), the shear planes were kept undistorted during the whole shear process. Only the interplanar distance was allowed to change in order to set the normal stress to a prescribed value. This approach was consistent with previous calculations [23, 116]. However, the computed Hellman–Feynman stress tensor was used here instead of total energy calculations. The other approach (relaxed-planes) utilized a full relaxation of the stress tensor (including possible in-plane stresses). The details of the related computational procedure can be found in [59]. In both approaches, the main attention was paid to the tensile region of normal stresses. The homogeneous shear was simulated using a one-atom simulation cell.

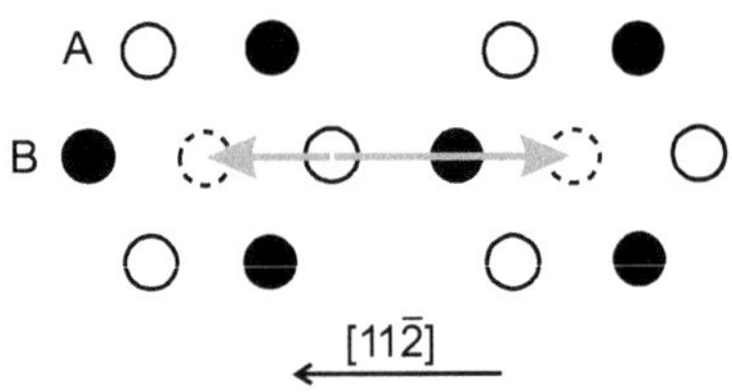

Figure 1.15 Two adjacent {111} A (*open circles*) and B (*solid circles*) planes in fcc crystals for illustration of the ⟨112⟩{111} shear system

The shear system studied is illustrated in Figure 1.15. For the sake of clarity, only two adjacent planes are displayed. When the upper plane A moves to the right, its atoms must overcome a high-energy barrier related to over-passing the atoms in the B plane. The final position of the selected atom in plane A is marked by the dashed circle. The corresponding structure has fcc symmetry of an opposite stacking order (with respect to the original state). The same state can be reached by moving the A plane to the left. In this case, the corresponding energy barriers as well as the related stresses are substantially lower. Although the plane shift cannot then continue the same path (because of the consequent higher energy barrier), the instant of approaching the shear strength is the first onset of instability. For the calculations of Hellman–Feynman stress tensor the Vienna *Ab initio* Simulation Package (VASP) was utilized. In the case of Ni, the projector augmented-wave potential was used along with the spin-polarized calculations taking the ferromagnetic ordering into account. The exchange-correlation energy was evaluated using either the local density approximation (Pt, Au) or the generalized-gradient approximation (Al, Ni, Cu, Ir). The functions $\tau_{is}(\sigma_n)$ were found to be almost linearly decreasing and could also be expressed by Equation 1.20. The regression parameters are collected in Table 1.5. When comparing the computed data with the previously published results [116] one can see good agreement in $\tau_{is,0}$ values while more significant differences can be found in b values. They are probably caused not only by a different assessment but also by a differently selected range of interpolated data with respect to normal stresses. Nevertheless, all the b values match the previous results within 15%.

Table 1.5 Regression parameters for {111} shear strength in both the rigid-planes and the relaxed-planes approaches

Element	Rigid-planes		Relaxed-planes		
	$\tau_{is,0}$ (GPa)	b	$\tau_{is,0}$ (GPa)	b	$\tau_{is,0}$[a] (GPa)
Al	3.12	0.238	3.07	0.319	2.84
Ni	5.64	0.139	5.05	0.123	5.05
Cu	3.01	0.117	2.43	0.080	2.16
Ir	17.1	0.223	17.3	0.249	
Pt	2.75	0.138	2.05	0.177	
Au	1.66	0.152	1.05	0.171	0.85

[a] [99]

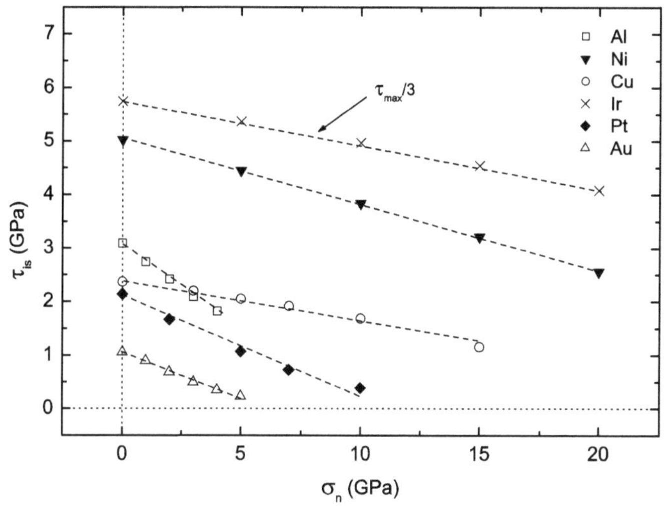

Figure 1.16 Theoretical shear strength as a function of normal stress in the relaxed-planes approach. *Dashed lines* represent linear regressions of the displayed data points

The results of relaxed-planes calculations are displayed in Figure 1.16. Again, the $\tau_{is}(\sigma_n)$ curves can be approximated by linear functions up to $\sigma_n = 20\,\text{GPa}$. The regression parameters are also included in Table 1.5. By comparing both approaches one can see that the full relaxation of stresses lowers the shear strength of all investigated crystals except for iridium.

Crystals of metals and diamond exhibit a nearly linear decrease in the ideal shear strength with increasing superimposed isotropic (hydrostatic) stress σ_h [55,119]. This behaviour seems to be closely related to the fact that, during a pure shear ($\sigma_h = 0$), the lattice volume of these crystals increases, thus minimizing the crystal energy. An application of the tensile isotropic stress works, therefore, as a shear driving force leading to a reduction of the ideal shear strength. On the other hand, the lattice volume of covalent crystals Si, Ge and SiC decreases during the pure shear so that the dependence $\tau_{is}(\sigma_h)$ exhibits an opposite trend [106]. This reasoning can be, to a large extent, also applied to the $\tau_{is}(\sigma_n)$ dependence. However, the ideal shear strength of Cu and Ni crystals is lowered by both tensile and compressive normal stresses [55,119]. This anomalous behaviour can be understood from the crystal relaxation during the shearing process. Under the pure shear deformation, these crystals prefer to extend the shear planes perpendicularly to the shear direction rather than increase their interplanar distance. Consequently, an application of additional compressive normal stresses contributes to further lateral extension that makes the shear easier [55,59,120].

1.1.4.3 Calculation of Tensile Strength from Shear Strength Data

The application of mechanical-stability criteria to the prediction of ideal tensile strength is computationally very time-consuming. These criteria can be interpreted approximately by assuming that a rupture of perfect crystals under uniaxial tension starts when reaching the shear strength in some convenient shear system [102]. An illustration of such a shear system in a crystal sample is given in Figure 1.17.

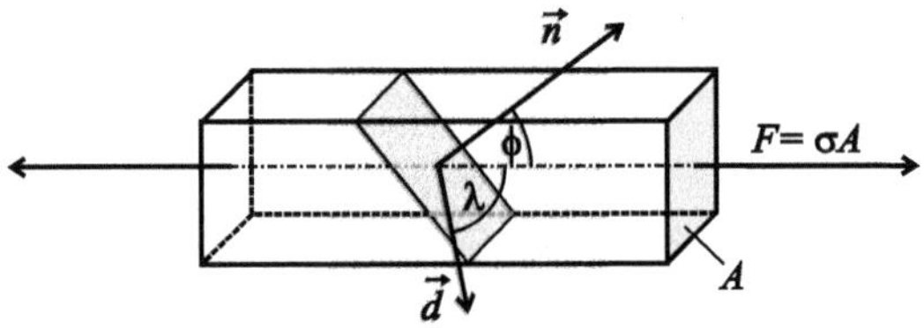

Figure 1.17 Illustration of a shear system in a crystal sample under tensile stress. The angles ϕ and λ are measured between the crystal axis and the normal **n** of the slip plane and the slip direction **d**, respectively. Reprinted with permission from Institute of Physics and IOP Publishing Ltd. (see page 265)

When the crystal is subjected to tensile stress, certain slip systems can be exposed to a combination of shear and tensile (normal to the shear plane) stresses. The displayed vectors **n** and **d** determine the vertical to the shear plane and the shear direction, respectively. The normal stress σ_n can be expressed by means of the tensile stress σ and the angle ϕ as

$$\sigma_n = \sigma \cos\phi. \tag{1.21}$$

Assuming that some shear instability can precede the volumetric instability, the ideal tensile strength σ_{iut} can be estimated from the corresponding theoretical shear strength τ_{is} using the relationship

$$\sigma_{iut,s} = \frac{\tau_{is}}{\cos\phi\cos\lambda}. \tag{1.22}$$

The relation is similar to the well known Schmid's law

$$\tau_c = \sigma_y \cos\phi\cos\lambda \tag{1.23}$$

that expresses the relationship between the critical resolved shear stress τ_c required to move dislocations across the slip plane and the yield stress σ_y in a crystal with dislocations. However, Equation 1.22 holds only for a perfect single crystal. Another significant difference between Equations 1.22 and 1.23 lies in the influence of the normal stress on the critical shear stress. Whilst, in Schmid's law, the critical shear stress does not depend on the normal stress σ_n (at least in fcc crystals), the significant influence of σ_n on the ideal shear strength has already been clearly demonstrated. The latter result can be used for a simple estimate of σ_{iut} for fcc crystals Al, Ni, Cu, Ir, Pt and Au.

In order to estimate the theoretical tensile strength σ_{iut}, Equations 1.20, 1.21 and 1.22 can be combined into the final form

$$\sigma_{iut,s} = \frac{\tau_{is,0}}{\cos\phi(\cos\lambda + b\cos\phi)} \tag{1.24}$$

when assuming $\sigma = \sigma_{iut,s}$ in Equation 1.21. Equation 1.24 can easily be used for an estimation of the ideal tensile strength from the ideal shear strength $\tau_{is,0}$.

The $\sigma_{iut,s}$ values obtained for uniaxial tension in [110], [11$\bar{1}$] and [100] directions are listed in Tables 1.6 and 1.7. Uniaxial tension was applied to the crystal in the most favourable representative of the family of symmetry-equivalent directions $\langle 110\rangle$, $\langle 111\rangle$ and $\langle 100\rangle$. The σ_{iut} values, that were collected from available literature, represent the corresponding values of the volumetric-instability stress (at the inflection point). It can be seen that the predicted $\sigma_{iut,s}$ values for [100] and [11$\bar{1}$] directions are substantially lower than the corresponding σ_{iut} values for all studied fcc crystals with the exception of Ir. On the other hand, the σ_{iut} values in [110] direction are so low that the predicted $\sigma_{iut,s}$ values are of a comparable magnitude. In the case of [100] direction, the predicted $\sigma_{iut,s}$ values for Cu, Ni and Al can be well compared with ideal strength σ^*_{iut} corresponding to the first onset of the mechanical instability as predicted in [78,98] or the phonon instability as reported in [101].

Table 1.6 The estimated theoretical tensile strengths $\sigma_{iut,s}$ in $\langle 110\rangle$ and $\langle 111\rangle$ directions (in GPa) along with the available literature data for σ_{iut} (volumetric instability)

Element	$\langle 110\rangle$ loading			$\langle 111\rangle$ loading		
	$\sigma_{iut,s}^{\text{rigid}}$ [GPa]	$\sigma_{iut,s}^{\text{relaxed}}$ [GPa]	σ_{iut} [GPa]	$\sigma_{iut,s}^{\text{rigid}}$ [GPa]	$\sigma_{iut,s}^{\text{relaxed}}$ [GPa]	σ_{iut} [GPa]
Al	5.0	4.5	4.2[a]	9.2	8.8	14.8[a]
Ni	10.0	9.1	11.7[a]	17.1	15.4	39.3[a]
Cu	5.5	4.6	5.5[a]	9.2	7.5	26.5[a]
Ir	27.6	27.1	26.5[b]	50.4	50.6	43.5[b]
Pt	4.9	3.5		8.3	6.1	30.0[b]
Au	2.9	1.8	2.8[a]	5.0	3.2	13.6[a]

[a] [78]
[b] [102]

Table 1.7 The estimated theoretical tensile strengths $\sigma_{iut,s}$ in $\langle 100\rangle$ direction along with the available literature data for the ideal strength σ_{iut} (volumetric-instability) and the ideal strength σ_{iut}^{*} corresponding to the first onset of the shear instability

Element	$\sigma_{iut,s}^{\text{rigid}}$ [GPa]	$\sigma_{iut,s}^{\text{relaxed}}$ [GPa]	σ_{iut}^{*} [GPa]	σ_{iut} [GPa]	
Al	9.9	9.0	11.1[a]	12.6[a]	
Ni	20.0	18.3	21.3[a]	39.0[a]	35.2[b]
Cu	11.0	9.3	9.8[a]	23.7[a]	24.1[b]
Ir	55.2	54.3		44.5[b]	
Pt	9.8	7.0		34.1[b]	
Au	5.8	3.6	10.0[a]	22.5[a]	18.6[b]

[a] [78]
[b] [98]

A good agreement of results predicted from shear strength data with those obtained by including the stability analysis also seems to be promising for an application of the proposed method to crystals other than fcc.

1.1.5 Nanocomposites

Composites represent a widely used successful way to improve mechanical characteristics of materials, in particular their elastic moduli, strength and fracture toughness. Resulting properties in real engineering macro-composites are functions of many parameters depending on the particular design and production technology. However, the situation is not so complicated for all mechanical characteristics. For example, the Young's modulus of long-fibre composites in the direction of the fibre can be sufficiently precisely assessed according to the following simple relationship:

$$E = \eta V_f E_f + (1 - V_f) E_m, \tag{1.25}$$

where E_f and E_m are Young's moduli of the fibre and the matrix, respectively, V_f is the volume fraction of fibres and η is the factor depending on the strength of cohesion of the fibre/matrix interface [121]. Obviously $\eta = 0$ when there is no cohesion at all, while $\eta = 1$ corresponds to the ideal binding between atoms of matrix and fibres. The value $\eta \approx 0.8$ is fulfilled for a majority of real composites of a standard quality. Similar simple linear relationships are also often used for other moduli as well as for the strength of composites.

Values of the equilibrium volume, elastic moduli and σ_{iut} of ideal nanocomposites were computed in [122–124] for different thicknesses of nanofibres from a single atom to several atomic distances. The fibres were made of W or Mo, whereas the matrix consisted of V or Nb atoms. In these first principles studies, the uniaxial tensile loading was applied parallel to the nanofibres. The main aim was to verify the validity of Equation 1.25 for elastic moduli and σ_{iut} in the ideal case of $\eta = 1$.

1.1.5.1 Calculation Method

A model of the nanocomposite was built up as a periodic repeating of $4 \times 4 \times 1$ bcc-based super-cell which is displayed in Figure 1.18 for the Nb–W composite. The crystal super-cell contained 32 atoms in both A (solid circles) and B (open circles) (001) planes. The grey solid circles in Figure 1.18 belong to other (adjacent) super-cells. The dashed contours define interfaces between the wires (W) and the matrix (Nb) in investigated lamina models of different concentrations (percentages) of wire atoms. In order to get a few more different concentrations, the systems Nb (wires) and W (matrices) were also studied. In this way, additional wire concentrations of 59.4%, 71.9% and 84.4% could be obtained.

The calculation procedure consisted of several steps. The first step was a computation of the total energy E_{tot} per atom as a function of the atomic volume V. Atomic positions within the cell had to be relaxed in order to minimize the interfacial stresses. The equilibrium volume per atom V_0 of each lamina model was obtained from the energy minimum and the lamina bulk modulus B could be determined according to Equation 1.18 ($U = E_{tot}$). Then the crystal system was elongated in the [001] direction to simulate uniaxial loading applied in the direction parallel to the lamina fibres. The system energy was minimized at any elongation by a full relaxation procedure regarding the atomic positions within the super-cell. The dependence of the total energy obtained on the relative elongation ε allowed authors to compute

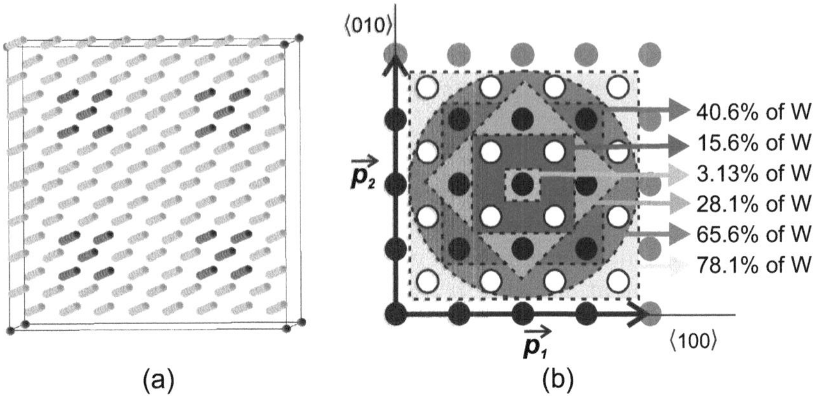

Figure 1.18 A sample of: (a) the nanocomposite, and (b) the super-cell

the values of the uniaxial stress σ applied to the system as well as the Young's modulus E_{001} value:

$$\sigma = \frac{1}{V}\frac{\mathrm{d}E_{tot}}{\mathrm{d}\varepsilon}$$

and

$$E_{001} = \frac{1}{V_0}\frac{\mathrm{d}^2 E_{tot}}{\mathrm{d}\varepsilon^2}. \qquad (1.26)$$

If no other instability precedes, the uniaxial stress reaches its maximum at the point of inflection of the E_{tot} function and the related stress σ_{iut} can be considered to be the IS of the nanocomposite under the conditions of uniaxial loading.

For electronic structure calculations the VASP code was utilized. The cut-off energy for the basis set was 290 eV and the exchange-correlation energy was evaluated using the generalized-gradient approximation of Perdew and Wang with Vosko, Wilk and Nusair interpolation [125]. The $3 \times 3 \times 12$ k-points mesh was used in all calculations. The solution was considered to be self-consistent when the energy difference of two consecutive iterations was smaller than 0.1 meV. Atomic positions within the super-cell were relaxed using the Hellman–Feynman stress tensor.

1.1.5.2 Computed Data for Composite Constituents

Although it could have been possible to compute the data for pure tungsten, vanadium and niobium using a primitive cell, the super-cell with all atomic positions occupied by the same kind of atoms was used in the calculations. This allowed us to check the results for the composites in a more reliable way.

Computed values of the equilibrium lattice parameter a_0, the bulk modulus B, the shear modulus G, the Young's modulus E_{001} and the Poisson's ratio ν for pure W, Nb, Mo and V are listed in Table 1.8 along with experimental data. The experimental values of a_0 and B were taken from [126]. Values of E_{001} and ν were computed from experimentally determined elastic moduli [82, 127] using Equations 1.8 and

$$\nu = \frac{C_{12}}{C_{11} + C_{12}}.$$

Table 1.8 Ground-state properties of Nb, W, Mo and V along with their ideal uniaxial strengths

Element		a_0 [Å]	B [GPa]	E_{001} [GPa]	ν	σ_{iut} [GPa]	
Nb	calc.	3.32	177	127	0.38	19.2[a]	18.8[b]
	exp.	3.30	170	152	0.35		
W	calc.	3.18	318	389	0.29	28.8[a]	28.9[c]
	exp.	3.16	323	417	0.28		
V	calc.	2.98	188	199	0.32	19.9[a]	
	exp.	3.03	162	151	0.34		
Mo	calc.	3.15	268	412	0.24	28.3[a]	28.8[b]
	exp.	3.15	272	394	0.26		

[a] [122]
[b] [49]
[c] [41]

The agreement between computed and experimental values of a_0 is excellent (within 1%). In spite of a most noticeable exception of the Young's modulus for Nb (underestimated by 16%), the agreement is very good also in the case of elastic moduli and the Poisson's ratio (mostly within 5%).

Computed values of the uniaxial stress σ_{iut} at the point of inflection on the $E_{tot}(V)$ curve for pure Nb, W, V and Mo are listed in Table 1.8 along with available literature data. The mutual agreement of values from different sources is very good.

1.1.5.3 Solutions for Composites

As an example, the dependence of the average atomic volume V_0 on the atomic concentration of W (nearly corresponding to the volume fraction V_f of tungsten fibres) in the niobium matrix is depicted in Figure 1.19 by solid squares. The dotted line interpolates between experimental values for pure Nb and W. As can be expected, the atomic volume linearly decreases with W concentration. Similar dependencies also hold for other studied composites.

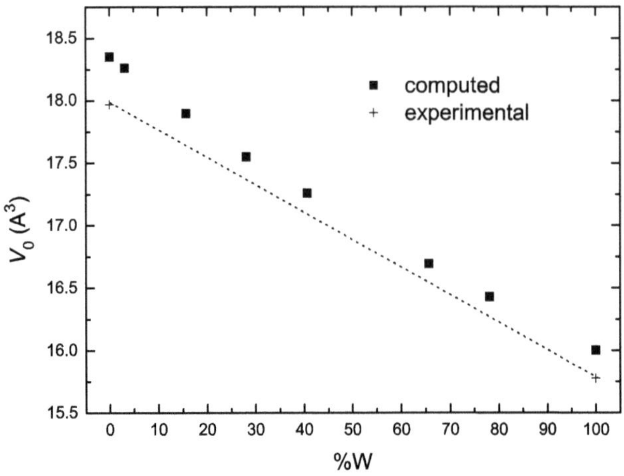

Figure 1.19 Equilibrium atomic volume as a function of W concentration

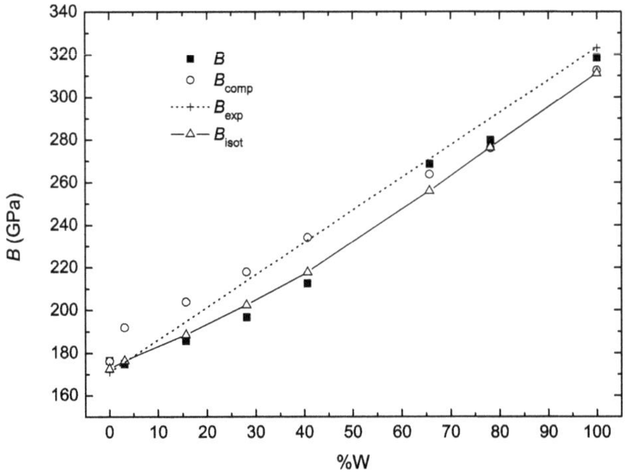

Figure 1.20 The composite bulk modulus as a function of the atomic concentration of tungsten fibres

The dependence of the composite bulk modulus B on the atomic concentration of W is depicted in Figure 1.20. Values obtained by means of Equation 1.26 are displayed by open circles. The triangles show values of B calculated from the Equation 1.27 (valid for an isotropic material)

$$B_{\text{isot}} = \frac{E_{001}}{3(1-2\nu)} \tag{1.27}$$

by using E_{001} values computed according to Equation 1.26. Both of them seem to follow Equation 1.25, where the bulk moduli are used instead of

E_{001}. The dotted line in Figure 1.20 follows Equation 1.25 for experimental values $B_f = 323\,\text{GPa}$ and $B_m = 170\,\text{GPa}$.

Values of the Young's modulus E_{001} of all investigated composites were also found to follow the linear mixture rule in a satisfactory manner. Consequently, all atomistic results confirm that deviations from Equation 1.25 observed for real composites are caused by their imperfections, in particular by a reduced interface cohesion.

While the computed quantities such as the equilibrium volume or elastic moduli obey the simple linear mixture rule, the IS of a pure reinforcing materials W and Mo can already be reached (or even exceeded) well below the 100% concentration of the fibres in the composite. Indeed, Figure 1.21 displays the σ_{ip} values as functions of the reinforcement concentration for all the investigated composites. The curves exhibit a simple linear dependence on the atomic concentration of fibre atoms up to about 60–80%. Above this concentration range they seem to reach a saturated value.

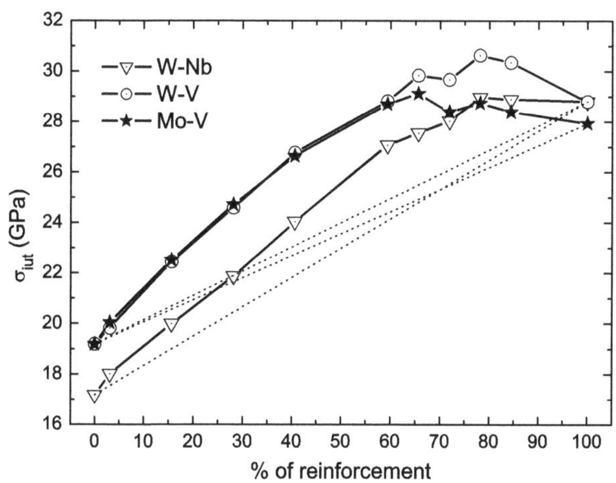

Figure 1.21 Theoretical tensile strength as a function of atomic concentration of reinforcing fibres for three composites. The fibres are made of W or Mo, whereas the matrix consists of V or Nb atoms

A possible explanation of this synergy effect can be the lattice mismatch owing to different Poisson's contractions of individual composite constituents. As a consequence, the fibres can be under transverse tensile or compressive stresses which can influence the uniaxial tensile strength of the whole composite (see Section 1.1.4). It was found that, in the case of cubic metals, the tensile strength increases under conditions of superimposed tensile transverse stresses. Therefore, a simple model has been developed to verify the above-mentioned explanation of the synergy effect in nanocomposites [128]. The model compares the data obtained for V–W nanocomposites with those achieved by modelling a deformation behaviour of two independent ideal crystals of pure W and V. In order to include the stress coupling effect, the individual crystals were loaded in a triaxial mode that simulated the stress state experienced by the fibre or the matrix in the composite. It was found that, during the deformation, the vanadium matrix was subjected to transverse (internal) tensile biaxial stress whereas the tungsten fibres experienced compressive stresses. By including these stresses into the simple proportional rule written in terms of stresses acting on the nanocomposite, the σ_{iut} values shift above the linear function interpolating strengths of W and V (the corresponding dotted line in Figure 1.21). However, such a shift does not fully reproduce the computed σ_{iut} values of the composite which means that either the predicted synergy effect is not only caused by the internal transverse stresses or the applied model is not sufficiently precise. Let us finally mention that the procedure simulating the deformation of the nanocomposite can also be improved by including a relaxation of all atomic positions within the bulk. These issues should be a matter of further analyses.

1.1.6 Influence of Crystal Defects and Temperature

All preceding results concerning the IS were obtained assuming a perfect crystal and an extremely low temperature ($T \rightarrow 0$). In real perfect crystals (whiskers), however, the presence of some imperfections is to be expected. At least, an equilibrium concentration of vacancies as well as a certain level of surface roughness are inevitable. Thus, it is worthwhile to assess possible effects of lattice defects, surface roughness and temperature on IS. For principal reasons, however, the strength of imperfect crystals cannot be identified as IS any more. Therefore, the term "strength" will simply be used instead of IS hereafter.

In general, the presence of lattice defects decreases the strength value. The only exception to this rule might be some special impurity dopants and alloying atoms. Since no grain (or subgrain) boundaries and secondary phase particles are expected in perfect single crystals, only point defects, dislocations, stacking faults, twins, free surfaces, cracks and phonons (temperature) are mentioned in the following brief summary.

1.1.6.1 Crystal Defects

Vacancies are inevitable in real crystals just near the zero Kelvin temperature. However, the effect of monovacancies on strength is negligible which simply follows from Equations 1.2 for σ_{iut}. Indeed, this classical formula for the tensile IS is equal to the Griffith criterion for a nanocrack of an atomic size – a vacancy. Larger clusters of vacancies (microcracks) may probably exist only at temperatures near the melting point that are far beyond the temperature range relevant for strength studies.

The effect of impurities and alloying atoms was studied in several works based on *ab initio* methods. Although Goodwin *et al.* [129] found that Ge and As impurities in aluminium raise the cohesive energy by up to 8%, no clear conclusions concerning the change in strength were made. Huang *et al.* [130] reported no effect on the σ_{iut} value of silicon in the case of p-type doping, unlike the 6% decrease in the case of very high levels of n-dopants. Both works used the pseudopotential approach combined with molecular dynamics. Song *et al.* [100] used the DVM method to realize that the 7% alloying of V, Cr, Fe and Mo slightly improves the value of σ_{iht}, unlike the 4% alloying that caused a slight decrease in σ_{iht}. Many authors (e.g., [131]) theoretically expect a slight decrease in τ_{is} owing to the impurity content. In summary, the influence of a low concentration of point defects on the strength seems to be very small.

Dislocation slip appears when shear stress in the slip plane parallel to the slip direction reaches the Peierls–Nabarro level. Since this level in metals is relatively very low, the strength might be dramatically reduced (by four orders of magnitude or even more). On the other hand, the P-N stress is extremely high in ceramic covalent crystals (C, Si, SiC, ZnS, Si_3N_4) as well as in complex ionic crystals ($MgAl_2O_4$, Al_2O_3, $Al_2O_3.MgO$). The dislocations are practically sessile at near zero temperatures and, therefore, the tensile strength might be reduced only by tens of percents owing to microcracks initiated by the stress relaxation around dislocations with long Burgers vectors.

As far as we know, no special studies regarding the effect of stacking faults (SF) on strength have been performed. Atoms on the stable SF plane lie in the local energy minimum and no stress is induced in the surrounding volume. The energy of the SF per atom is about two orders of magnitude lower than that of the free surface. We may, therefore, deduce that the influence of SFs on strength can be neglected. This also refers, most probably, to twins as was shown for the NiTi crystal from first principles [132].

The effect of a perfectly flat free surface on the strength can also be considered to be negligible. However, small imperfections like scratches or dimples act as stress concentrators (micronotches). Their maximum effect on strength can be roughly estimated by a factor $1+2(l/\rho)^{1/2}$, where l is the notch depth and ρ is the curvature of the notch root ($\rho > 0$) [133]. Therefore, sharp notches can significantly reduce the strength value.

According to the Griffith law for perfectly brittle materials, an atomically sharp crack of length a causes a drop in the tensile strength value by a factor of $(2a_0/(\pi a))^{1/2}$ (a_0 is the lattice parameter). However, only ceramics, semiconductors and, most probably, molybdenum and tungsten can be considered to be intrinsically brittle crystals at absolute zero temperature (see Section 1.2). In all other metals, dislocation emission precedes unstable crack growth. This process increases the effective surface energy and blunts the crack tip (see Chapter 2). Consequently, the strength drop must be much less than that predicted by the Griffith law.

1.1.6.2 Temperature

The variation of strength with temperature is generally attributed to the role of phonons in deformation and fracture processes. Although the Frenkel formula at Equation 1.1 simply suggests that the variation should be the same as that of the shear modulus, the Orowan–Polanyi relation at Equation 1.2 brings a complication with the temperature dependence of the surface energy. From a historical point of view, two sufficiently relevant methods were applied to predict the temperature influence: the Einstein model of harmonic oscillators combined with the elastic instability criterion and the model of dislocation nucleation supported by phonon fluctuations [1]. The first approach predicted a drop in the uniaxial strength within the range of units to tens of percents when changing the temperature from 0 K to 1000 K. The results depend on the type of utilized interatomic potentials (short- or long-range). However, this drop was found to be very close to that predicted when considering only the change in Young's modulus. The second approach deals with the maximum energy of up to $50\,kT$ (k is the Boltzmann constant) that the thermal fluctuations can supply at any temperature. Typically, the results reveal tens of percents decrease in the strength within that temperature range. An analysis of a possible thermal crack initiation has shown that relatively small amounts of available thermal energy are unlikely to affect the process very much up to 1200 K (in comparison with the total amount of strain energy conversion into surface energy). It should be emphasized, however, that the most physically justified approaches to the temperature problem are complex analyses of phonon spectra or the quantum (*ab initio*) MD simulations. However, recently available studies of that kind are too rare and incomplete to allow general conclusions.

Nevertheless, one can assume that the temperature change in the relevant elastic modulus might be considered to be an acceptable lower-band approximation to the strength-temperature dependence.

1.2 Intrinsic Brittleness and Ductility

The intrinsic tendency of perfect crystals to brittleness or ductility can be quantitatively assessed by considering crack stability conditions. Indeed, when a cracked crystal is subjected to a tensile loading, there are basically two possibilities for its behaviour – either an unstable cleavage fracture is observed (brittleness) or a dislocation emission stabilizes the crack tip by plastic blunting (ductility). Physical solutions of that problem are usually constructed by means of two classical approaches introduced by Kelly *et al.* [23] and Rice and Thompson [134]. The former deals with the ratio of tensile to shear ideal strengths σ_{iut}/τ_{is} of an investigated crystal, whereas the latter analyses a related mechanism of dislocation emission. The principles of both concepts, which lead to essentially equivalent results, are outlined in the following subsection along with a description of more general approaches. Further parts are devoted to an extended calibration of crystal brittleness/ductility performed by Rice *et al.* [135] and, in particular, to that developed by us.

1.2.1 Fundamentals

1.2.1.1 Classical Criteria

The classical understanding of intrinsic brittle/ductile behaviour was introduced by Kelly, Tyson and Cottrell (KTC) [23]. According to this concept, the crystal is considered to be intrinsically ductile when τ_{is} is exceeded earlier than the σ_{iut} at the crack tip. This corresponds to the following condition of intrinsic ductility:

$$\frac{\sigma_1 \tau_{is}}{\sigma_{iut} \tau_r} < 1, \tag{1.28}$$

where σ_1 is the maximum principal stress and τ_r is the resolved shear stress on the active slip plane adjacent to the crack tip. However, only slip in crystallographic planes containing the whole crack front can effectively produce blunting. Let us consider the blunting configuration depicted in Figure 1.22 in a special case of mode I loading (the shear stress intensity factors $K_{II} = K_{III} = 0$). Then, Equation 1.28 can be rewritten as

$$\frac{\sigma_{iut}}{\tau_{is}} > \frac{2[1 + \sin(\Theta/2)]}{\sin\Theta \cos\Phi}, \tag{1.29}$$

where Θ is the angle between the crack plane and the intersecting slip plane and Φ is the angle between the Burgers vector and the direction perpendicular to the crack front [136].

The following reasoning leads to a derivation of Equation 1.29. The stress tensor at the crack tip under the condition of plane strain [137] reads

$$\tau_{ij} = \begin{pmatrix} A\left[1 - \sin\left(\frac{\Theta}{2}\right)\sin\left(\frac{3\Theta}{2}\right)\right] & A\sin\left(\frac{\Theta}{2}\right)\cos\left(\frac{3\Theta}{2}\right) & 0 \\ A\sin\left(\frac{\Theta}{2}\right)\cos\left(\frac{3\Theta}{2}\right) & A\left[1 + \sin\left(\frac{\Theta}{2}\right)\sin\left(\frac{3\Theta}{2}\right)\right] & 0 \\ 0 & 0 & 2A\nu \end{pmatrix},$$

where

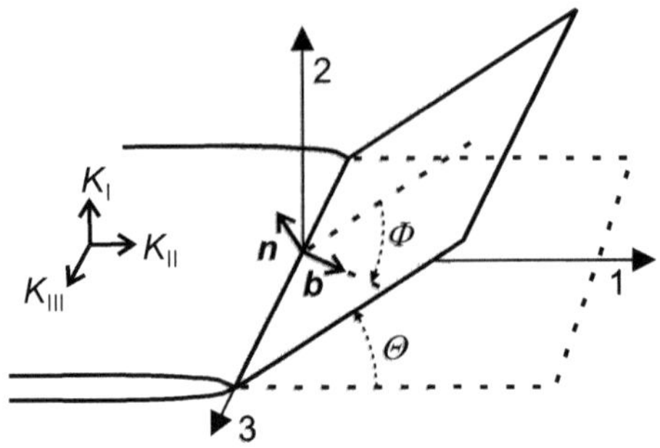

Figure 1.22 Scheme of the blunting crack-slip plane configuration

$$A = \frac{K_I}{\sqrt{2\pi r}} \cos\left(\frac{\Theta}{2}\right)$$

and

$$\sigma_1 = A\left[1 + \sin\left(\frac{\Theta}{2}\right)\right]. \tag{1.30}$$

Regarding Figure 1.22, the coordinates of the stress vector $\mathbf{T}$ in the slip plane can be expressed by means of the stress tensor τ_{ij} and the normal $\mathbf{n} = (-\sin\Theta, \cos\Theta, 0)$ as $T_i = \tau_{ij} n_j$ which gives

$$\mathbf{T} = (T_1, T_2, T_3) = \left(-\frac{A}{2}\sin\Theta, A\cos^2\left(\frac{\Theta}{2}\right), 0\right).$$

The magnitude of the resolved shear stress τ_r depends on τ_{ij} and $\mathbf{b} = (\cos\Theta\cos\Phi, \sin\Theta\cos\Phi, \sin\Phi)$ as

$$\tau_r = \mathbf{T}\cdot\mathbf{b} = \frac{A}{2}\sin\Theta\cos\Phi. \tag{1.31}$$

By substituting Equations 1.30 and 1.31 into the criterion at Equation 1.28 one finally obtains Equation 1.29.

Another criterion based on the ability of the material to emit dislocations from the crack front was proposed by Rice and Thomson (RT) [134]. There are two principal forces acting on the emitted dislocation in the adjacent slip plane. The first is the repulsive force that is induced by external loading and the second is the attractive mirror force induced by a presence of free surfaces (crack flanks). In the frame of the elastic approximation and in terms of the K_I factor, the resulting force component within the slip plane can be expressed as follows:

$$f_d(r) = \frac{b_e}{2\sqrt{2\pi r}} K_I \sin\Theta\cos\frac{\Theta}{2} - \frac{G}{4\pi r}\left(b_s^2 + \frac{b_e^2}{1-\nu}\right), \tag{1.32}$$

where b_e and b_s are magnitudes of edge and screw components of the Burgers vector, respectively, and G is the shear modulus. The function at Equation 1.32 is schematically plotted in Figure 1.23. One can see that for small distances of the dislocation from the crack front ($r < r_0$, where $f_d(r_0) = 0$) the attractive force is dominating, whereas for greater distances the repulsive interaction prevails. This means that the elastic theory always predicts a mechanical instability, i.e., the intrinsic brittleness. However, the elastic approximation fails for $r \leq r_c$, where r_c is the size of the dislocation core. In this case the Burgers vector of the emitted dislocation is expected to be proportional to r and, consequently, the repulsive force remains dominating over the whole distance. In other words, the condition $r_0 \leq r_c$ means a spontaneous emission of dislocations and, with regard to Equation 1.32 and $r_0 = r_c$, the stress intensity factor corresponding to this emission can be written as

$$K_{Ie} = \frac{G}{\sqrt{2\pi r_c}} \frac{b_s^2 + b_h^2/(1-\nu)}{b_h \sin\Theta \cos(\Theta/2)}.$$

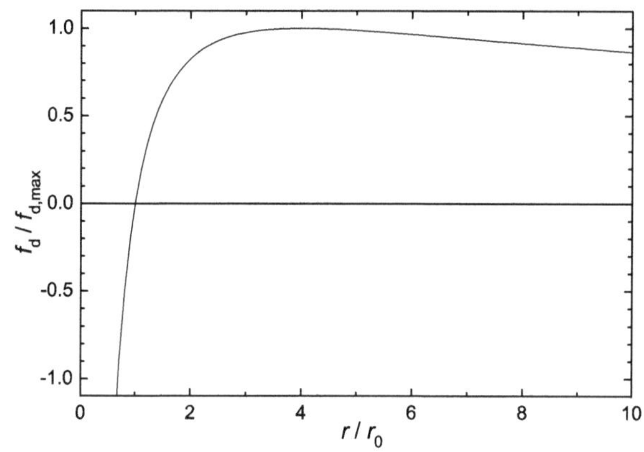

Figure 1.23 The dependence of the resulting force on the relative distance for a dislocation emitted from the crack front

Because the Griffith criterion for brittle fracture reads

$$K_{Ic} = \sqrt{\frac{2G\gamma}{1-\nu}},$$

where γ is the fracture energy, the combined criterion for intrinsic ductility becomes

$$K_{Ie} < K_{Ic}. \tag{1.33}$$

Note that no thermal activation was assumed to help the dislocation emission in the RT concept which corresponds to absolute zero temperature. This means that Equation 1.33 overestimates the tendency of crystals to intrinsic brittleness when applied to higher temperatures.

At zero Kelvin, the criteria KTC and RT are essentially equivalent and yield similar results (see later). Indeed, when one assumes remote mode I loading, the simplest IS estimates at Equations 1.1 and 1.2 and an averaged angle $\bar{\Theta}$ [138], both criteria can be rearranged to

$$\frac{Gb}{10\gamma} < 1.$$

1.2.1.2 Beyond the Classical Criteria

The efficiency of both criteria in their classical forms is restricted in many directions. First, there is only one selected configuration of crack-slip planes. In the KTC criterion, moreover, the Orowan estimate of σ_{iut} contains an uncertain γ-value and it does not reflect the real triaxial state of stress at the crack tip. The Frenkel formula for τ_{is} does not take the normal/shear stress coupling into account. Second, the RT criterion contains a purely defined r_c-value and stands only for the mode I case. Therefore, more generalized and accurate approaches are needed in order to obtain a plausible calibration of crystals with respect to their intrinsic brittle/ductile response.

The RT-criterion was upgraded by application of the Peierls–Nabarro model to the process of the dislocation nucleation as well as by replacing the K_I-concept by that of the effective crack driving force G_{eff} [135, 139] (see also Appendix B). This leads to the following condition for dislocation nucleation:

$$\begin{aligned} &\left[f_I(\Theta)K_I + f_{II}K_{II}\right]\cos\Phi + f_{III}(\Theta)K_{III}\sin\Phi = \\ &= \sqrt{\frac{2G}{1-\nu}\left[\cos^2\Phi + (1-\nu)\sin^2\Phi\right]\gamma_b}, \end{aligned} \tag{1.34}$$

where K_I, K_{II} and K_{III} are stress intensity factors (SIFs) for modes I, II and III, f_I, f_{II}, f_{III} are related angular functions and γ_b is the energy barrier for the dislocation nucleation. In the case of a pure opening mode I Equation 1.34 reduces to

$$G_I = 8\frac{1 + (1-\nu)\tan^2\Phi}{(1+\cos\Theta)\sin^2\Theta}\gamma_b,$$

where $G_I = (1-\nu)K_I^2/2G$.

The intrinsic ductility means that, during the loading, Equation 1.34 is satisfied earlier than the generalized Griffith condition $G_{eff} = 2\gamma$. This leads to

$$\frac{\gamma}{\gamma_b} > \frac{4[1 + x^2 + z^2/(1-\nu)][1 + (1-\nu)\tan^2 \Phi]}{(1 + \cos\Theta)[\sin\Theta + (3\cos\Theta - 1)x + 2z\tan\Phi]^2}, \qquad (1.35)$$

where $x = K_{II}/K_I$ and $z = K_{III}/K_I$. In the case of the pure opening mode ($x = z = 0$) one obtains

$$\frac{\gamma}{\gamma_b} > \frac{4[1 + (1-\nu)\tan^2 \Phi]}{(1 + \cos\Theta)\sin^2\Theta} = N_{\mathrm{RT}}. \qquad (1.36)$$

The KTC-criterion can be improved by inserting more sophisticated estimates of IS values based on atomistic models [25, 136]. The Orowan estimate of σ_{iut} can be replaced by values of σ_{iht} based on the Morse potential [36]. These values reflect well the stress state of almost isotropic tension ($\sigma_1 = \sigma_2 = 1.6\sigma_3$) at the crack tip and, fortunately, they are also sufficiently precise as proved by recent *ab initio* calculations (see Section 1.1.3). The Frenkel formula for τ_{is} is quite satisfactory but the atomistic results are still more precise [31]. The normal/shear stress coupling can be reflected by inserting both linear and parabolic functions $\tau_{is}(\sigma_n)$ which, again, was verified by means of *ab initio* computations (see Section 1.1.4). Moreover, both the crack plane and the front positions that correspond to the many possible combinations of Miller indices related to all possible blunting slip systems are to be analyzed in order to obtain a complex picture of the intrinsic response. When taking these corrections into account, Equation 1.29 (for $\sigma_{iut} \to \sigma_{iht}$) and Equation 1.36 constitute very useful stress and energy criteria. They enable us to order crystals of pure elements and compounds according to their tendency to exhibit intrinsic brittleness (or ductility).

1.2.2 Calibration of Crystals

1.2.2.1 Models Based on Dislocation Emission

When considering Equation 1.36 and calculating the values of $\Gamma_{\mathrm{RT}} = N_{\mathrm{RT}}\gamma_b/\gamma$ for various crystals of elements and compounds, one can order the crystals with respect to their intrinsic brittleness/ductility. The values $\Gamma_{\mathrm{RT}} \geq 1$ predict intrinsic brittleness, whereas $\Gamma_{\mathrm{RT}} < 1$ means the intrinsic ductility. Such an ordering is called the calibration of the intrinsic brittleness/ductility of crystals. Rice *et al.* [135] performed such a calibration of selected metallic and ceramic elements by assuming angles Θ and Φ associated with one characteristic blunting configuration of the crack plane-slip system for each crystal. The results are displayed in Table 1.9. One can see that all the diamond crystals and metallic crystals of bcc structure (except for alkali metals) are intrinsically brittle since their Γ_{RT} values are distinctly greater than 1. The extreme behaviour of alkali metals is not very surprising. Indeed, their fcc and bcc structure modifications have approximately the same energy and, moreover, crystals of Na and Li keep closed-packed structures even up to about 50 K [140]. On the other hand, all the fcc crystals (except for iridium) are intrinsically ductile, since their Γ_{RT} values lie close to 1. Note that this result agrees well with engineering experience of imperfect crystals and polycrystals of analyzed materials at room temperature. Here, due to the effect of thermal activation, all the Γ_{RT} values must be lower to promote the ductility. In this way, perfect bcc crystals might also cross over to the ductile state as is well known for engineering materials. When applying such a simple model of dislocation emission, however, it is not to be expected that the brittle/ductile boundary will be exactly 1. Anyway, more important information is offered just by the order of crystals according to their brittleness/ductility behaviour.

The influence of the loading mode on brittle/ductile behaviour of perfect crystals can be assessed by means of Equation 1.35 [136]. Let us consider, for example, bcc structures, where the crack lies in the $\{100\}$ plane with an adjacent slip system $\{110\}\langle 1\bar{1}1\rangle$ and the associated angles are $\Theta = 45°$ and $\Phi = 35.3°$. For a pure mode I ($x = z = 0, \nu = 0.3$) one obtains $\gamma/\gamma_b > 6.3$. Already a very small ratio of shear modes such as $K_{II} = K_{III} = 0.1K_I$ ($x = z = 0.1$) results in $\gamma/\gamma_b > 3.5$, i.e., to a considerable transfer of the intrinsic response towards ductility. The same behaviour is also exhibited by perfect crystals of fcc metals. On the other hand, the crystals with diamond structure remain intrinsically brittle even under high applied shear modes. Similar behaviour is typical also for single crystals containing dislocations. However, this is not necessarily the case for cracked polycrystals owing to the effect of friction produced by the crack-flank roughness (see Chapter 2 for more details).

1.2.2.2 Models Based on Ideal Strength

These models mostly utilize Equation 1.29 in terms of an extended analysis of the influence of the crack plane position in the crystal lattice. Crack plane and front directions $\{hkl\}$ and $\langle uvw\rangle$ related to all possible combinations of low Miller and direction indices ($0 \leq h, k, l, u, v, w \leq 4$) along with all possible blunting slip systems (the crack front forms an intersection of crack and slip planes) were investigated in our works [25, 136]. The slip configurations for fcc, bcc, diamond, B1 and B2 structures are shown in Table 1.10. The developed computer code contained procedures that were able to select various possible slip systems within the crystallographically equivalent set and construct associated blunting crack-slip configurations. The left-hand side of Equation 1.29 depends solely on the crystallography. Indeed, the atomistic simulations [31] have shown that the shear IS can be simply expressed as

Table 1.9 Values of the parameters Γ_{RT} and $\bar{\Gamma}_{\mathrm{KTC}}$ for investigated crystal structures

Crystal	Structure	Γ_{RT}	$\bar{\Gamma}_{\mathrm{KTC}}$	${}^{p}\bar{\Gamma}^{\varepsilon}_{\mathrm{KTC,min}}$	${}^{p}\bar{\Gamma}^{\sigma}_{\mathrm{KTC,min}}$	${}^{l}\bar{\Gamma}^{\varepsilon}_{\mathrm{KTC,min}}$	${}^{l}\bar{\Gamma}^{\sigma}_{\mathrm{KTC,min}}$
NaCl	B1		10.2	1.0	1.0	1.5	1.2
CsCl	B2		7.5	1.0	1.0	1.6	1.2
CsBr	B2		6.9	1.0	1.0	1.6	1.2
CsJ	B2		6.7	1.0	1.0	1.6	1.2
Mo	bcc	5.2	6.3	1.0	0.9	1.4	1.0
W	bcc	4.7	6.1	1.0	0.9	1.4	1.0
C	diam	4.5	6.0	1.0	1.0	1.3	1.2
LiF	B1	2.6	5.8	1.0	1.0	1.6	1.2
MgO	B1	2.9	5.0	1.0	1.0	1.5	1.2
Ir	fcc	3.6	4.2	0.9	0.7	1.1	0.7
Ge	diam	3.4	4.1	1.0	0.9	1.3	0.9
Fe	bcc	2.4	4.0	1.0	0.7	1.3	0.8
Si	diam	2.7	3.8	1.0	0.8	1.3	0.9
V	bcc	1.9	3.5	1.0	0.7	1.3	0.7
Ta	bcc	2.0	3.5	1.0	0.7	1.3	0.7
Nb	bcc	1.7	2.7	1.0	0.5	1.2	0.5
Ni	fcc	1.7	2.3	0.7	0.4	0.8	0.4
Al	fcc	1.2	2.1	0.8	0.4	0.9	0.4
Th	fcc		1.9	0.6	0.3	0.6	0.3
Pt	fcc	1.3	1.6	0.8	0.3	0.9	0.3
Cu	fcc	1.2	1.6	0.6	0.3	0.7	0.3
Ag	fcc	1.1	1.5	0.7	0.3	0.7	0.3
Na	bcc	1.2	1.4	0.7	0.3	0.7	0.3
Rb	bcc		1.3	0.7	0.3	0.7	0.3
K	bcc	1.3	1.2	0.8	0.3	0.8	0.3
Li	bcc	0.7	1.1	0.6	0.2	0.6	0.2
Au	fcc	0.9	0.9	0.7	0.2	0.7	0.2
Pb	fcc	0.8	0.9	0.6	0.2	0.6	0.2

$\tau_{is} = \kappa G$, where κ is the parameter characteristic for the particular structure and the slip system (see Table 1.3). Consequently, Equation 1.29 can be rearranged as

$$\frac{\sigma_{iht}}{G} > \frac{2\chi[1 + \sin(\Theta/2)]}{\sin\Theta \cos\Phi} = N_{\mathrm{KTC}}, \tag{1.37}$$

where the exchange $\sigma_{iut} \to \sigma_{iht}$ was also performed. The left-hand side of Equation 1.37 is a ratio of material characteristics and the right-hand side (N_{KTC}) depends only on crystallography. Thus, another computer procedure was developed to calculate values N_{KTC} for all selected crack-slip configurations in all investigated structures. There is a wide range of N_{KTC} values for individual crystal structures as well as for each crystal. In the latter case, the set of blunting crystallographic configurations, labelled by N_{KTC} values, can be divided into brittle and ductile parts by the characteristic value σ_{iht}/G. This means that the intrinsic response of each crystal substantially depends on the crack-plane position towards the basic crystallographic system. Nevertheless, one can determine a global characteristic $\bar{\Gamma}_{\mathrm{KTC}} = \bar{N}_{\mathrm{KTC}}G/\sigma_{iht}$, where $\bar{N}_{\mathrm{KTC}}$ is the mean value of N_{KTC}, averaged over all investigated configurations in a particular crystal.

Table 1.10 Slip configurations used in the model based on the ideal strength

Lattice	bcc		fcc		Diamond	B1	B2
Plane	{112}	{110}	{111}	{111}	{111}	{110}	{110}
Direction	⟨111⟩	⟨111⟩	⟨110⟩	⟨112⟩	⟨110⟩	⟨110⟩	⟨100⟩

All the studied crystal elements and compounds are ordered according to the global parameter $\bar{\Gamma}_{\mathrm{KTC}}$ in Table 1.9. One can clearly see that the ordering of crystals is very similar to that determined by using the parameter Γ_{RT} (shown in the same table). All ionic and covalent ceramic crystals can be considered to be intrinsically brittle, unlike those of fcc structure and the alkali metals. However, values of Γ_{RT} and of $\bar{\Gamma}_{\mathrm{KTC}}$ for practically all crystals are higher than 1 which infers their intrinsic brittleness at 0 K. In the case of RT criterion, the thermal activation at the room temperature is expected to produce a significant reduction of all values so that the alkali metals and fcc metals become ductile. In the case of KTC criterion, however, the ratio of ideal strengths in Equation 1.29 should not depend too much on temperature. The main reason for too high values must be sought in the strong dependence of τ_{is} on the normal stress component σ_n (see Section 1.1.3) which was not taken into account. This dependence was considered in both parabolic and linear approximations as

$$\tau_{is}(\sigma_n) = \tau_{is0}\left(1 - \sigma_n/\sigma_{iht}\right)^2,$$

$$\tau_{is}(\sigma_n) = \tau_{is0}\left(1 - k\sigma_n/\sigma_{iht}\right), \quad k \in \langle 0, 1 \rangle.$$

Moreover, both the plane strain ($\sigma_1 = \sigma_2 = 1.6\sigma_3$) and plane stress $\sigma_1 = \sigma_2, \sigma_3 = 0$) conditions at the crack tip were studied. The computer procedure was completed by a routine that could select minimum values $\bar{\Gamma}_{\mathrm{KTC,min}}$ of all $\bar{\Gamma}_{\mathrm{KTC}}$ parameters related to investigated crack-slip configurations in a particular crystal. When the condition $\bar{\Gamma}_{\mathrm{KTC,min}} \geq 1$ is fulfilled, no blunting of the crack tip by shear can appear and, consequently, the crystal can be assumed to be definitely brittle. The computed parameters $\bar{\Gamma}_{\mathrm{KTC,min}}$, obtained for both the parabolic and the linear ($k = 0.5$) approximation, are also presented in Table 1.9. The upper indices ε (σ) stand for the plane strain (plane stress) case and the upper indices l (p) refer to the linear (parabolic) approximation of the dependence τ_{is} *vs* σ_n. The results show that, in the plane strain state, all ceramic and bcc crystals exhibit intrinsic brittleness at absolute zero temperature. All fcc crystals including the bcc alkali metals indicate a ductile behaviour. It should be noted, however, that $\bar{\Gamma}_{\mathrm{KTC,min}}$ values slightly lower than 1 provide only a necessary but not sufficient condition for intrinsic ductility. In that case, indeed, a high dominance of brittle crack-slip configurations might be expected. Therefore, the borderline brittleness/ductility should rather correspond to $\bar{\Gamma}_{\mathrm{KTC,min}} \approx 0.9$. In the plane stress case, all the ionic ceramics (including covalent diamond) are predicted to behave in a brittle manner. However, crystals of Mo, W, Ge and Si also lie close to the brittle/ductile borderline. All other metals can be assumed as definitely intrinsically ductile. These results already seem to reflect plausibly the intrinsic response of crystals at 0 K. Moreover, the plane stress state prefers the ductile behaviour which is also to be expected.

In summary, the following conclusions with respect to the intrinsic brittleness/ductility of perfect crystals can be derived:

1. Diamond and all ionic ceramics are intrinsically brittle. This also holds for sufficiently large perfect crystals of Mo, W, Si and Ge (plane strain).
2. The perfect bcc crystals, except for alkali metals, lie close to the brittle/ductile borderline at absolute zero temperature. They are expected to exhibit a transition to ductile behaviour at higher temperatures similarly to the behaviour of real bcc crystals and polycrystals.
3. All perfect crystals of fcc metals, including the bcc alkalis, are intrinsically ductile in the whole temperature range.

The transient behaviour of the perfect body-centred Fe crystal was also recently verified by means of molecular dynamics simulations [141]. It should be emphasized that the brittleness/ductility behaviour of perfect crystals is in good quantitative agreement with long-term experimental experience with engineering single crystals and polycrystals containing defects. This means that this kind of behaviour is, at least partially, predetermined by the intrinsic properties of the crystal lattice.

1.3 Multiscale Model of Nanoindentation Test

Indentation tests represent a very effective tool to explore mechanical properties of materials. Besides hardness, many other characteristics such as strength, yield stress, fatigue limit and plastic strain can be approximately assessed from these tests. The nanoindentation test provides useful information about very local mechanical characteristics of the material's microstructure. However, this test also became a very promising experimental method for identification of the ideal shear strength (e.g., [142]). Due to a very small penetration depth, the stressed volume beneath the sharp nanoindenter may be free of preexisting dislocations. Moreover, this volume is usually constricted to one grain (single crystal) even in polycrystalline materials. During the nanoindentation, therefore, the increasing local shear stress can reach the ideal shear strength value and, consequently, dislocations can nucleate in the ideal crystal lattice. Such an onset of local plastic deformation could be detected as a pop-in on the load-displacement curve by several authors [15, 143–145]. However, a physically plausible model of the nanoindentation test must be developed in order to determine the value of τ_{is} related to the pop-in load.

During the last 10 years, several models of the nanoindentation test have been published (e.g., [142, 146, 147]). These models utilized either the simple Hertzian solution of the stress distribution under a spherical indenter or more sophisticated multiscale approaches. The latter models can be built by using different concepts (see Preface). Hereafter, a multiscale approach that couples an indentation model based on continuum mechanics (finite elements) with material characteristics determined on both the mesoscopic (crystallography) and atomistic (*ab initio*) levels is introduced. In order to reflect sufficiently the physics of the indentation process, a plausible model of that kind must comply with several fundamental relationships and boundary conditions: (1) the *ab initio* calculations of pure ideal shear strength are to be generalized in order to reproduce correctly the stress state under the indenter – in particular, the influence of the normal stress σ_n on τ_{is} must be considered; (2) both the nonlinearity and the anisotropy of the elastic response of the ideal crystal are to be taken into account; (3) the evolution of the resolved shear stress and the normal stress on all appropriate crystallographic planes must be evaluated during the simulation; (4) a 3D finite element model of the nanoindentation process must be developed since the real stress state is quite different from that obtained from 2D models; (5) an analysis of friction forces between the substrate and the indenter should be also included.

To our knowledge, the first multiscale model including all the above-mentioned corrections was published by Krenn *et al.* [142]. The authors gave a very good interpretation of the pop-in effects in tungsten and molybdenum single crystals in terms of reaching the τ_{is} value. Unfortunately, a detailed description of all utilized computational procedures was not given in that paper. Therefore, the reported results cannot be checked and the methodology remains questionable. In our first work concerning the nanoindentation in the polycrystalline copper [146], only the first two corrections were incorporated into the model. In more recent papers [145, 148] practically all the necessary corrections were already taken into account. Hereafter, the utilized computational procedures and the results of these models for copper and nickel will be briefly outlined.

1.3.1 Description of Submodels

The three-dimensional isotropic FEM model of the nanoindentation tests was performed by using the finite element code ANSYS. A frictionless sphero-conical indenter with a radius of 0.2 μm (0.5 μm) in agreement with related experiments for Ni (Cu) was pressed into a 5 µm thick substrate disc with a radius of 10 µm (see Figure 1.24). The values of the Young's modulus and Poisson's ratio of the diamond indenter were taken as $E = 1141$ GPa and $\nu = 0.07$, respectively. In the vicinity of the interface, the sphere was discretized with elements approximately 0.2 nm (0.5 nm) wide and the mesh of the substrate was refined from 0.4 mm (1 mm) at the outer edge of the disc to 0.2 nm (0.5 nm) directly beneath the indenter.

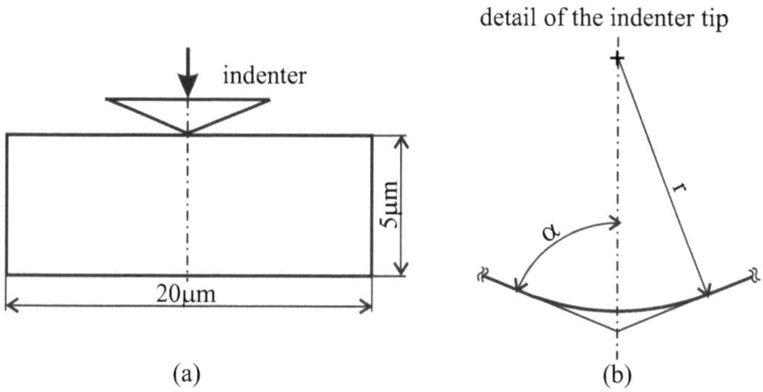

Figure 1.24 The geometrical configuration of the model: (a) cross-section of the nanoindenter and the substrate disc, and (b) scheme of the tip of the sphero-conical indenter with the semi-angle and the tip radius. Reprinted with permission from Trans Tech Publications Inc. (see page 265)

Owing to the rotational symmetry of the indentation model, only a 2D section could be analyzed as depicted in Figure 1.25. To model a contact area between the indenter tip and the specimen, the potential contact surfaces during the deformation had to be identified via a target (indenter) and the contact (substrate) elements in terms of so-called contact pairs. The elements of the substrate were, unlike those of the indenter, constrained against penetration into the opposite surface. The tangential contact stiffness based on current contact normal pressure and maximum allowable elastic slip could be updated in the frame of the ANSYS code.

The Coulomb model was utilized in the analysis involving friction. In this model, an equivalent shear stress was defined as a fraction of the contact pressure corresponding to the onset of sliding between the two surfaces. A special sticking/sliding calculation procedure determined the transitions from sticking to sliding and *vice versa* (see [148] for details).

The nonlinearity and, to a certain extent, the anisotropy of the elastic response was taken into account by a multilinear approximation of tension-compression stress-strain curves for Cu and Ni crystals. These curves were calculated from first principles in accordance with the experimental [001] indentation direction. The multilinear approximations were utilized in the ANSYS code procedure as an equivalent stress-strain dependence.

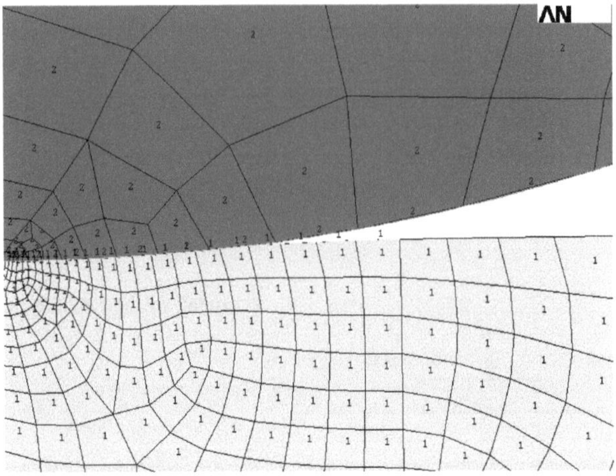

Figure 1.25 The finite element network near the contact boundary. Reprinted with permission from Trans Tech Publications Inc. (see page 265)

The function $\tau_{is}(\sigma_n)$ for the set of $\{111\}$ slip planes was also computed by means of the *ab initio* approach. Parameters of this linear dependence (see Equation 1.19) for Ni and Cu crystals can be found in Table 1.5.

The global calculation procedure simulated, step by step, the penetration of the indenter into the crystal. In order to identify the appropriate crystallographic plane in which the condition for the dislocation emission was first reached, the activity in the shear systems $\langle 112\rangle\{111\}$ was continuously tracked in stepwise calculations. The stress tensor transformation was performed to obtain all crystallographic systems related to the cylindrical symmetry. This enabled us to compute the values of τ and σ_n as functions of the rotation angle ϕ. Consequently, the maximum of the ratio $\theta_i(\phi) = \tau_i(\phi)/\tau_{is,i}(\phi)$ was searched for the entire circle going through each mesh node. The highest value of that ratio for all nodes and angles was denoted $\theta_{\max}$. Thus, the values of $\theta_{\max}$ could be related to each individual deformation step characterized by the penetration depth h.

1.3.2 Simulation of Pop-in Effects

The results obtained by using frictionless analysis were found to be practically identical with those obtained by using the model involving friction. This finding was in agreement with the conclusion already reported by Krenn *et al.* [142].

The dependence of $\theta_{\max}$ on the penetration depth h is depicted in Figure 1.26 for the nickel crystal. When the value of $\theta_{\max}$ slightly exceeds 1, the condition for the dislocation emission can be considered to be safely fulfilled. This nearly corresponds to the depth $h = 5.5\,\mathrm{nm}$ and the related indentation load $F = 35\,\mu\mathrm{N}$. The associated experimental pop-in values were found to be $h_{\mathrm{exp}} = 4.0 - 4.5$ nm and $F_{\mathrm{exp}} = 40 - 48\,\mu\mathrm{N}$. The following values referred to the copper crystal: $h = 12$ nm ($h_{\mathrm{exp}} = 10 - 20$ nm) and $F = 100\,\mu\mathrm{N}$ ($F_{\mathrm{exp}} = 150 - 270\,\mu\mathrm{N}$).

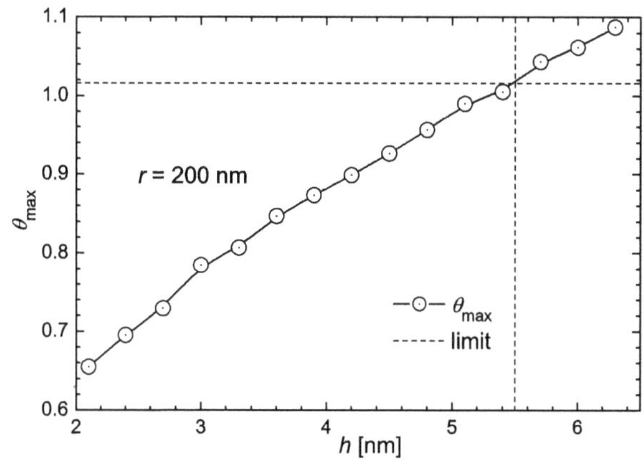

Figure 1.26 The dependence of the parameter $\theta_{\max}$ on the penetration depth h. The predicted moment of the first emission of dislocation loops corresponds to exceeding the critical value $\theta_{\max} \approx 1$ for $h \approx 5.5\,\mathrm{nm}$

In spite of some differences in experimental and theoretical values, the results reveal that nanoindentation may serve as a very good tool for the measurement of ideal shear strength. One should note that the deviations within the range of tens of percents are highly acceptable when compared to those in the range of hundreds of percents obtained by means of other experimental methods (see Table 1.2).

Chapter 2
Brittle and Ductile Fracture

This chapter is devoted to damage and fracture micromechanisms operating in the case when monotonically increasing forces are applied to engineering materials and components. According to the amount of plastic deformation involved in these processes, the fracture events can be categorized as brittle, quasi-brittle or ductile.

Brittle fracture is typical for ceramic materials, where plastic deformation is strongly limited across extended ranges of deformation rates and temperatures. In polycrystalline ceramics the reasons lie in a high Peierls–Nabarro stress of dislocations due to strong and directional covalent bonds (this holds also for some ionic compounds), and in less than five independent slip systems in ionic crystals (e.g., [149]). In amorphous ceramics it is simply because of a lack of any dislocations and, simultaneously, strong covalent and ionic interatomic bonds. Metallic materials or polymers exhibit brittle fracture only under conditions of extremely high deformation rates, very low temperatures or extreme impurity concentrations at grain boundaries. In the case of a strong corrosion assistance, brittle fracture can also occur at very small loading rates or even at a constant loading (stress corrosion cracking). A typical micromechanism of brittle fracture is so-called cleavage, where the atoms are gradually separated by tearing along the fracture plane in a very fast way (comparable to the speed of sound). During the last 50 years, the resistance to unstable crack initiation and growth, i.e., the fracture toughness, became a very efficient measure of brittleness or ductility of materials. In the case of cleavage, this quantity can be simply understood in a multiscale context. The macroscopic (continuum) linear–elastic fracture mechanics (LEFM) developed by Griffith and Irwin brought to light an important relationship between the crack driving force G (the energy drop related to unit area of a new surface) and the stress intensity factor K_I as

$$G = \frac{1-\nu^2}{E} K_I^2.$$

This relation holds for a straight front of an ideally flat crack under conditions of both the remote mode I loading and the plane strain. The energy necessary for creation of new fracture surfaces can be supplied from the elastic energy drop of the cracked solid and/or from the work done by external forces (or the drop in the associated potential energy). Thus, at the moment of unstable fracture, the Griffith criterion gives $G_c \approx 2\gamma$, where γ is the surface (or fracture) energy that represents a resistance to cleavage. Consequently

$$\gamma \approx \frac{1-\nu^2}{2E} K_{Ic}^2. \tag{2.1}$$

However, the surface energy can be expressed also in terms of the cohesive (bonding) energy needed to break down an ideal crystal or an amorphous solid into individual atoms. The bonding energy of a surface atom is a half of that associated with an internal atom [150] and, because of two fracture surfaces, one can simply write

$$\gamma = \frac{U}{4S}, \tag{2.2}$$

where U is the cohesive energy assigned to one atom and S is the area per atom on the fracture surface. With regard to Equations 2.1 and 2.2 it reads

$$K_{Ic} \approx \left(\frac{EU}{2S}\right)^{1/2}. \tag{2.3}$$

Values of U can be calculated either *ab initio* or by using semi-empirical interatomic potentials (see the previous chapter), and they can also be experimentally determined as twice the sublimation energy. For most metallic and ceramic crystals, values of U and S are in units of eV/atom and $10^{-19}\,\mathrm{m}^2$, respectively. Thus, according to Equation 2.3, values of fracture toughness in the case of an ideal brittle fracture are as low as $K_{Ic} \in (0.5, 1)\,\mathrm{MPa\,m}^{1/2}$. This range represents a lower-bound physical benchmark for the fracture toughness of engineering materials, and it corresponds well to experimental results achieved in tests with classical ceramic materials such as glasses or porcelain. Similar considerations can also be applied to classical ceramic materials that do not contain macroscopic pre-cracks. Indeed, some pores or microcracks are always present in such materials.

In advanced ceramic materials for engineering applications, however, the level of fracture toughness is substantially enhanced. This can be achieved by microstructurally induced crack tortuosity combined with the presence of many small particles (or even microcracks) around the crack front. In this way the crack tip becomes shielded from the external stress supply and the stress intensity factor at the crack tip reduces. Both the theoretical background and the practical example of that technology are discussed in Section 2.1 in more details. Another method, commonly utilized for an additional improvement of fracture toughness of ceramics, is the distribution of supplied energy to damage mechanisms other than pure cleavage. This can be succeeded, for example, by an enforcement of phase transformations in the vicinity of the advancing crack front [149].

In cracked metallic solids, however, the measured values of K_{Ic} are at least an order of magnitude higher than the lower-bound benchmark. This holds even for ferrite (bcc Fe) at very low temperatures, where almost microscopically smooth cleavage fractures along $\{001\}$ planes appear (note that the $\langle 001 \rangle$ direction in Fe is associated with the lowest ideal tensile strength). The value of related fracture energy was experimentally found to be about $14\,\mathrm{Jm}^{-2}$ [149]. This means that the energy supplied for the unstable fracture is also considered here for the development of localized plastic deformation around the crack tip. Hence, the general thermodynamic criterion for unstable crack growth [19] can be written in the Griffith–Orowan form

$$\frac{1-\nu^2}{E}K^2 \geq 2\gamma + w_p(K,\gamma), \tag{2.4}$$

where $w_p(K,\gamma)$ is the plastic work needed for building the plastic zone at the crack tip. While this work can be neglected in the case of brittle fracture, it is of the same order of magnitude as 2γ in the case of quasi-brittle fracture in metals. Note that the crack tip emission of dislocations in metals already occurs at very low K values in units of $\mathrm{MPa\,m}^{1/2}$ (see Section 3.2 for more details). The dislocations emitted from the crack tip generate an opposite stress intensity factor so that the crack tip becomes shielded from increasing external (remote) loading. The plastic work consumption proceeds until the moment when the sum of external and internal stress intensity factors at the crack tip (the local K-factor) exceeds the critical value necessary for separating atoms to produce new surfaces in an unstable (cleavage) manner [151,152]. This is mathematically expressed in Equation 2.4 so that the plastic work $w_p(K,\gamma)$ is written as a function of both γ and K. Thus, the moment of cleavage fracture is somewhat delayed and, as reported by many authors [153–155], a short stage of stable crack growth often precedes the unstable propagation. The microstructurally induced heterogeneity in the resistance to both the unstable crack growth (γ) and the dislocation emission can, sometimes, produce a series of elementary advances and arrests of the crack tip.

Many quasi-brittle fractures in practice occur as a consequence of pre-existing corrosion dimples, large inclusions or fatigue cracks. However, the localized plastic deformation at favourable sites in the bulk also enables the creation of microcracks as nucleators of the quasi-brittle fracture in solids which do not contain any preliminary defects. At phase or grain boundaries it can be accomplished by many different and well known micromechanisms conditioned by the existence of high stress concentrations in front of dislocation pile-ups. Let us briefly mention another mechanism of crack initiation in bcc metals first introduced by Cottrell [156]. When two edge dislocation pile-ups are driven by the applied stress σ and meet on different $\{110\}$ glide planes in the grain interior, their interaction results in the nucleation of a [001] sessile dislocation. This dislocation can be considered to be a wedge in the {001} cleavage plane. Interaction of n dislocations of Burgers vector b then creates a microcrack with flank opening nb. The work $W = \sigma n^2 b^2$ done by the force σnb acting at the front of n dislocations along the distance nb must be equal to the energy $2\gamma nb$ for the creation of new crack surfaces. This gives the microscopic criterion for quasi-brittle fracture as

$$\sigma_c nb = 2\gamma_s, \tag{2.5}$$

where σ_c is the critical (fracture) stress. Assuming the relation connecting the number of dislocations with the grain size d in terms of the Hall–Petch relation, Equation 2.5 can be rearranged to

$$\left(\sigma_0\sqrt{d} + k_y\right) k_y = \beta G\gamma_s, \tag{2.6}$$

where σ_0 is the yield stress, k_y constant in the Hall–Petch relation (temperature dependent), β the temperature independent constant and G the shear modulus (weakly temperature dependent). Thus, the right-hand side of Equation 2.6 is practically independent of temperature. If the left-hand side is equal to or higher than the right-hand side, the brittle (or quasi-brittle) fracture initiates just at the moment of reaching the yield stress. In an opposite case, the ductile failure occurs after some deformation hardening period. Both the high deformation rate and the low temperature enhance σ_0 as well as k_y, thereby giving rise to quasi-brittle fracture. The same is caused by a large grain size. Thus, the criterion at Equation 2.6 correctly predicts the experimentally observed fracture behaviour. Note that this simple model for single-phase bcc metals is of a two-level type, since the Hall–Petch relation can be easily interpreted by combined atomistic-dislocation considerations [149].

In Section 2.2 a statistical approach to geometrical shielding effects occurring in multi-phase engineering materials is outlined. This two-level concept can be used to give quantitative interpretation of some rather surprising results obtained when measuring the fracture toughness and the absorbed impact energy (notch toughness) of some metallic materials. Examples of such interpretation are documented for ultra-high-strength low-alloyed (UHSLA) steels and Fe-V-P alloys.

Unlike brittle or quasi-brittle fracture, the ductile fracture starts with a rather long period of stable crack or void growth due to the bulk plastic deformation. In the case of pre-cracked solids this means that the surface energy 2γ becomes negligible when compared to the plastic term $w_p(K, \gamma)$ in Equation 2.4, and this criterion loses its sense. Therefore, instead of stress-based criteria (fracture stress, critical stress intensity factor) the deformation-based criteria are more appropriate for a quantitative description of ductile fracture. In the first stage of ductile fracture, microvoids (micropores) nucleate preferentially at the interface between the matrix and secondary phase particles.

The physical reasons are clear: high interfacial energy (low fracture energy), the incompatibility strains (dislocation pile-ups) and the mosaic stresses induced by a difference in thermal dilatations of the matrix and inclusions. Nucleated voids experience their stable growth controlled by the plastic deformation. In the tensile test, for example, the voids become cylindrically prolonged by uniaxial deformation up to the moment when the ultimate strength is reached. Beyond that limit they also expand in transverse directions under the triaxial state of stress inside the volume of developing macroscopic neck.

Although the bulk ductile fracture occurs only very exceptionally in engineering practice, the research of that process is important for forging technologies. Besides the two-scale analysis of plastic deformation, some models of void coalescence during the tensile test are outlined in the last section of this chapter. It should be emphasized that the damage process inside the crack-tip plastic zone of many metallic materials can also be described in terms of the ductile fracture mechanism (e.g., [157]). Therefore, an analytical model that enables a prediction of fracture toughness values by means of more easily measurable ductile characteristics is also presented.

2.1 Brittle Fracture

From the historical point of view, brittle fracture proved to be one of the most frequent and dangerous failures occurring in engineering practice. Besides the well known brittleness of utility ceramics and glasses, metallic materials may also exhibit intrinsically brittle properties dependent on temperature; there exists a critical temperature, the so-called ductile-brittle transition temperature (DBTT) under which the material is brittle, while it is ductile above that temperature. This holds particularly for bcc metals, in which cores of screw dislocation are split into sessile configurations [4,158]. They remain immobile at low temperatures so that,under such conditions, cleavage is a dominant fracture mechanism. However, a steep exponential increase of ductility appears when approaching the DBTT owing to thermal activation helping to increase the mobility of screw segments. Improper application of a material below this temperature can have catastrophic consequences, such as, for example, the sinking of the RMS Titanic nearly one hundred years ago. The material of Titanic, although representing the best-grade steel at that time, was characterized by coarsed grain and high level of inclusions so that DBTT was higher than 32°C. No wonder this ship was catastrophically destroyed by brittle fracture during its impact with the iceberg at the water temperature of −2°C [159].

However, brittleness is often induced by other effects such as flawed material processing or segregation of deleterious impurities at grain boundaries. Grain boundary segregation can result in a local enrichment of thin but continuous interfacial layers throughout the polycrystalline material with concentrations as much as several orders of magnitude higher than that in the grain interior [160]. The most dangerous impurities segregating in bcc iron and steels are phosphorus, tin and antimony. For example, the disintegration of the rotor at the Hinkley Point Power Station turbine generator in 1969 was caused by 50% of phosphorus segregated at grain boundaries of the 3Cr1/2Mo low-alloy steel containing a few tenths of a percent of phosphorus in the bulk [161].

Brittle intercrystalline (intergranular) decohesion caused by impurity segregation exhibits relatively high microroughness of fracture surfaces. Moreover, the secondary cracks identifying the splitting of the main crack front are often observed preferentially at triple points. Both these phenomena lead to the so-called geometrically induced shielding (GIS) of the crack tip that has a favourable effect on decreasing the local stress intensity factor, thereby increasing the fracture toughness. This kind of shielding is one of the so-called extrinsic components of fracture toughness that can be considered as a possible toughening mechanism in the research and technology of advanced materials.

In the next subsections, the theory of GIS and its practical application to an improvement of fracture toughness of brittle materials is outlined.

2.1.1 Geometrically Induced Crack Tip Shielding

Crack front interactions with secondary–phase particles or grain (phase) boundaries in the matrix structure cause deflections of the crack front from the straight growth direction resulting in the microscopic tortuosity of cracks. As already mentioned, such waviness combined with crack branching (splitting) is a natural property of intergranular cracks in metals as well as ceramics. In general, the tortuosity induces a local mixed-mode I+II+III at the crack front even when only a pure remote mode I loading is applied. In order to describe the crack stability under mixed-mode loading, various LEFM-based criteria were proposed (see, e.g., [162–164]). Several of the most frequently used mixed-mode criteria can be found in Appendix B, where conditions of their validity are also briefly described. When selecting a suitable criterion one should note that an unstable brittle fracture in metallic materials is usually preceded by a stable corrosion and/or fatigue crack growth to some critical crack size. During such growth the crack always turns perpendicularly to the direction of maximal principal stress, i.e., to the opening mode I loading. This physically corresponds to minimization of both the crack closure (see Chapter 3 for more details) and the friction so that the rough crack flanks behind the tortuous crack front do not experience any significant sliding contact. Because the crack-wake friction is responsible for somewhat higher fracture toughness values measured under remote sliding modes II and III when compared to those under mode I [164], one can consider an approximate equality $K_{Ic} \approx K_{IIc} \approx K_{IIIc}$ along tortuous crack fronts of remote mode I cracks. Moreover, first unstable pop-ins at these fronts follow, most probably, the local planes of already pre-cracked facets. Consequently, the simplest stability criterion

$$G_{eff} = G_I + G_{II} + G_{III},$$

can be accepted, where G_{eff} is the effective crack driving force. An almost equivalent relation is often used in terms of stress intensity factors:

$$K_{eff} = \sqrt{K_I^2 + K_{II}^2 + \frac{1}{1-\nu}K_{III}^2}. \tag{2.7}$$

For example, in the case of a long straight crack with an elementary kinked tip, it simply reads

$$K_{eff} = \cos^2(\theta/2)K_I, \tag{2.8}$$

where θ is the kink angle. One can clearly see that $K_{eff} < K_I$ for $\theta > 0$. This inequality generally holds for any spatially complex crack front. Hence, the local stress intensity K_{eff} at such a front is always lower than the remote K_I-factor applied to a straight (smooth) crack of the same macroscopic length. The geometrically induced shielding (GIS) effect belongs, according to Ritchie [165], to so-called extrinsic shielding mechanisms. The resistance to crack propagation in fracture and fatigue has, in general, many components that can be divided into two main categories: intrinsic and extrinsic toughening. The first mechanism represents the inherent matrix resistance in terms of the atomic bond strength or the global rigidity, strength and ductility. Appropriate modifications to both the chemical composition and the heat treatment are typical technological ways to improve the intrinsic fracture toughness. On the other hand, processes like kinking, meandering or branching of the crack front, induced mostly by microstructural heterogeneities, belong typically to the extrinsic toughening mechanisms. They reduce the crack driving force and, apparently, increase the intrinsic resistance to crack growth. Thus, the measured fracture toughness can be expressed as a sum of the intrinsic toughness and extrinsic components:

$$K_{Ic} = K_{Ici} + \sum K_{Ice}. \tag{2.9}$$

The standardized procedure for calculation of K_{Ic}-values [166] assumes a planar crack with a straight front and, therefore, does not take the extrinsic shielding effect associated with the crack microgeometry into account. Hence, surprisingly high K_{Ic}-values might be measured, particularly for materials with coarse microstructures and highly tortuous cracks. General expressions for GIS contributions in both brittle and quasi-brittle fracture were derived in [167, 168] by following the approach first introduced by Faber and Evans [169]. In the case of brittle fracture

$$K_{Ici} = \left(\frac{\bar{g}_{eff,r}}{R_A} \right)^{1/2} K_{Ic}, \tag{2.10}$$

where K_{Ic} and K_{Ici} are respectively the measured (nominal) and intrinsic values of fracture toughness, $\bar{g}_{eff,r}{}^{1/2} = \bar{k}_{eff,r}$ is the mean effective k-factor for the tortuous crack front, normalized to the remote K_I ($k_{eff,r} = K_{eff}/K_I$), and R_A is the area roughness of the fracture surface. Equation 2.10 can be derived by the following simple reasoning.

Let us consider a cracked body of a thickness B with an intrinsic resistance G_{Ici} against the crack growth under remote mode I loading. The coordinate system x, y, z is related to the crack front in the usual manner (Figure 2.1). The straight crack front with no geometrical shielding (GIS) represents a trivial case. Here, obviously, the measured fracture toughness value G_{Ic} (or K_{Ic}) is equal to its intrinsic value, i.e., $G_{Ic} \equiv G_{Ici}$ (or $K_{Ic} \equiv K_{Ici}$).

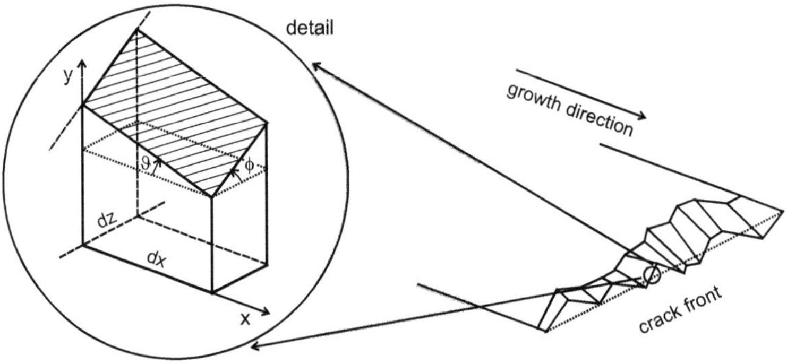

Figure 2.1 Scheme of the tortuous crack front and its segment. Reprinted with permission from John Wiley & Sons, Inc. (see page 265)

When the crack front is microscopically tortuous, a variable local mixed-mode 1+2+3 characterized by g_{eff} or k_{eff} values is present generally at each site along the crack front. During the external loading under increasing remote value G_I, the proportionality $g_{eff} \sim G_I$ or $k_{eff} \sim K_I$ must be valid. Thus, the ratio $g_{eff,r} = g_{eff}/G_I$ can be introduced as independent of G_I but dependent on the crack front tortuosity. Let G_{uI} be the remote crack driving force at the moment of an unstable elementary extension $\mathrm{d}x$ of the crack front. This value is equal to the conventionally measured (nominal) fracture toughness G_{Ic}. Then the nominal elementary energy release rate due to the creation of a new crack surface area $\mathrm{d}x\mathrm{d}z$ is equal to $G_{uI}\mathrm{d}x\mathrm{d}z$. However, the actual (local) elementary energy release rate at the tortuous crack front is

$$g_{eff}\mathrm{d}x\mathrm{d}z = g_{eff,r}G_{uI}\mathrm{d}x\mathrm{d}z.$$

Consequently, the total energy available for the creation of a new surface area $B\mathrm{d}x$ along the crack front can be written as

$$\mathrm{d}W = G_{uI}\mathrm{d}x \int_0^B g_{eff,r}\mathrm{d}z. \tag{2.11}$$

As follows from Figure 2.1, however, the real new elementary surface area $\mathrm{d}S = R_A B\mathrm{d}x$ is greater than $B\mathrm{d}x$ since

$$R_A = \frac{1}{B}\int_0^B \frac{\mathrm{d}z}{\cos\phi(z)\cos\vartheta(z)}. \tag{2.12}$$

In Equation 2.12, R_A is the roughness of the fracture surface and $\mathrm{d}x\mathrm{d}z/(\cos\phi\cos\vartheta)$ is the area of the hatched rectangle in Figure 2.1. Because G_{Ici} is the intrinsic resistance to crack growth, the total fracture energy must be

$$\mathrm{d}W = G_{Ici}\mathrm{d}S = G_{Ici}\, R_A\, B\, \mathrm{d}x. \tag{2.13}$$

Combining Equations 2.11 and 2.13 and denoting

$$\bar{g}_{eff,r} = \frac{1}{B}\int_0^B g_{eff,r}\mathrm{d}z,$$

one obtains

$$G_{uI} \equiv G_{Ic} = \frac{R_A}{\bar{g}_{eff,r}}\, G_{Ici}. \tag{2.14}$$

In general, $G_{Ic} \geq G_{Ici}$ since $\bar{g}_{eff,r} \leq 1$ and $R_A \geq 1$. Therefore, the nominally measured fracture toughness G_{Ic} is usually higher than the intrinsic (real) matrix resistance G_{Ici}. According to the relation $G_{Ic}/G_{Ici} = (K_{Ic}/K_{Ici})^2$, Equation 2.14 can be eventually rewritten to obtain Equation 2.10.

Values of $\bar{g}_{eff,r}$ and R_A must be estimated by using numerical (or approximate analytical) models of the real tortuous crack front combined with appropriate experimental methods for fracture surface roughness determination. In Sections 2.1.2 and 2.1.3, the so-called pyramidal- and particle-induced models are presented. In the context of 2D crack models, the tortuosity is usually described by a double- or even single-kink geometry and $R_A = 1/\cos\theta$ is assumed. In the 2D single kink approximation at Equation 2.8, the crack front is assumed to be straight ($R_A = 1$). Consequently, Equation 2.10 takes the following form:

$$K_{Ici} = \cos^2(\theta/2) K_{Ic}.$$

Besides both the kinking and the meandering, the crack branching can also take place especially in the case of intergranular fracture. This process causes further reduction of SIF ahead of the crack tip and, therefore, Equation 2.10 is to be further modified. According to [170], the crack branching reduces the local SIF approximately to one half of its original magnitude. Let us denote A_b the area fraction of the fracture surface influenced by crack branching. When accepting a linear mixed rule, Equation 2.10 can be then modified as

$$K_{Ici} = \left(\left(\frac{\bar{g}_{eff,r}}{R_A} \right)^{1/2} (1 - A_b) + 0.5 A_b \right) K_{Ic}. \tag{2.15}$$

The area A_b can be determined by measuring the number of secondary cracks (branches) occurring on fracture profiles prepared by polishing metallographical samples perpendicular to the fracture surface [171] (see also Section 3.2). Twice the sum of projected lengths of branches into the main crack path divided by the true crack length yields a plausible estimate of A_b.

When omitting the crack branching and considering Equations 2.9 and 2.10, the extrinsic GIS component of fracture toughness can be simply expressed as $K_{Ice} = (1 - \sqrt{\bar{g}_{eff,r}/R_A})K_{Ic}$. Brittle fracture in metallic materials occurs only when a pure cleavage or intergranular decohesion takes place. In these cases the extrinsic components other than geometrical (such as zone shielding or bridging) can be neglected. In the particular case of cleavage fracture (bcc metals at very low temperatures) one usually observes that $R_A < 1.2$ and $\bar{g}_{eff,r} > 0.9$. This means that GIS is rather insignificant. On the other hand, the extrinsic component K_{Ice} might be very high when the intergranular fracture cannot be avoided (strong corrosion or hydrogen assistance, grain-boundary segregation of impurities and tempering embrittlement of high-strength steels). In that case, however, the favourable effect of the extrinsic component is usually totally destroyed by an extreme reduction of the intrinsic component K_{Ici}. Nevertheless, one can still improve the fracture toughness of both metals and ceramics by increasing the extrinsic (shielding) component without the loss of general quality in mechanical properties (see Sections 2.1.2, 2.2.2 and 3.2.6).

2.1.2 Pyramidal Model of Tortuous Crack Front

A plausible assessment of the GIS effect is possible only when the following steps can be realized:

1. building of a realistic model of the crack front based on a 3D determination of fracture surface roughness;
2. calculation of local normalized stress intensity factors k_{1r}, k_{2r} and k_{3r} along the crack front;
3. calculation of the effective stress intensity factor $k_{eff,r}$.

The first step can be achieved by the 3D reconstruction of fracture morphology. The second problem can be solved, for example, by using the software package FRANC3D based on the boundary element method [172]. The third step is solvable by standard mathematics. A nearly exact numerical solution by means of the FRANC3D code is, however, usually extremely time consuming. Therefore, a simple pyramidal model of the crack front was proposed for approximate analytical estimations [168, 173]. This model is based on a pyramid-like periodic approximation of the tortuous crack front, each element of which is characterized by the twist angle Φ and the highest tilt angle Θ_m towards the macroscopic crack plane; see Figure 2.2.

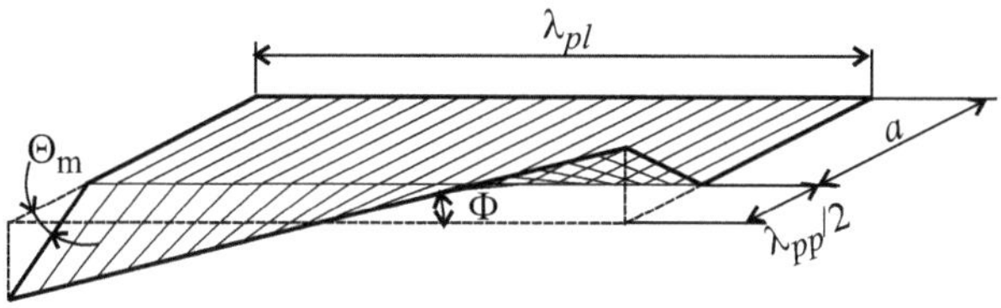

Figure 2.2 A periodic element of the pyramidal model of tortuous crack front

The profile roughness R_L (measured along the crack front) and the periodicity λ_{pl} (λ_{pp}) measured parallel (perpendicular) to the crack front are associated with the angles Φ and Θ_m by the following simple equations:

$$\lambda_{pp} \tan \Theta_m = \lambda_{pl} \tan \Phi, \qquad R_L = \cos^{-1} \Phi. \tag{2.16}$$

The characteristic periodicities λ_{pl} and λ_{pp} can be determined either by the Fourier analysis of roughness profiles measured at appropriate locations on the fracture surface, or simply identified with a characteristic microstructural periodicity, e.g., with the mean grain size. The effective stress intensity factor $k_{eff,r}$ (normalized to the remote K_I factor) at each point of the pyramidal front can be calculated by using Equation 2.7 with the following approximate analytical expressions for local stress intensity factors:

$$\begin{aligned}
k_{1r} &= \cos\left(\frac{\Theta}{2}\right)\left[2\nu \sin^2 \Phi + \cos^2\left(\frac{\Theta}{2}\right)\cos^2 \Phi\right], \\
k_{2r} &= \sin\left(\frac{\Theta}{2}\right)\cos^2\left(\frac{\Theta}{2}\right), \\
k_{3r} &= \cos\left(\frac{\Theta}{2}\right)\sin \Phi \cos \Phi \left[2\nu - \cos^2\left(\frac{\Theta}{2}\right)\right].
\end{aligned} \tag{2.17}$$

The results calculated according to Equation 2.17 are sufficiently accurate provided that $\lambda_{pp} \ll 2a$, where a is the pre-crack length. The global effective factor $\bar{k}_{eff,r}$, averaged for the periodic crack front geometry composed of identical pyramidal elements, can then be computed as

$$\bar{k}_{eff,r} = \frac{\pi - 2}{2\Theta_m(2R_L + \pi - 4)} \int_{-\Theta_m}^{\Theta_m} \left(k_{1r}^2 + k_{2r}^2 + \frac{k_{3r}^2}{1-\nu}\right) d\Theta. \tag{2.18}$$

Comparison of results obtained by means of the pyramidal model and the FRANC3D code revealed that, in the whole range of both the surface roughness and the roughness periodicity typical for real intergranular surfaces, the difference lies within the 10% of error band [168]. Although the pyramidal model yields very promising results predominantly in the case of intercrystalline fracture (see Section 2.2.2), it can also be quite successfully applied to other brittle fracture modes, as shown in the next section.

2.1.3 Fracture Toughness of Particle Reinforced Glass Composite

Traditional ceramic materials such as glass or porcelain possess amorphous microstructures. An absence of crystallographically conditioned dislocations makes these materials extremely brittle. However, the very low intrinsic fracture toughness of glass in the range $K_{Ici} \in (0.5, 1)\,\mathrm{MPa\,m^{1/2}}$ may be improved, for example, by reinforcing with second constituents with high modulus, high strength and/or high ductility in the form of fibres, whiskers, platelets or particulates embedded into the matrix [174, 175]. A successful example of ceramic platelet reinforcement of glass is the borosilicate glass/Al_2O_3 platelet composite that was first introduced by Boccaccini *et al.* [176]. Based on this system, environmentally friendly and cost-effective materials can be produced as alumina platelets for the building industry or as abrasives for the polishing industry. The enhancement in fracture toughness can be ascribed here to four concurrent phenomena [176–180]: the Young's modulus increment resulting from the platelets addition (the intrinsic component), the presence of a compressive residual stress in the glass matrix, the crack tip shielding produced by platelets and the crack deflection mechanism (extrinsic components). The shielding effect is a result of local mixed-mode I+II+III induced by rigid particles surrounding the crack tip. The crack deflection is forced particularly by a necessity to bypass rigid particles when searching the direction of the highest crack driving force (compare Section 2.2.1). This leads to a zig-zag crack propagation in between the platelets (crack tortuosity) and a reduction of the crack driving force in comparison to that of the straight crack. This must be associated with an enhanced microroughness of fracture surfaces. A direct correlation between the roughness of the fracture surface and the fracture toughness of dispersion reinforced ceramic and glass composites has been suggested and experimentally proved [180–182]. Because these systems provide an excellent possibility to verify theoretical GIS models, quantitative assessments of all the above-mentioned intrinsic and extrinsic effects have been performed [183,184]. Moreover, extended experimental analysis of fracture toughness, fracture surface roughness and microstructure was performed on samples made of borosilicate glass containing different volume fractions of alumina platelets.

2.1.3.1 Experimental Procedure and Results

The experimental glass matrix composite was fabricated via powder technology and hot pressing. Alumina platelets of a hexagonal shape, with major axes between $5 - 25\,\mu m$ and axial ratio of 0.2, were used. The commercially available borosilicate glass was selected for the composite matrix. The microstructure of specimens containing 0, 5, 10, 15 and 30 vol.% of platelets [176] consisted of a dense glass matrix with a more or less homogeneous distribution of platelets. A strong bond between the matrix and the platelets was confirmed by transmission electron microscopy [185]. Upon cooling from the processing temperature, the thermal expansion mismatch between matrix and reinforcement induces tangential compressive and radial tensile residual stress in the matrix around the particles. Fracture toughness values were obtained using test pieces of a standard cross-section ($3 \times 4\,mm^2$) with the chevron notch machined by an ultra thin diamond blade. A Zwick/Roell electromechanical testing machine was utilized for the three-point bending test with a span of 20 mm. Scanning electron microscopy (SEM) was used for the fractographic analyses of fracture surfaces. Roughness parameters were measured by the optical profilometer MicroProf FRT based on a chromatic aberration of its lens. The device works with vertical resolution of 3 nm and lateral resolution of about $1\,\mu m$. A three-dimensional reconstruction of surface topography was performed by means of the software Mark III. The surface roughness was quantified by the average area roughness, R_A, defined on the basis of the ISO 4278 norm as the arithmetic mean of the deviations of the roughness profile from the central line. The profile roughness R_L, defined in a standard manner as the true profile length divided by its projected length, was also determined. The profiles obtained from 3D fracture surface morphology quantification were subjected to Fourier analysis in order to determine the characteristic periodicities λ_{pp} and λ_{pl}. The measured values for all specimens are displayed in Table 2.1.

Note that values of λ_{pp} are an order of magnitude lower than the double-length of the pre-crack ($2a = 4\,mm$) which ensures a reasonable validity of the pyramidal model.

Dependencies of both the relative area roughness $R_r = R_A(X\%)/R_A(0\%)$ and the average fracture toughness on different alumina platelet volume contents (X%) are shown in Figure 2.3. It is seen that both curves increase linearly with increasing content of alumina platelets in the matrix approx-

Table 2.1 Characteristics of the pyramidal model related to tortuous cracks in measured specimens

Al_2O_3 (vol.%)	R_L	λ_{pp} [μm]	λ_{pl} [μm]	Θ_m	$\bar{k}_{eff,r}$
0	1.011	373	114	0.0455	0.983
5	1.053	412	171	0.3178	0.924
10	1.199	102	32	0.2040	0.763
15	1.115	341	170	0.2410	0.719
30	1.229	102	128	0.7311	0.714

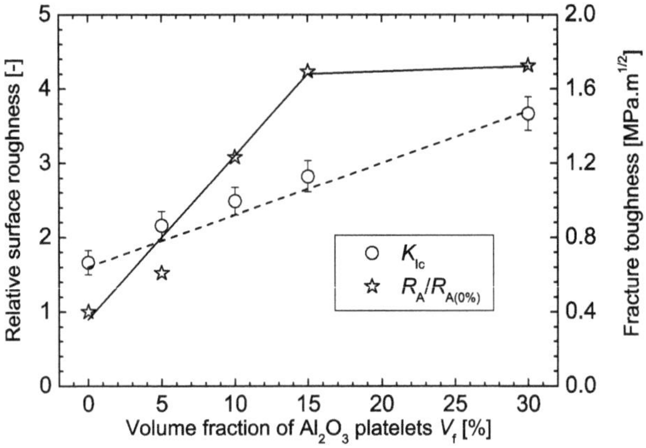

Figure 2.3 Dependence of the relative surface roughness and the fracture toughness on the volume content of alumina platelets in the glass matrix. Reprinted with permission from Elsevier B.V. (see page 265)

imately up to $X = 15\%$. The increase in roughness is, unlike that of the fracture toughness, effectively stopped at higher platelet contents. This also means that other mechanism(s) should be acting to counteract the loss of effectiveness of crack deflection here. Typical examples of reconstructed fracture surfaces obtained from the profilometric measurement for both 0 and 30 vol% of alumina platelets are depicted in Figure 2.4. It is evident that the fracture surface roughness was significantly increased when alumina platelets were incorporated into the borosilicate glass matrix.

At the highest volume fraction of alumina platelets (30 vol%), however, platelet clusters are already observed as shown in Figure 2.5. It seems to be plausible that the crack front interacts with the whole cluster rather than with all its individual platelets. Thus, some particles inside clusters do not directly contribute to the crack front deflection (the surface roughness).

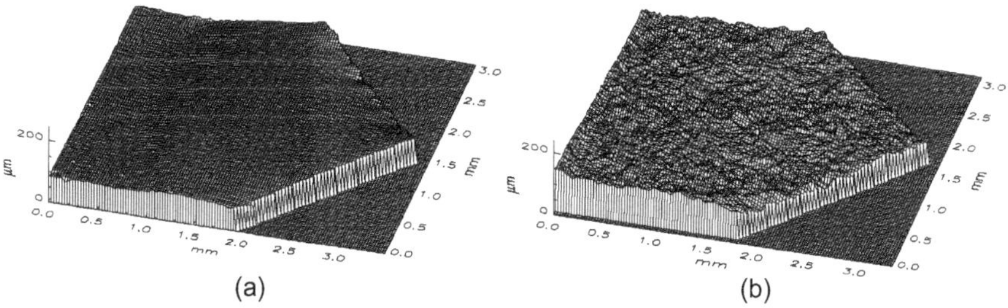

Figure 2.4 Three-dimensional reconstructed fracture surfaces for 0 and 30 vol% of alumina platelets. Reprinted with permission from Elsevier B.V. (see page 265)

Figure 2.5 Clusters of platelets in the sample with 30% reinforcement volume. Reprinted with permission from Elsevier B.V. (see page 265)

2.1.3.2 Theoretical Assessment of the Shielding Effect

Besides the roughness-induced shielding (RIS), the crack tip shielding caused by surrounding rigid particles has to also be considered. This effect can be approximately assessed according to results reported in [178, 183]. In these works, the shielding effect produced by rigid circular particles was analyzed in the frame of the 2D ANSYS model based on the finite element method. The presence of such inclusions generally induces the mixed-mode I+II at the tip of the straight crack.

The rigid particles possessed 20 times higher Young's modulus than the matrix. Particles of different sizes (diameter $d = 6, 12, 30, 60, 120, 240\,\mu\text{m}$), spaced by $l = 30, 60, 90, 120\,\mu m$, were considered. Moreover, geometrically identical particles of negligibly small moduli (holes) were also studied for comparative reasons. Note that the range $d/l \in (0.1, 3)$ corresponds to the particle volume fraction $f_p \in (0.04, 25)\%$. Several thousands of possible positions of the crack tip were analyzed within an investigated area in between a pair of spherical particles, according to the scheme in Figure 2.6. This area was long enough to incorporate the influence of further neighbouring pairs of particles (behind and ahead of the investigated area, dashed lined) which were not explicitly considered in the analysis.

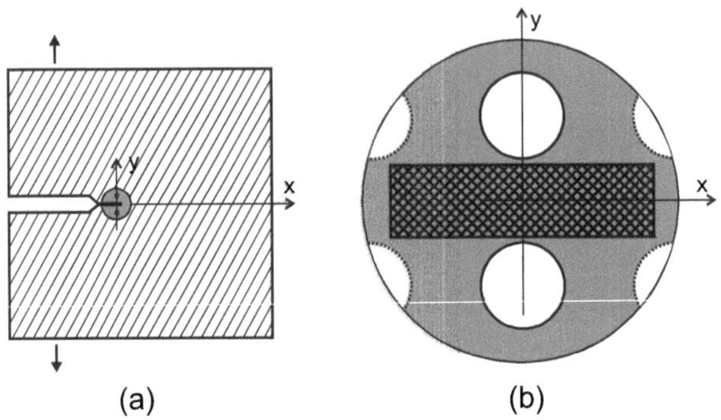

Figure 2.6 The scheme related to the model of particle-induced crack tip shielding: (a) position of the investigated region in the testing sample, and (b) detail of the region (*black rectangle*) and circular particles

This enabled us to generalize the results to a periodic square network of particles by multiplication of normalized effective SIFs in the points which lie within both the left-hand and the right-hand parts of the investigated region and are associated owing to the translation periodicity.

The effective SIF ($k_{eff,r} = \sqrt{k_{I,r}^2 + k_{II,r}^2}$) was used to assess the effective crack driving force. Averaged values of $\bar{k}_{eff}$ as functions of the ratio d/l for all analyzed types of particles are displayed in Figure 2.7. One can see that the rigid inclusions start to produce some shielding after reaching the critical value $(d/l)_c = 0.2$ (or $f_{pc} = 0.5\%$). Then the normalized effective k-factor rather slowly drops to the value of 0.9 that corresponds to $(d/l) = 3$ (or $f_p = 25\%$). Practically the same decrease was identified in the case of holes. Here, however, the drop was shifted to a higher critical value $(d/l)_c = 1$ (or $f_{pc} = 6.5\%$). Despite this rather slight difference, the shielding effect of both rigid particles and holes appeared to be similar.

In order to asses the RIS, the pyramidal model was applied in the first approximation by using the roughness characteristics from Table 2.1. The dependence of the relative fracture toughness $K_{Ic}(X\%)/K_{Ic}(0\%)$ (where $K_{Ic}(0\%) = K_{Ici} = K_{Icm}$), calculated using Equations 2.16, 2.17 and 2.18, on the volume fraction of Al_2O_3 platelets is plotted by the dashed line in

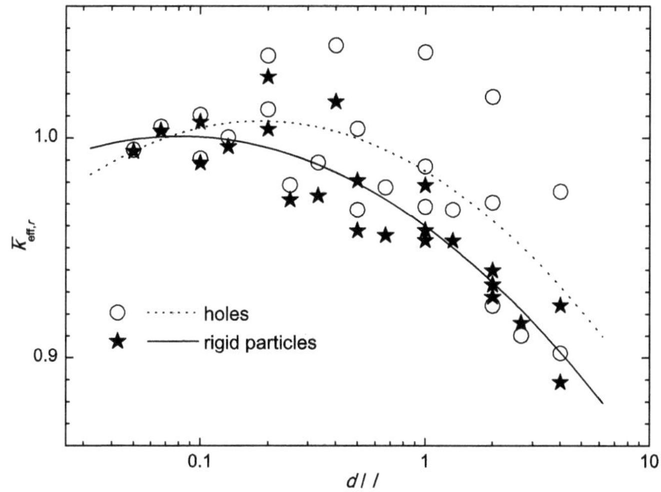

Figure 2.7 Averaged values of the effective stress intensity factor as a function of the particle size/spacing ratio for all analyzed types of particles

Figure 2.8 along with the experimental data. As expected, the maximal predicted relative increase of 40% in the fracture toughness cannot fully explain the real improvement of 120% that was achieved by the 30% volume fraction of platelets.

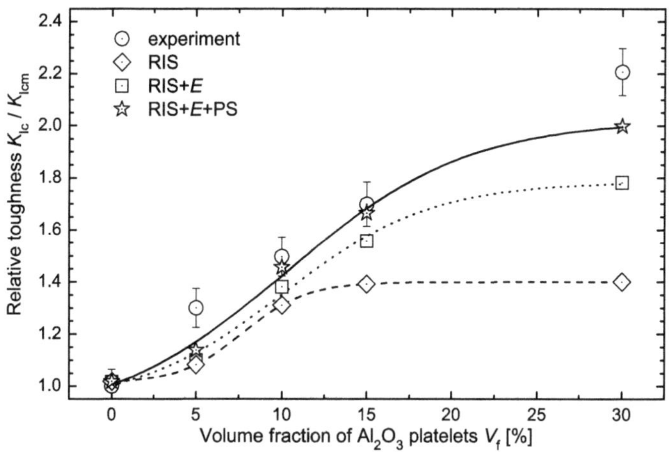

Figure 2.8 Theoretical curves of the relative fracture toughness as functions of the percentage of Al_2O_3 particles in comparison with experimental data. The *full line* shows the theoretical prediction including all considered corrections

Indeed, one must also consider the decrease in the crack driving force related to both the increase in Young's modulus and the shielding induced by platelets [177]. The increase in fracture toughness due to the increase in Young's modulus can be calculated from

$$\frac{K_{Ic}}{K_{Icm}} = \sqrt{\frac{E}{E_m}\frac{G_c}{G_{Icm}}}, \tag{2.19}$$

where E is Young's modulus of the composite and $E_m = E(0\%)$ is Young's modulus of the matrix, the values of which are given in Table 2.2 [176]. The improved prediction (RIS+E) including this effect is shown by the dotted line in Figure 2.8.

Because the difference in shielding produced by rigid particles and holes was found to be negligible, one can use the result plotted in Figure 2.7 as a further correction of the theoretical curve. By considering the relevant volume fractions of experimental samples (the value $\bar{k}_{eff,r} = 0.9$ was used for $X = 30\%$), the final theoretical prediction (RIS+E+PS) is shown by the full line in Figure 2.8.

Table 2.2 Young's moduli of borosilicate glass matrix composite containing different volume fractions of Al_2O_3 platelets

Platelets content [vol.%]	E [GPa]
0	63
5	65
10	70
15	79
30	102

One can see that the agreement between theory and experiment is reasonable.

A somewhat more complex and exact model was proposed by Kotoul *et al.* [184]. This model considered several additional toughening mechanisms, such as compressive residual stresses or crack front trapping at platelets that may be effective in these composites. Moreover, it could explain the experimental fact that the fracture roughness ceased to increase from about 15vol% of Al_2O_3. The model follows from the theory of particle-induced crack deflection that was developed by Faber and Evans. However, some errors in the expression for the strain energy release rate, appearing in their original paper [169], had to be corrected. After relevant modification, the following equation for the normalized effective crack driving force was obtained:

$$g_{eff,r} = \cos^2\frac{\lambda}{2}\left(2\nu\sin^2\phi + \cos^2\frac{\lambda}{2}\right)^2\cos^4\phi + \\ + \cos^2\phi\sin^2\frac{\lambda}{2}\cos^4\frac{\lambda}{2} + \\ + \frac{\cos^2\frac{\lambda}{2}\sin^2\phi\cos^2\phi}{1-\nu}\left(2\nu - \cos^2\frac{\lambda}{2}\right)^2. \tag{2.20}$$

Here Θ is a tilt angle and Φ is a twist angle of crack front elements induced by their interaction with platelets. Note that Equation 2.20 possesses the required limiting properties, i.e.,

$$\lim_{\phi\to\pi/2} g_{eff,r} \to 0, \qquad \lim_{\phi\to 0} g_{eff,r} \to \cos^4\frac{\theta}{2}$$

which was not the case in the originally derived expressions in [169]. Details concerning the calculation of the averaged crack driving force $\bar{g}_{eff,r}$ can be found elsewhere [184]. This solution also involves the contribution of the change in Young's modulus according to Equation 2.19. The theoretical prediction was in very good agreement with experimental data. This result revealed that, most probably, the contributions of residual stresses as well as crack trapping could be negligible. Indeed, as shown in [184], the presence of high residual radial tensile stresses along the platelet circumference leads to crack front propagation around the particle to relieve these stresses (no crack trapping). Simultaneously, however, the segment of the crack front propagating through the matrix is shortened and the corresponding twist angle decreases which results in flattening of the crack front in the matrix. This raises the energy release rate $\bar{g}_{eff,r}$ and makes it easier for the crack propagation in the tangential compressive stress field within the matrix. As a result, the net toughening increment given by $\bar{g}_{eff,r}$ remains unchanged and the residual stress contribution also does not take any effect.

Taking the above-mentioned considerations into account, the peculiar occurrence of a plateau in the plot of fracture surface roughness as a function of platelet volume fraction (Figure 2.3) can also be elucidated. There are two contributions to the surface roughness related to (1) crack propagation around the platelets and (2) crack propagation within the matrix. The former grows with increasing platelet concentration. The latter decreases with increasing platelets concentration because the fracture surface in the matrix flattens. Moreover, due to clustering of platelets, their vicinity becomes less effective at deflecting cracks and, as a result, the corresponding contribution to the surface roughness further decreases. Beyond about 15% volume fraction of platelets, the positive and the negative contributions to the surface roughness mutually compensate and the increase in surface roughness stops.

In summary, one can say that the models based on coupled shielding effects are able to elucidate quantitatively the increase in fracture toughness caused by particle reinforcement of glass-based ceramics.

2.2 Quasi-brittle Fracture

Microscopically smooth cleavage cracks, observed in ferrite at very low temperatures, possess the surface energy of $14\,\mathrm{Jm}^{-2}$ which is much higher than that of about $1\,\mathrm{Jm}^{-2}$ related to the lower-bound benchmark for ideal cleavage cracks. A satisfactory explanation gives the cleavage mechanism based on alternative short-range dislocation slip proposed by Knott [186]. A pair of $(1/2)\langle 111\rangle(1\bar{1}0)$ and $(1/2)\langle 1\bar{1}\bar{1}\rangle(110)$ edge dislocations are emitted from the cleavage crack front [010] to a distance of several Burgers vectors to produce an elementary crack advance in [100] direction. After that advance the dislocations immediately disappear on newly created surfaces as a consequence of attractive mirror forces. In this way, microscopically flat fracture surfaces with neither slip markings nor dislocations in their vicinity might be produced. The fracture energy associated with that process can be assessed as $12\,\mathrm{Jm}^{-2}$ which is close to the above-mentioned experimental value.

However, this kind of perfectly smooth cleavage crack is the exception rather than the rule. The cleavage fracture in metallic materials is, even at low temperatures, usually accompanied by clear microscopic traces of local plastic deformation. The so-called steps, tongues and fishbones are produced by interaction of the crack front with dislocation tangles or twins [187]. The most typical morphological features are so-called river markings that are created by shear connections of steps originating at grain boundaries with non-zero twist component. They are enforced by a necessity of gradual re-initiation of the cleavage crack when penetrating to a twisted cleavage plane in the adjacent grain (see Figure 2.9). In metallic materials, therefore, at least a small plastic zone at the crack tip is always to be expected. Even in the case of intergranular fracture along strongly weakened grain boundaries, at least several percent of cleavage and/or dimple fracture facets are always present on the fracture surface. This experimental observation was proven by a theoretical model [188] showing that a pure intergranular crack front would, during its advance, require higher tortuosity and more spatial geometrical complexity [189]. This would demand a steadily increasing fracture energy to a unit projective area. Consequently, at rather early crack growth stages, the cleavage of some of the largest grains becomes more energetically favourable than their intergranular decohesion. Thus, the fracture morphology of quasi-brittle cracks consists of a mixture of intergranular facets (microscopically tortuous) and transgranular cleavage or dimple facets (microscopically nearly straight). The intergranular, cleavage and dimple fractures represent brittle, quasi-brittle and ductile components of the fracture process, respectively. Thus, in spite of a macroscopically brittle appearance, the fracture processes in metallic materials are to be considered as quasi-brittle.

It is well known that LEFM can be successfully used for the description of quasi-brittle fracture only when the plastic zone size

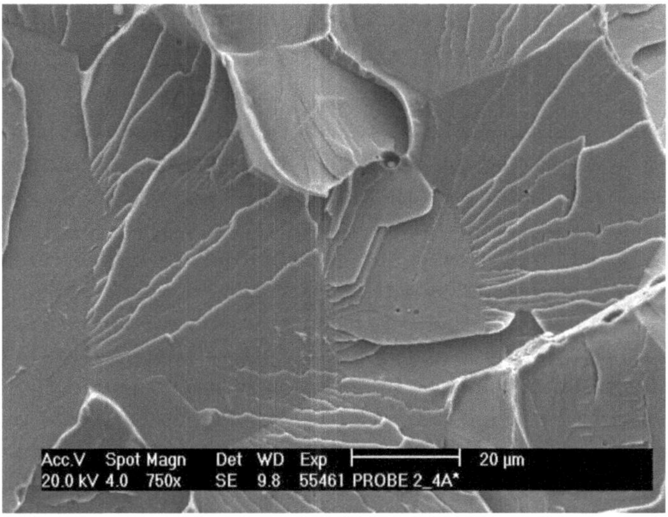

Figure 2.9 A typical picture of river markings on cleavage facets (mild steel)

$$r_p = \frac{1}{\xi}\left(\frac{K}{\sigma_y}\right)^2, \tag{2.21}$$

where $\xi \approx 2\pi$, is very small in comparison to the crack length a and the thickness B of the component. Under such conditions of small-scale yielding and plane strain the laboratory tests of fracture toughness K_{Ic} remain more or less invariant with respect to both the specimen shape and the crack location. Indeed, the elastic-plastic stress-strain behaviour of even very ductile materials becomes quasi-brittle inside the small plastic zone (see Section 2.3.2). When also taking the constraint effect into account, which means comparable values of T-stress (the second term in the Williams expansion) depending on the ligament w between the crack tip and the free surface of the specimen and the component, the K_{Ic} values obtained in the laboratory tests can be more safely transferred to large engineering components [190, 191]. However, the plastic zone size is also an important quantity with respect to the microscopic mode of crack propagation that is affected by materials microstructure. In that respect, naturally, it should be related to a characteristic microstructural distance d, e.g., to the grain (particle) size or interparticle (interphase) spacing. Such a parameter is called the size ratio. Similarly to the brittle fracture case shown in Section 2.1, the microscopic mode of crack propagation might also have a strong impact on the resistance to unstable crack growth in the case of quasi-brittle failures. In this connection, the size ratio plays a very important role. This will be demonstrated in the next subsections in a theoretical as well as an experimental manner.

2.2.1 Statistical Approach to Geometrical Shielding Based on Size Ratio Effect

A lot of experimental evidence about the special role of the size ratio $S_R = d/r_p$ was collected during the last three decades of fracture and fatigue research. Hitherto, only the mean values d_m or $S_{Rm} = d_m/r_p$ were usually taken into consideration. Numerous experiments [192–194] revealed that the crack path is particularly influenced by microstructure (grain boundaries, phase boundaries, precipitates, inclusions) when the plastic zone size is comparable to or smaller than the mean characteristic microstructural distance, i.e., when $S_{Rm} \geq 1$. Microscopically tortuous, intergranular or crystallographic fracture morphology is predominantly observed under these circumstances. On the other hand, the crack path becomes insensitive to microstructure when the plastic zone size embraces many microstructural elements, i.e., when $S_{Rm} \ll 1$. In this case, ductile dimples or non-crystallographic quasi-cleavage are observed on rather smooth and straight transgranular fracture surfaces. In fatigue, many authors [195–197] reported the maximum percentage of intergranular facets at those sites of the fracture surface, where the cyclic plastic zone size was exactly equal to the mean grain size ($S_{Rm} \approx 1$).

The theory of yield stress gradient effects in inhomogeneous materials (e.g., [198]) or the discrete dislocation models in fatigue (e.g., [199]) yield a general basis for elucidation of the above-mentioned phenomena. According to the former theory, the interaction between the crack tip and the near-inhomogeneous region becomes significant only when $S_{Rm} \geq 1$. Following this interaction, the crack will either circumvent high-strength heterogeneity or deflect to the low-strength one. Similarly, the discrete dislocation models reproduce well the strong interaction of crack tip with grain boundaries when the crack tip plasticity is constrained within one or a small number of grains. Once the plastic zone size becomes much higher than the mean grain size, the interaction effects decay and the crack growth rate starts to be quite unaffected by the microstructure. The maximum of intergranular morphology at $S_{Rm} \approx 1$ in fatigue can be explained by considering intergranular (or interfacial) misfit strains [168]. If the above condition holds, grains of the mean size closely adjacent to the fatigue crack front become, unlike their neighbours (more distant to the crack front), cyclically plastically deformed. The related cyclic mismatch stresses at boundaries between neighbouring grains can preferentially lead to intergranular (or interphase) relaxation cracking. Due to only slightly asymmetric Weibull (or log-normal) probability density of the grain size in metallic materials [200–202], the grain sizes close to the average are the most probable ones. Thus, the large strain mismatch causing intergranular fracture becomes most probable just when the condition $S_{Rm} \approx 1$ becomes fulfilled at the advancing fatigue crack front.

In most engineering materials, the scatter of both the grain size and the particle spacing causes variation of the parameter d within more than two orders of magnitude along the crack front. On the other hand, a sharp decrease of the stress with distance from the crack front leads to localization of the plastic deformation within a narrow zone of approximately constant width along the whole crack front. Therefore, the plastic zone size follows the Hall–Petch rule in terms of the mean grain size and the yield stress [203], and remains rather invariable along the crack front with the exception of plane stress regions close to free specimen surfaces. As a consequence, the size ratio S_R follows the grain size distribution and varies in a wide range along the crack front inside the specimen. In other words, there are many sites at the crack front where $S_{Rm} \ll 1$ or $S_{Rm} \geq 1$ is to be expected. However, only the latter locations contribute to the microscopic tortuosity that induces GIS, while the effect of the former ones can be neglected. Therefore, the statistical approach seems to be the most relevant way to describe the GIS effect in quasi-brittle materials.

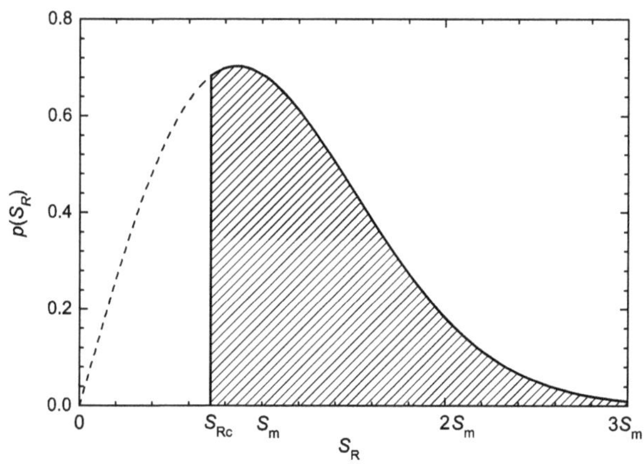

Figure 2.10 Probability density function $p(S_R)$ in terms of the size ratio S_R. The ratio of *hatched and white areas* under the curve determines the probability of finding a geometrically shielded element at the crack front

The basic idea of the statistical approach, first introduced in [200], lies in an assumption that the microstructural elements can be simply divided into two main categories of low and high S_R. It means that the low S_R part of the probability density function does not influence the shielding phenomenon that is controlled by the high S_R part. This is clearly seen in Figure 2.10, where the area under the probability curve is divided into two parts. Only elements falling into the hatched part contribute to the shielding effect at the crack front. The S_R value determining the boundary between both parts is denoted by S_{Rc}. The two-part concept is introduced for the sake of both clarity and simplicity because, in fact, some transient range must exist. The S_{Rc} value is expected to vary, for example, with the impurity concentration at grain boundaries in a particular material: Because of generally higher probability of intergranular cracking, the higher the impurity concentration the lower S_{Rc} value is to be assumed. Therefore, this value is to be considered as a free (fitting) parameter in the statistical analysis. Nevertheless, it should lie somewhere within the transient range $S_{Rc} \in (0.2, 1.0)$.

Denoting the probability density function $p(S_R)$, the relative length of the crack front contributing to GIS can be expressed as

$$\eta = \int_{S_{Rc}}^{\infty} S_R p(S_R) \mathrm{d}S_R.$$

In other words, the parameter η means the probability of finding a shielded element at the crack front. This statistical parameter is suggested to be a suitable measure of the GIS efficiency and will be used in further analysis. For a particular material, the value of η can be calculated when determining both the yield stress σ_y from the tensile test (in order to estimate r_p according to Equation 2.21) and the statistical distribution of d from the metallographic sample. Fortunately, the two-parameter (ξ, S_{Rm}) Weibull distribution, where $\xi \approx 2.2$, can be successfully used for all engineering materials [200, 201]. Consequently, one can easily show [11, 204] that the parameter η can be estimated on the basis of the yield stress and the mean grain size only:

$$\eta = \exp\left[-\left(\frac{0.886\, r_p}{d_m}\right)^{2.2}\right]. \tag{2.22}$$

The portion p_t of tortuous intergranular or crystallographic morphology of the fracture surface can be experimentally determined by using SEM. Indeed, such morphology is clearly different from that of rather straight transgranular facets (quasi-cleavage or ductile dimples). The tortuous part can also be distinguished by topographical methods utilizing various roughness parameters (see Section 3.1 in more detail). Note that, according to the Cavallery principle [201], the area-based value of p_t is equal to the tortuous portion of crack front. As shown above, however, this portion can also be assessed by setting $p_t = \eta$. Let us assume the proportional rule for a mix of the tortuous and straight morphology, and note that the probability of finding the tortuous element of the crack front (which produces shielding) is equal to η. Then the following modification of Equation 2.15 can be utilized:

$$K_{Ici} = \left[1 - \eta + \eta\left(\left(\frac{\bar{g}_{eff,r}}{R_A}\right)^{1/2}(1 - A_b) + 0.5A_b\right)\right] K_{Ic}. \tag{2.23}$$

Obviously, Equation 2.23 reduces to Equation 2.15 for $\eta = 1$ (pure brittle mode, GIS along the whole crack front), to Equation 2.10 for $\eta = 1$ and $A_b = 0$ (GIS along the whole crack front, no branching) or to the identity

$K_{Ici} = K_{Ic}$ for $\eta = 0$ (pure quasi-cleavage and/or ductile mode, no GIS). The term in the square brackets is called the static shielding factor [168,171].

2.2.2 Anomalous Fracture Behaviour of Ultra-high-strength Steels

An interesting application example of the statistical approach is a quantitative interpretation of the peculiar fracture behaviour of ultra-high-strength low-alloy (UHSLA) steels. During the 1970s and 1980s, many authors (e.g., [193, 205–208]) reported an unexpected increase in fracture toughness with increasing prior austenite grain size (or austenitizing temperature). This is clearly documented by the experimental data of various authors in Figure 2.11, where the mean prior austenite grain size is varied over a wide range $d_m \in \langle 5, 265 \rangle$ µm. On the other hand, the impact absorbed energy, as expected, dramatically decreased as shown in Figure 2.12 for the US military 300M steel (AISI 4340 with enhanced silicon content). Such contradictory behaviour is rather exceptional since values of fracture toughness and absorbed energy are usually well correlated [209].

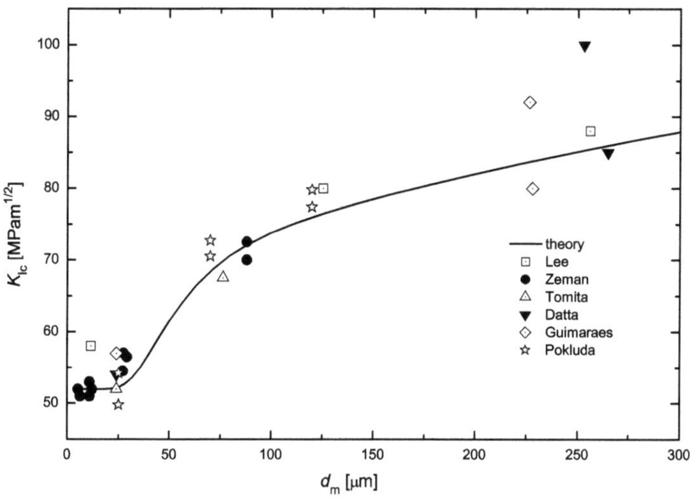

Figure 2.11 Fracture toughness of UHSLA steels as a function of the mean prior austenite grain size d_m. Reprinted with permission from John Wiley & Sons, Inc. (see page 265)

This anomalous behaviour could be explained in a satisfactory manner only after recognizing the difference in fracture modes between fine- and coarse-grained steel grades [210]. While the fine-grained samples exhibited a transgranular dimple fracture morphology, the coarse-grained ones fractured

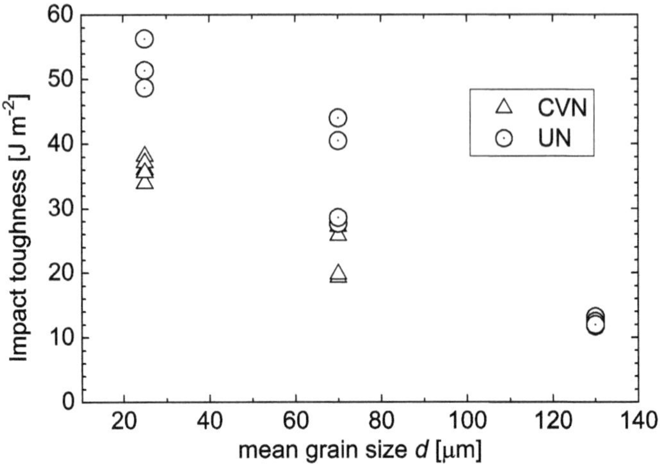

Figure 2.12 Experimental data of the absorbed impact energy in dependence on the mean prior austenite grain size

by intergranular decohesion along prior austenite grain boundaries as documented in Figure 2.13 for the 300M steel. Thus, during the measurement of fracture toughness K_{Ic}, the short stable stage of intergranular crack growth in coarse grained specimens was accompanied by a high level of GIS at the fatigue pre-crack tip. However, the first models [210–212] that attempted to estimate quantitatively the GIS level were only partially successful. The reasons were their two-dimensional and deterministic characters. Only the subsequent, statistically based approaches [168, 213] have led to very reasonable agreement between the theory and experiment. The statistical parameter was determined using Equation 2.22 and the fracture toughness K_{Ic} was assessed according to Equations 2.18 and 2.23 ($A_b = 0$). Nearly the same value of the yield stress was measured for all microstructures since this value is controlled by the martensitic matrix (not by the prior austenite grain size). Thus, $\sigma_y = 1500\,\text{MPa}$ and $S_{Rc} = 0.5$ were considered in the calculations of GIS. The computed values of η are plotted as a function of the mean austenite grain size d_m in Figure 2.14. This function reproduces very well the fraction of intergranular morphology of samples with different mean grain size. In particular, specimens having $d_m < 20\,\mu\text{m}$ fractured in a pure transgranular dimple mode, whereas for those of $d_m > 150\,\mu\text{m}$ the fracture was fully intergranular.

The characteristic periodicities λ_{pl} and λ_{pp} in the pyramidal model were identified with the mean austenite grain size and $R_A = 1.6$ was used as a typical value for intergranular fracture surfaces [214]. The identity $K_{Ici} = 52\,\text{MPa}\,\text{m}^{1/2}$ that corresponds to the fracture toughness of steels with finest grains (no GIS) was accepted for all steel grades. Indeed, owing to the extreme purity of UHSLA steels, the intrinsic fracture toughness associated with both

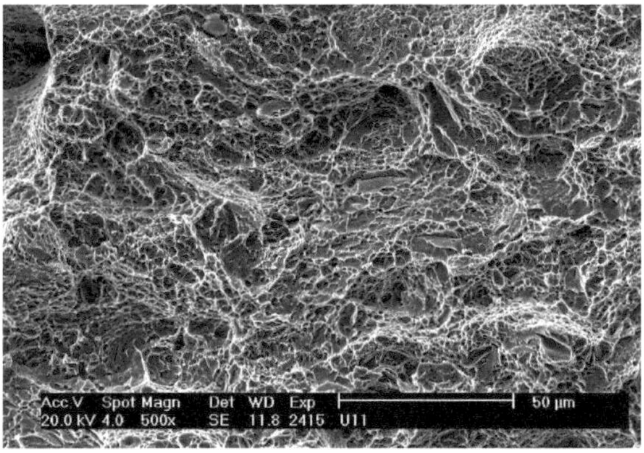

(a)

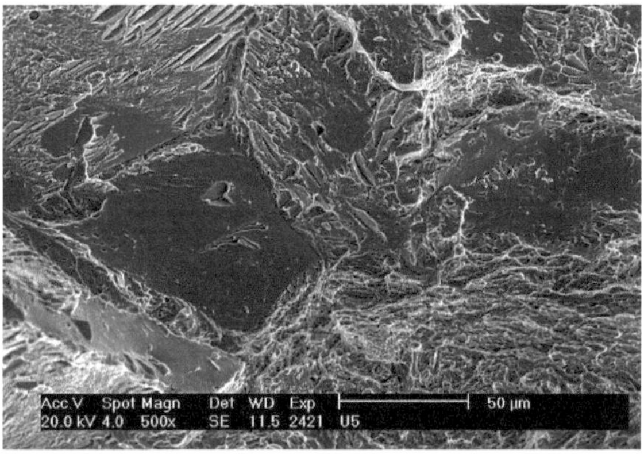

(b)

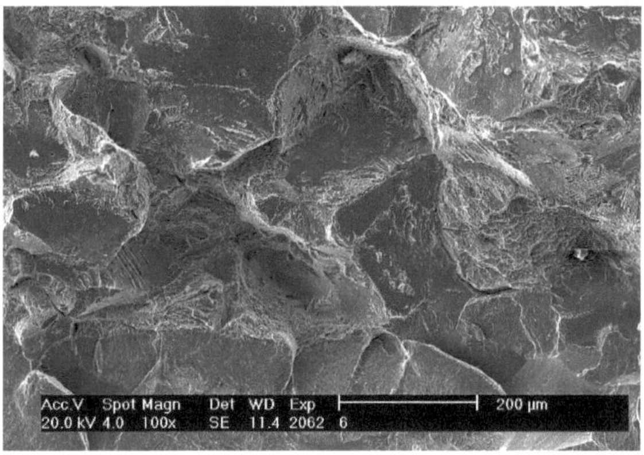

(c)

Figure 2.13 Fracture surface morphology of three grades of 300M steel: (a) transgranular ($d_m = 20\,\mu\mathrm{m}$), (b) mixed trans-intergranular ($d_m = 70\,\mu\mathrm{m}$), and (c) intergranular ($d_m = 120\,\mu\mathrm{m}$). Reprinted with permission from John Wiley & Sons, Inc. (see page 265)

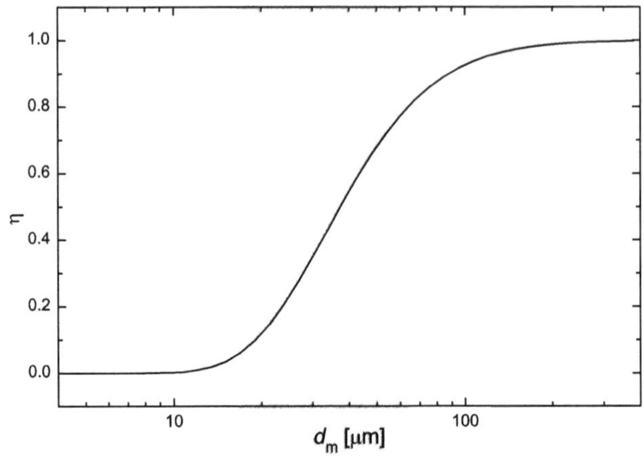

Figure 2.14 Statistical parameter η as a function of the mean grain size d_m

the intergranular decohesion and the fine dimple fracture must be nearly equal. Finally, the dependence of the K_{Ic} values on d_m could be predicted by using the pyramidal approximation according to Equations 2.16, 2.17, 2.18 and 2.23. The result is shown in Figure 2.11 by the full curve. The agreement between theory and experiment is very satisfactory in spite of the fact that only two fitting parameters S_{Rc}, K_{Ici} of physically plausible values were used in this analysis ($A_b = 0$). For the coarsest grades, however, the theoretical curve lies slightly below the experimental data that exhibit the highest scatter. This can be explained by the fact that branching of some of the intergranular crack front segments was not taken into account in the theoretical analysis. As can be seen from Equation 2.23, the branching term might be responsible for the slight difference between theory and experiment in the pure intergranular region.

Let us emphasize that the fracture mechanism in K_{Ic} specimens was quite different from that in the notched specimens used in the impact tests for the measurement of absorbed energy. In the latter case the crack had to be first initiated at the notch root. Therefore, the initiation energy represented a substantial portion of the total absorbed energy. This energy could be well correlated with an extent of shear (stretch) zones adjacent to notches that were clearly identified on the fracture surfaces of all specimens. These zones indicate areas, where the crack was initiated by ductile fracture accompanied by long-range shear coalescence of microvoids with the notch root (see Section 2.3). The shear zones in the coarse-grained specimens were found to be very narrow (and the related energy consumption very low) in comparison to those in fine-grained samples, as documented in Figures 2.15 and 2.16. This phenomenon can be understood in terms of a capability of local micro-crack initiation at inclusions or grain (phase) boundaries. Indeed, the local microdefect (microvoid) nucleation assisted by high local stresses ahead of long dislocation pile-ups in large grains must be much easier, thus reducing the average fracture strain within the notch plastic zone.

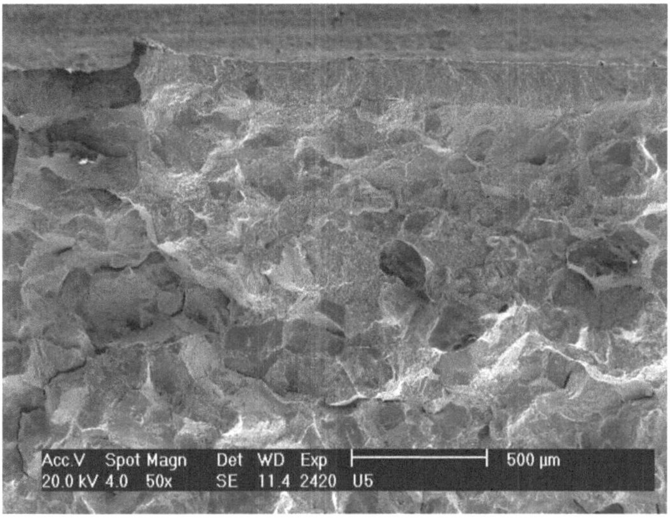

Figure 2.15 The narrow crack-initiation shear zone at the notch (*at the top*) in the coarse-grained sample

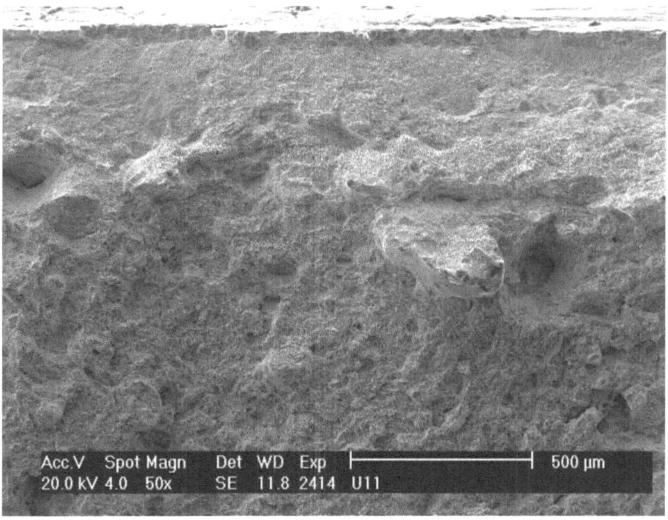

Figure 2.16 The wide crack-initiation shear zone at the notch (*at the top*) in the fine-grained sample

It should be finally noted that an improvement in fracture toughness caused by microstructure coarsening as discussed above is usually associated with a significant decrease in most other mechanical properties [168]. Nevertheless, the increase in an extrinsic component is not necessarily always accompanied by such a detrimental effect. For example, duplex ferrite-austenite microstructures possess extremely high components of GIS while exhibiting other very good mechanical properties (see Section 3.2).

2.2.3 Mixed Intergranular and Cleavage Fracture of Phosphorus-doped Fe-2.3%V Alloy

It is well known that grain boundary segregation of phosphorus, tin and antimony can lead to intergranular fracture. However, finding a quantitative relationship between the segregation level and fracture toughness (or surface energy) constitutes a very difficult task. Therefore, an extended study of fracture behaviour and grain boundary chemistry in an Fe-3%Si based alloy, which contained traces of phosphorus, has recently been performed [215]. Even though this study provided us with unambiguous results, the data may have suffered from the fact that both silicon and phosphorus segregate to the grain boundaries and reduce the cohesion of the material [160]. Additionally, due to repulsive interaction between phosphorus and silicon, a complex segregation behaviour occurs (i.e., enrichment with phosphorus but depletion of silicon) [216]. Many straightforward results in this respect could be obtained by measurement of phosphorus segregation and fracture behaviour in phosphorus-doped iron or in a ferrous alloy containing an element indifferent to phosphorus segregation. Therefore, the phosphorus-doped Fe-2wt%V base alloy was selected [217]. Here, the austenite γ-phase was fully avoided and the system remained bcc up to the congruent melting point. This made it possible to grow bicrystals directly from the melt. Moreover, vanadium does not affect the grain boundary cohesion significantly and, in addition, its segregation is expected to be rather low. From this point of view, one may consider this alloy as a pseudobinary Fe-P system.

Polycrystals of an Fe-2.3wt%V-0.12wt%P alloy were prepared by hot rolling of the vacuum cast master alloy between 1070 K and 1370 K annealing at 973 K for 1 h so that an average grain size of 0.2 mm was achieved. The notched samples for fracture testing were annealed for interfacial segregation at 1073 K (24 h), 973 K (48 h), 873 K (72 h) and 773 K (168 h). Annealed samples were deformed in three-point bending at room temperature using the Zwick Z 020 testing machine. Because the validity of small-scale yielding was not fulfilled for all samples, the fracture toughness K_{Jc} was evaluated according to ASTM E399-72 procedure. The data are listed in Table 2.3. It should be emphasized that force-displacement curves of all specimens exhibited a short nonlinear part indicating a dislocation-assisted stable crack growth before the fast unstable tearing. However, this initiation stage of such quasi-brittle fracture became negligible in cases of prevailing portions of intergranular morphology on the fracture surface.

Table 2.3 Experimental data on chemical composition of grain boundaries and on fracture behaviour of polycrystalline Fe-2.3wt%V-0.12wt%P alloy

T [K]	X_P^{Φ} [at%]	X_V^{Φ} [at%]	K_{Jc} [MPa m$^{1/2}$]	η	R_L^*	K_{Jic} [MPa m$^{1/2}$]
773	25.3 ± 6.9	14.4 ± 9.2	24.5 ± 7.5	0.89 ± 0.07	4.8 ± 2.2	7.9 ± 3.6
873	17.6 ± 8.6	9.2 ± 8.0	49.0 ± 9.4	0.47 ± 0.14	4.6 ± 1.6	31.8 ± 10.3
973	19.4 ± 6.4	5.6 ± 5.6	66.4 ± 16.6	0.74 ± 0.07	5.7 ± 2.2	28.8 ± 13.7
1073	20.6 ± 6.7	4.4 ± 2.0	38.8 ± 14.6	0.77 ± 0.03	5.8 ± 0.1	14.6 ± 6.5

K_{Jc} – the fracture toughness, T – the annealing temperature, η – the portion of intergranular fracture, R_L^* – the linear (profile) roughness corrected to the portion of transcrystalline cleavage, K_{Jic} – the intrinsic fracture toughness, X_P and X_V – the grain boundary concentrations of phosphorus and vanadium, respectively

Fracture surfaces of the samples were inspected by Philips XL-30 scanning electron microscope and the portions of intergranular fracture, η, were determined (see Table 2.3). The values correspond to a close vicinity of the fatigue crack front, i.e., to the initiation stage of the quasi-brittle fracture process. As a rule, the η -values in the final region of unstable fracture were found to be higher.

The composition of the grain boundaries was studied by Auger electron spectroscopy (AES) using a Microlab 310F VG-Scientific facility equipped with a field emission gun. Notched cylindrical samples of 5 mm diameter and 30 mm length were fractured *in situ* by impact at about $-120°$C, and subsequently analyzed in ultra-high vacuum of 5×10^{-8} Pa. To determine the phosphorus and vanadium grain boundary concentrations, several tens of AES analyses per sample were done at many intergranular and transgranular facets. Typical results of the AES analysis are shown in Figures 2.17 and 2.18. In Figure 2.17(a), the intergranular facets and the analyzed points are depicted along with related Auger spectra, where clear peaks of P, V and Fe are visible (Figure 2.17(b)). In Figure 2.18, an example of transgranular facets and associated spectra are presented. Note that no evidence of phosphorus was detected on these facets. The chemical composition of the grain boundary monolayer (see Table 2.3) was obtained from the derivative Auger peak-to-peak heights measured at the fracture surface according to the method published in [216].

AES measurements revealed that annealing at different temperatures results in segregation of both P and V. The temperature dependence of their grain boundary concentrations is shown in Figure 2.19. While the concentra-

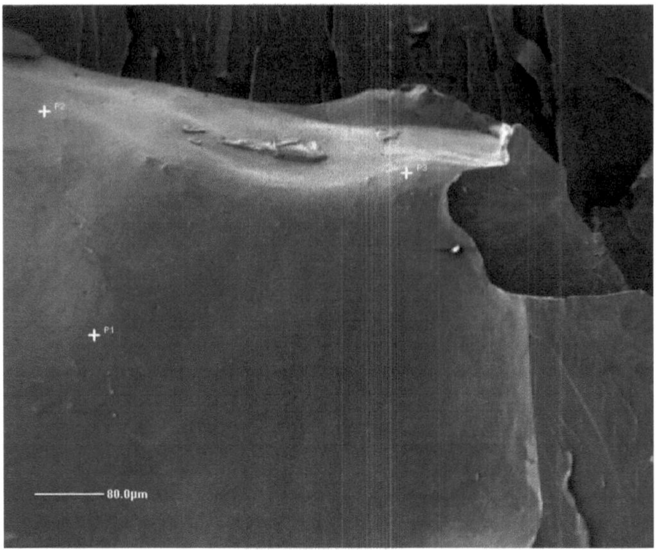

(a)

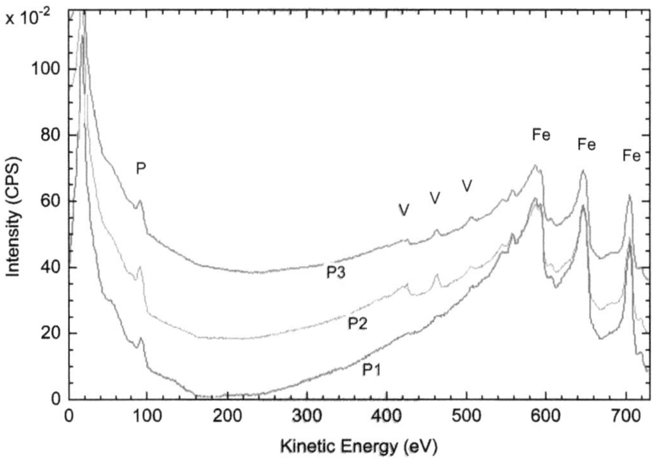

(b)

Figure 2.17 (a) Intergranular facets with analyzed points, and (b) related Auger spectra with clear peaks of P, V and Fe

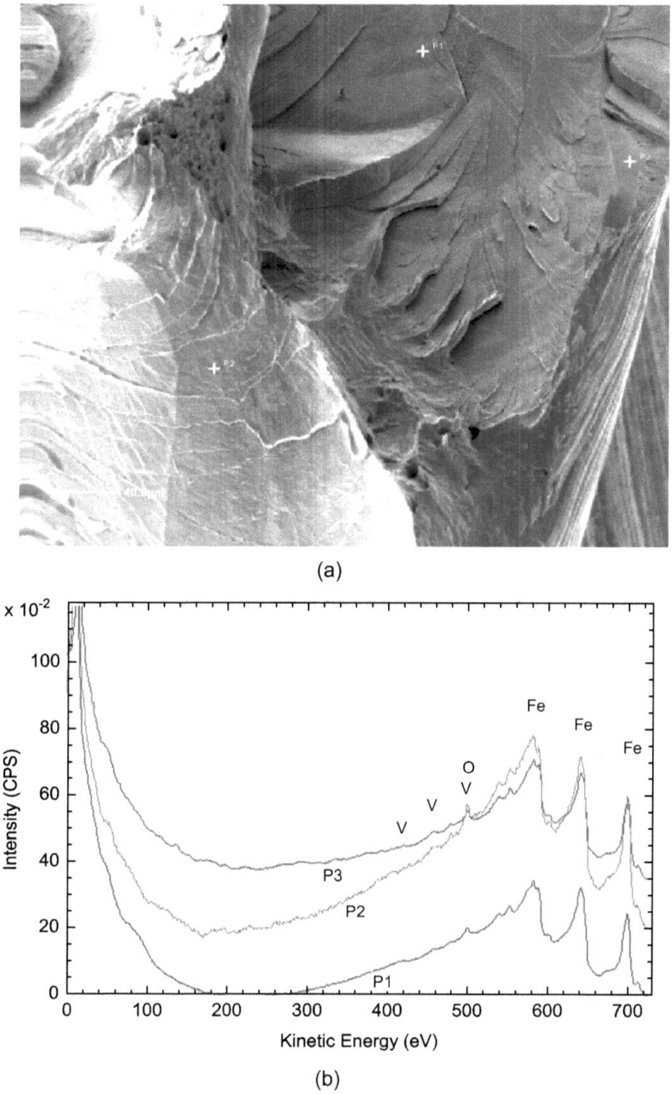

Figure 2.18 (a) Transgranular facets with analyzed points, and (b) related Auger spectra without peaks of P

tion of vanadium decreases with increasing temperature, phosphorus does not exhibit monotonous dependence showing a minimum of segregation at 873 K, followed by a rather surprising increase with a further raising of temperature. This behaviour clearly suggests that the segregation at lower temperatures is far from equilibrium. This was also proven by comparison of the results with prediction of the grain boundary composition in Fe-P and Fe-V binary systems [218] (see the averaged experimental points and the theoretical lines in Figure 2.19). Only the data for 1073 K represent, most probably, the equilibrium grain boundary composition. In all cases, the scatter of the data predominantly reflects the heterogeneity of grain boundary segregation.

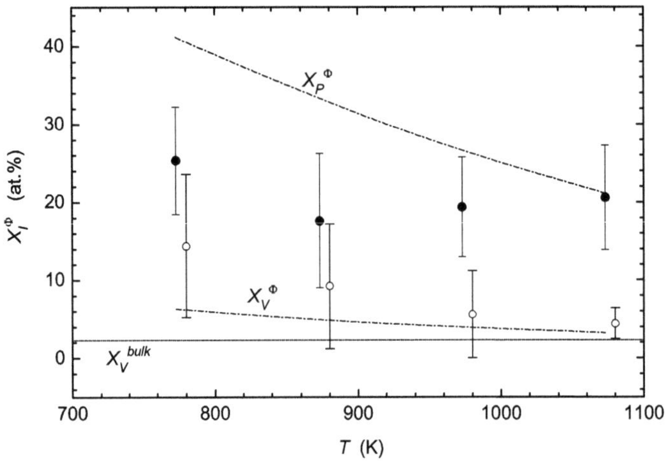

Figure 2.19 Concentrations X_I^Φ of phosphorus (*solid circles*) and vanadium (*open circles*) at grain boundaries of polycrystalline Fe-2.3wt%V-0.12wt%P alloy at different temperatures. The *dashed-dotted lines* show predicted equilibrium concentrations of P and V at general grain boundaries. The *dotted line* depicts the vanadium bulk concentration

Analysis of the fracture surfaces revealed that the fracture process was predominantly intergranular but accompanied by various amounts of transcrystalline cleavage (Figure 2.20). Because of the very large mean grain size of 200 μm, transgranular cleavage was observed rather than the ductile dimple morphology, as one can predict from Equation 2.6.

The portion of intercrystalline fracture morphology, η, changes with annealing temperature of the sample, i.e., with the level of interfacial segregation. In Figure 2.21, the curves of the probability density *vs* phosphorus concentration are plotted, where $1/N$ is the increment of the probability density and N is the number of measured AES data at each temperature. It is apparent that the experimental data are well correlated by the Gaussian cumulative distribution function

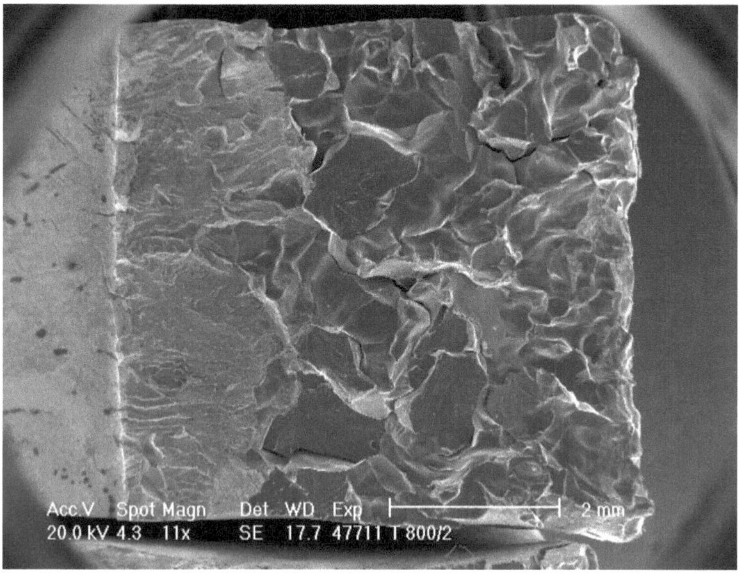

Figure 2.20 Fracture surfaces of polycrystalline Fe-2.3%V-0.12%P alloy annealed at 773 K and 1073 K, broken at room temperature. Typical morphology of prevailing intercrystalline brittle fracture ($\eta = 0.73$) with several transcrystalline cleavage facets near the fatigue pre-crack front is apparent in the middle of the fracture surface. On the *left-hand side* the notch and the pre-crack are clearly visible. Reprinted with permission from Elsevier B.V. (see page 265)

$$F(X_P) = \int_{-\infty}^{X_P} f(\xi)\,\mathrm{d}\xi$$

for all annealing temperatures, where

$$f(\xi) = \frac{1}{\sigma\sqrt{2\pi}} \exp\left[-\frac{1}{2}\left(\frac{\xi-\mu}{\sigma}\right)^2\right].$$

Here, $f(\xi)$ is the probability density function, μ is the mean value and σ is the standard deviation. It is assumed that transcrystalline cleavage occurs when $X_P < X_{Pc}$ while intercrystalline fracture dominates for $X_P > X_{Pc}$, where X_{Pc} is some critical concentration of phosphorus. Thus, the portion of transcrystalline fracture should be proportional to $F(X_{Pc})$. Portions of transcrystalline and intercrystalline fracture related to the critical value $X_{Pc} = 17$ at% are plotted in Figure 2.21 by dashed lines. The curves of a predicted portion of intergranular fracture for various selected X_{Pc} in the range of 5–20 at% P (spline curves based on calculated points for individual annealing temperatures) are shown in Figure 2.22, where they are also compared with the experimental values of η. It is apparent that the theoretical prediction for

$X_{Pc} = 17\,\mathrm{at\%}$ P yields the best fit to averaged experimental data independently of annealing temperature. This means that, at room temperature, all grain boundaries containing more than about 17 at% of phosphorus fractured in an intergranular manner. Note that the intergranular fracture at low temperature of −120°C occurs at 4–6 at% P (see Figure 2.21) which is in excellent agreement with the recent result reported by Wu *et al.* [219].

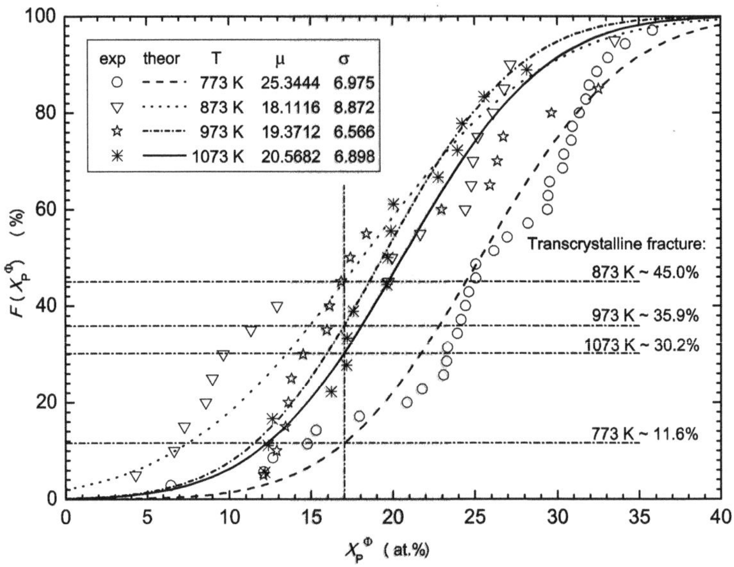

Figure 2.21 Distribution curves of the grain boundary concentration of phosphorus, X_P^{Φ} for samples annealed at various temperatures. Experimental points are correlated with Gaussian curves with characteristic parameters μ and σ. The *vertical line* marks the critical phosphorus concentration $X_{Pc} = 17\mathrm{at\%}$. Portions of transcrystalline fracture determined by X_{Pc} for individual annealing temperatures are given at *horizontal lines*

The presence of intergranularly fractured facets induces an extreme microscopic tortuosity of the crack front and, therefore, a strong GIS occurs [168]. This explains surprisingly high values of fracture toughness measured according to the ASTM standard. These values were in the range $20-90\,\mathrm{MPa\,m^{1/2}}$. To obtain the values of both the intrinsic fracture toughness and the fracture energy at intercrystalline facets, the GIS effect has to be subtracted from the measured values of K_{Jc}. A 3D analysis of fracture surface profiles was performed close to the fatigue pre-crack front (the initiation site of intercrystalline fracture). A 3D reconstruction using the program code Mark III enabled the evaluation of the profile roughness R_L. An example of the tortuous profile is shown in Figure 2.23. The coordinate z represents the height of the profile and the coordinate x runs along the crack front. Because of a high portion of transgranular fracture, the values of R_L had to be corrected to

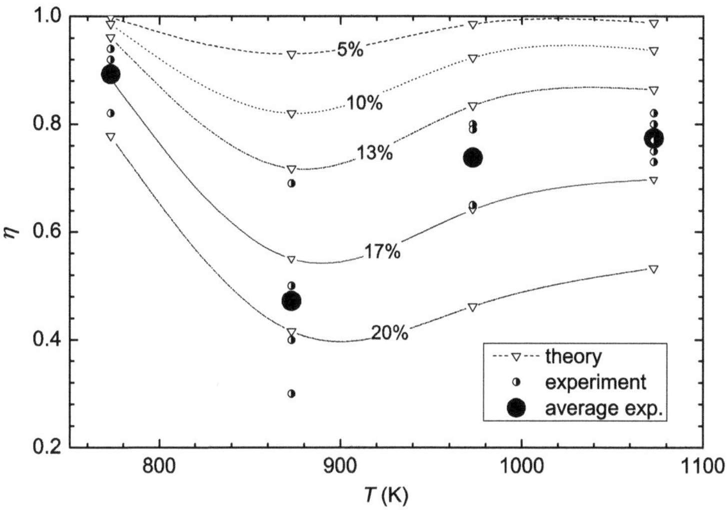

Figure 2.22 Comparison of calculated ratios of intercrystalline fracture morphology for selected values of X_{Pc} (5–20 at%) with experimental data (*semi-solid points*). The averages of these values for individual temperatures are also shown (*large filled circles*). Reprinted with permission from Elsevier B.V. (see page 265)

the portion $1 - \eta$ of the transcrystalline morphology according to the relation $R_L^* = (R_L - 1 + \eta)\eta$. Corrected values were used to determine the intrinsic fracture toughness, K_{Jic}, in the frame of the pyramidal model.

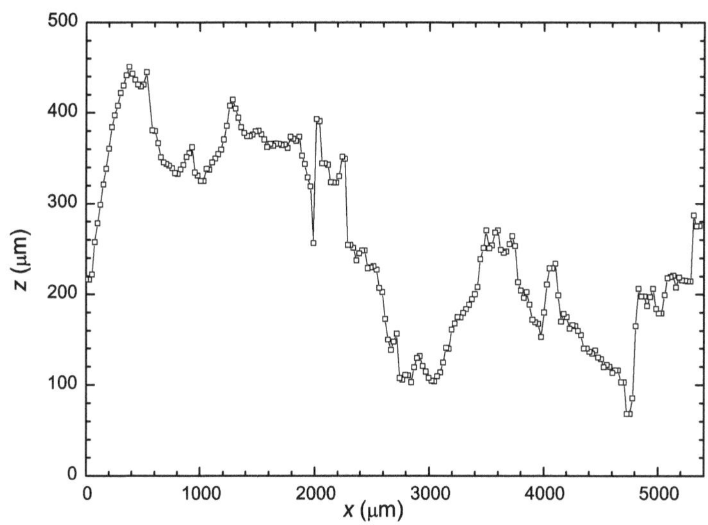

Figure 2.23 An example of fracture profile of intercrystalline morphology near the fatigue pre-crack front

The calculated values of K_{Jic} are plotted as a function of η in Figure 2.24. An extrapolation of this dependence to 100 % of intergranular fracture ($\eta = 1$) enables us to obtain the value of K_{Jic} that would correspond to a pure intergranular fracture. As expected, the data are significantly scattered due to a pronounced anisotropy of phosphorus grain boundary segregation in the range of $1 - 35$ at%. Despite this scatter, the dependence can be linearly well correlated to obtain $K_{Jic} = 3.8\,\mathrm{MPa\,m^{1/2}}$. By considering this value in the relation

$$\gamma_i = \frac{K_{Jic}^2(1-\nu)^2}{2E}$$

one obtains the mean value of the surface energy on intercrystalline facets, $\gamma_i = 19\,\mathrm{Jm^{-2}}$. This value is well comparable with the surface energy of cleavage facets in ferrite at very low temperatures. It is also in agreement with an estimate of 20 $\mathrm{Jm^{-2}}$ found previously for an Fe-Si-P alloy [220]. Thus, more than 17 at% of phosphorus causes a grain boundary embrittlement to this limit level even at room temperature.

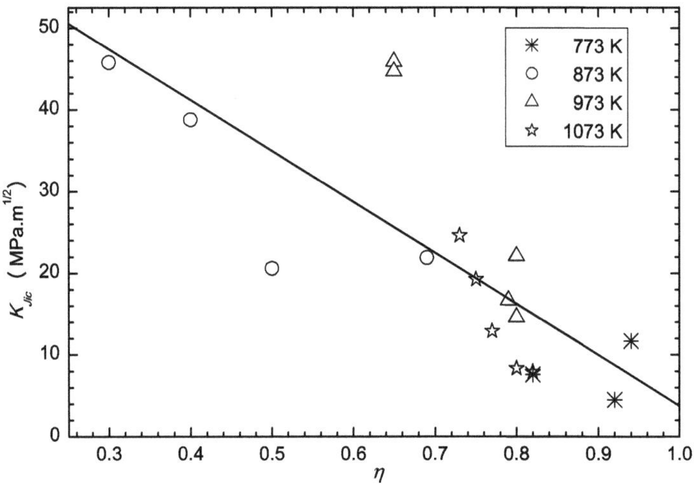

Figure 2.24 Plot of experimental values of the intrinsic fracture toughness, K_{Jic}, *vs* the ratio of intercrystalline brittle fracture, η, for all annealing temperatures. Reprinted with permission from Elsevier B.V. (see page 265)

The concentration of phosphorus at some grain boundaries along the crack front was significantly less than the critical value of 17 at%. Consequently, the related grains failed by the dislocation-assisted transgranular cleavage fracture under intrinsic stress intensity factors as high as about 70 $\mathrm{MPa\,m^{1/2}}$given by an extrapolation of experimental data in Figure 2.24 to 100% of cleavage fracture ($\eta = 0$). Such relatively high values were found to be not only due to the blunted tip of the fatigue pre-crack. The dislocation arrangements within the plastic zone both ahead of the crack front and in the crack wake produced additional crack tip shielding [11] (see also Section 3.2). Due to the large scatter of phosphorus concentration at grain boundaries, however, there is a competitive mechanism of intergranular decohesion operating along supercritically segregated (and suitably oriented) grain boundaries of all samples already at K_{Jic}-values of an order of units of MPa m$^{1/2}$, i.e., just at the onset of the short crack initiation stage. Such isolated intergranular ledges create local peaks of K-factor along the crack front. These peaks enable an easier initiation of further intergranular or cleavage facets so that the mixed intergranular/cleavage fracture can spread across the whole crack front during the short initiation stage. Thus, the subsequent unstable rupture could sometimes happen under intrinsic stress intensity factors of an order of units of MPa m$^{1/2}$.

In the following stage of fast unstable fracture, the stress intensity factor and crack growth rate rapidly increase and allow an immediate intergranular fracture at less segregated grain boundaries adjacent to the crack front. This means that the portion of intergranular morphology increases with the distance from the fatigue pre-crack front. This is clearly seen in Figure 2.20, where almost a pure intergranular morphology appears near the right edge of the fracture surface corresponding to the last stage of unstable fracture.

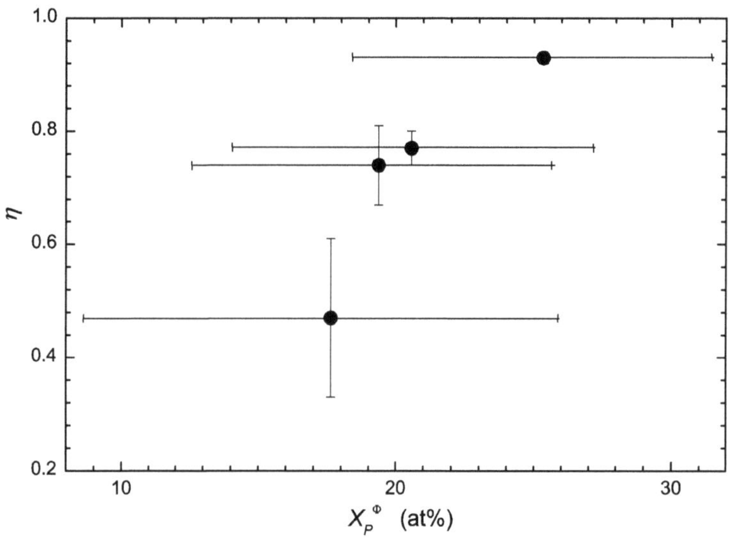

Figure 2.25 Cumulative plot of experimental ratio of the intercrystalline brittle fracture η *vs* grain boundary concentration X_P^Φ for all annealing temperatures. Reprinted with permission from Elsevier B.V. (see page 265)

As follows from the cumulative plot of η *vs* X_P^Φ in Figure 2.25, the ratio of the intercrystalline brittle fracture increases, although not necessarily linearly, with increasing grain boundary concentration of phosphorus. This fact suggests that the grain boundary concentration of phosphorus rather than the annealing temperature controls the fracture mode. This is a consequence of the fact that the segregation was not allowed to reach an equilibrium.

In summary, the results of the analysis revealed that, at room temperature, the surface energy of grain boundaries segregated by phosphorus is very low and comparable with the surface energy of cleavage facets at very low temperatures. However, the crack-tip shielding induced by intergranular fracture significantly increases the fracture toughness of iron to values measured in steels of a high purity. This means that, similarly to the above-mentioned case of the ultra-high-strength steel, the results of fracture toughness tests performed according to the standard ASTM procedure need not necessarily reflect the inherent brittleness of materials.

2.3 Ductile Fracture

The micromechanism of ductile fracture consists of void generation from secondary phase particles and their growth and coalescence. The classical McClintock's model [221] considers isolated, initially cylindrical spaced voids uniformly distributed in a plastic solid. The aim of this model was to determine the dependence of void dimensions on both the axial plastic strain ε_p and the superimposed respective radial and tangential stresses σ_2 and σ_3. The analysis led to a rather complicated dependence of the radius of voids on the triaxiality parameter $\kappa = \sigma_m/\sigma_i$, where $\sigma_m = 1/3(\sigma_1 + \sigma_2 + \sigma_3)$ is the mean hydrostatic stress and $\sigma_i = \frac{\sqrt{2}}{2}\left[(\sigma_1 - \sigma_2)^2 + (\sigma_1 - \sigma_3)^2 + (\sigma_2 - \sigma_3)^2\right]^{1/2}$ is the von Mises stress intensity. More recent relationships (e.g., [222–224]) confirmed that the void growth can be generally described as

$$R = R_0 \exp\{f(\kappa)\varepsilon_p\}\,,$$

where $f(\kappa)$ is an appropriate function of the stress triaxiality, and R and R_0 are running and initial sizes of voids, respectively. It should be noted that, in the vicinity of free surfaces of growing voids, the movement of dislocations is accelerated by image stresses and stress concentration. Consequently, a localization of plastic deformation must take place.

The axially symmetric tensile test can be assumed as a simple and efficient example of the ductile fracture process. In the initial stages of plastic deformation up to the ultimate strength, both the density and the volume fraction of voids remain small and the voids grow preferentially in one direction only. Therefore, the localization of plastic deformation does not play any important role and the global deformation process proceeds in a uniform manner. Just before reaching the strength limit, however, the density of voids in one of the volume elements becomes critical and starts a local void coalescence.

The related pronounced localization of plastic deformation causes a global plastic instability and the macroscopic homogeneity of plastic deformation is lost. This results in necking and a development of triaxial stress state in the adjacent bulk region. In the region of necking, the Bridgman solution

$$\frac{\sigma_2}{\sigma_i} = \frac{\sigma_3}{\sigma_i} = \ln(1 + r/2R) \tag{2.24}$$

can be applied, where r/R is the ratio of the radius of the cross-section to the radius of the specimen profile in its narrowest part. This means that the tensile triaxiality develops inside the neck region. If we simply assume that, just before the onset of necking, first void nuclei of a radius a_0 appear by decohesion of particle-matrix interfaces, the McClintock's growth relations can be simply approximated as

$$a_1 \approx a_0 \exp(\lambda_1 \varepsilon_p), \tag{2.25}$$

$$a_2 \approx a_0 \exp(\lambda_2 \varepsilon_p), \tag{2.26}$$

where a_1 and a_2 are the characteristic sizes of growing elliptical voids. The parameters λ_1 and λ_2 vary from the onset of necking to the final fracture in the range of $\lambda_1 \in \langle 0.8, 2\rangle$ and $\lambda_2 \in \langle 0, 1\rangle$, respectively [225]. When the voids nucleate only by partial decohesion of the particle-matrix interface or by broken carbides, Equations 2.25 and 2.26 should be modified as

$$a_i \approx a_0 [\exp(2\lambda_i \varepsilon_p) - 1], \tag{2.27}$$

where $i = 1, 2$ [226]. Thus, from the moment of plastic instability on, the voids expand into all space directions, and the kinetics of plastic deformation within the necked region is predominantly influenced by their presence. In Section 2.3.1, the dislocation based model of plastic deformation during the tensile test is presented and compared with experimental data [225]. This model clearly demonstrates the role of voids growing during the plastic instability stage.

Originally, the ductile fracture process was assumed to be finished by a continuous void coalescence due to local necking of inter-void ligaments until their total contraction [221, 227]. Based on the growth relations, the strain to fracture ε_f could be expressed as

$$\varepsilon'_f = \frac{(1-n)\ln(\Lambda/2a_0)}{\sinh[(1-n)(\sigma_2 + \sigma_3)/(2\sigma_i\sqrt{3})]}, \tag{2.28}$$

where n is the hardening exponent ($\sigma = \sigma_0 \varepsilon^n$) and Λ is the initial mean intervoid distance. The symbol ε'_f used in Equation 2.28 instead of ε_f means that the fracture strain according to this equation does not include the value of the uniform strain related to ultimate strength. Because $\Lambda/(2a_0) \approx f_v^{-1/3}$, where f_v is the volume fraction of secondary phase particles, this equation correctly reflects qualitatively the experimental findings that ε_f decreases with both increasing the volume fraction of particles and stress triaxiality (if $\sigma_2 = \sigma_3 = 0$ then $\varepsilon_f \to \infty$). However, further research revealed that the fracture strain calculated according to Equation 2.28 significantly exceeds experimental data. One of the main reasons is the fact that the coalescence process does not proceed in a stable manner but it is accomplished by rapid unstable collapse after the intervoid ligament has reached a critical width [228].

Because the triaxiality is highest within the narrowest part of the specimen on its central axis, the first noticeable coalescence of voids usually appears there. Thus, the process of final fracture starts at the moment when a penny-shaped central crack has been developed. As soon as the crack front approaches the external traction-free surfaces, the stress state in the remaining ligament changes from the triaxial to the plane stress (biaxial) one. This causes a sudden increase in the maximum shear stress on the plane inclined by approximately 45° to that of the crack. Therefore, the final rupture occurs by shear along this plane and the creation of shear lips. The fracture surfaces of cone-shaped broken parts become a crater-like macromorphology (see Figure 2.26). Dimples in the central part reflect the process of void growth and coalescence, as can be clearly seen from Figure 2.27. Inside the dimples, the nucleating secondary phase particles can often be detected. In Section 2.3.2, some models of void coalescence are presented and compared with experimental data on tensile fracture strain of various steels. These models give a physical basis of an empirically determined dependence of the ductile fracture strain on the state of stress that serves as a critical strain assessment when applying various forging technologies (see Section 2.3.3).

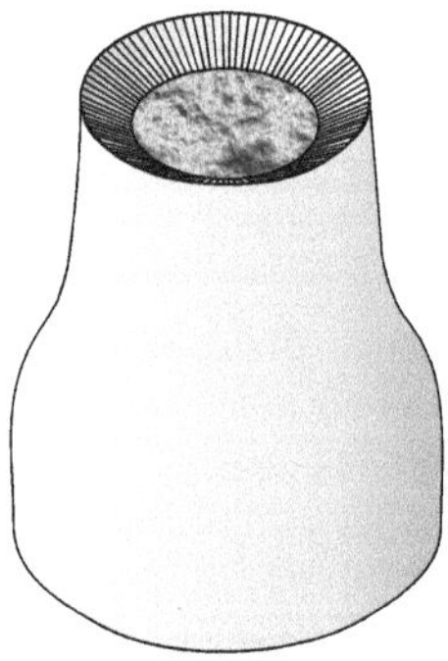

Figure 2.26 The scheme of crater-like morphology typical for one part of a ductile sample fractured during the tensile test

At present, numerical models based on constitutive models of porous solids [223, 229, 230] are widely used to simulate the ductile fracture processes of various materials and components under different loading conditions. These

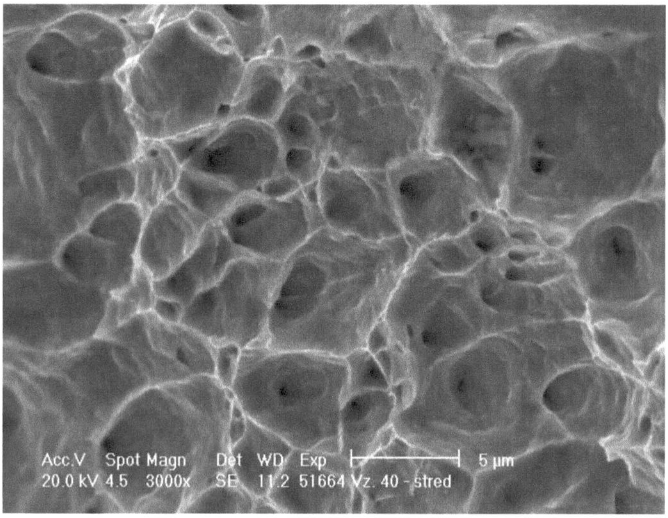

Figure 2.27 A typical SEM picture of ductile dimples (aluminium alloy)

semi-empirical models use a modified yield function and plastic potential, including the "porosity" terms as additional internal variables responsible for "softening" of the material. The models need to be implanted into the finite element codes and calibrated by the experimental stress-strain curve of the particular material in order to obtain at least two empirical parameters. A more detailed description of these models and their applications lies beyond the scope of this book.

The last subsection is devoted to a simple method, based on the above-mentioned theoretical concepts of ductile fracture, that can be used for the prediction of fracture toughness values of high-strength metallic materials. This method avoids expensive fracture toughness tests when we need only approximate K_{Ic} values.

2.3.1 Kinetics of Plastic Deformation During Uniaxial Ductile Fracture: Modelling and Experiment

Whereas the classical works [221,222,231] presented void growth as a result of bulk plastic deformation, Lui and Le Mai [232] and Jonas and Baudelet [233] have shown an active role of voids in the creation of plastic instability during the tensile test. Later on, detailed measurements revealed a significant acceleration of plastic strain rate in the smallest cross-section of the neck during the tensile test performed under conditions of constant crosshead speed [234]. In order to explain this behaviour, Staněk and Pokluda assumed an active role of voids inside the neck in the promotion of dislocation movement [225,235]. They used the well known equation connecting the macroscopic plastic flow with the rate of dislocation movement proposed by Gilman [236] as

$$\dot{\varepsilon}_p = \mu b v_c(\rho_0 + M\varepsilon_p) f_m \exp\left[-(D + H\varepsilon_p)/\sigma_s\right], \tag{2.29}$$

where $\mu \approx 0.5$, b is the magnitude of the Burgers vector, v_c the limit speed of dislocations, ρ_0 is the initial dislocation density, M the coefficient of dislocation multiplication, f_m the fraction of mobile dislocations, D the effective friction stress, σ_s the mean effective true stress and H the coefficient of dislocation trapping. In engineering materials, the fraction f_m can vary in a very wide range in dependence on ε_p and, in particular, on ρ_0. However, Equation 2.29 provides a good description of plastic flow rate even when assuming $f_m \neq f_m(\varepsilon_p)$ [235]. Equation 2.29 can be used for a reasonable simulation of the kinetics of plastic deformation up to the ultimate strength, i.e., within the region of uniform deformation. However, it must be substantially modified to allow a description of localized plastic deformation inside the neck due to the influence of void growth.

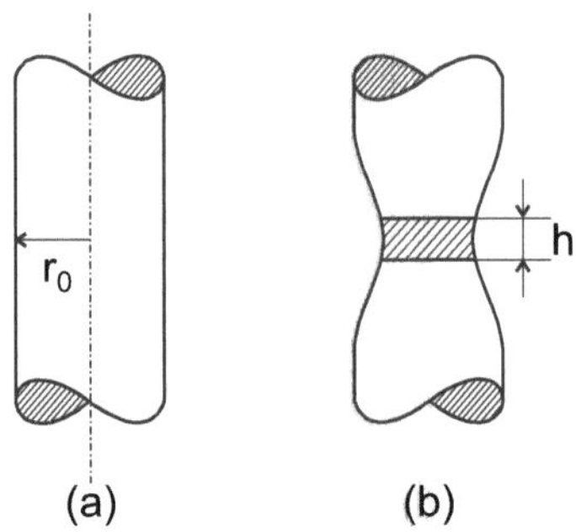

Figure 2.28 Scheme of: (a) the initial dimension of the sample, and (b) the axial dimension h of the bulk element within the neck that experiences the highest localized plastic deformation during the tensile test

Both experiments and finite element calculations [234,237] revealed that an axial dimension $h \leq r_0$ (r_0 is the initial specimen radius) of the bulk element of highest localized macroscopic plastic deformation in the thinnest part of the neck remains nearly constant during the tensile test – see Figure 2.28. This active volume changes with the tensile strain as

$$V = V_0 \exp(-\varepsilon_p), \tag{2.30}$$

where $V_0 = \pi r_0^2 h$.

As mentioned before, the microvoids promote the dislocation mobility by increasing free surfaces in the bulk and inducing local strain concentrations. This evokes an idea of small volumes adjacent to microvoids, where all dislocations can be considered to be mobile. Consequently, the active volume V of the specimen can be divided into three parts – see Figure 2.29: V_1 is the volume unaffected by voids, V_2 is the sum of small volumes containing only mobile dislocations and V_3 is the total volume of voids.

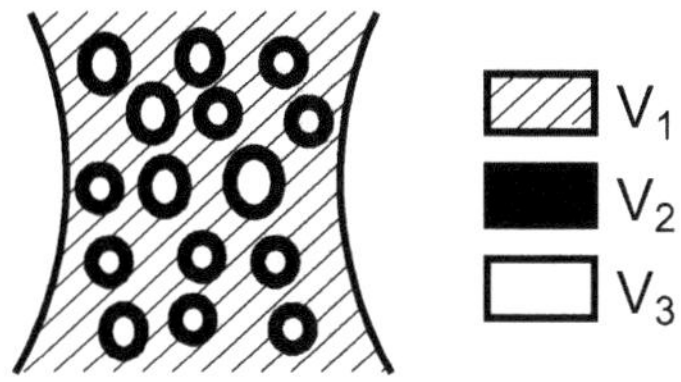

Figure 2.29 The scheme of active volumes within the neck. V_1 is the volume unaffected by voids, V_2 is the sum of small volumes containing solely mobile dislocations and V_3 is the total volume of voids

The relation $V_1 \gg V_2 \approx V_3$ holds during the whole deformation, almost up to the final fracture. Let us denote l_{im} $(i = 1, 2)$ and ρ_{im} the respective total length and density of mobile dislocations in the relevant volumes and ρ the mean total density of dislocations. Then the mean relative density f_m of mobile dislocations in the active volume can be expressed as

$$f_m = \frac{l_{1m} + l_{2m}}{\rho(V_1 + V_2)} = \frac{\rho_{1m}}{\rho}\frac{V_1}{V_1 + V_2} + \frac{\rho_{2m}}{\rho}\frac{V_2}{V_1 + V_2}.$$

Since

$$\rho_{2m} = \rho; \qquad \frac{\rho_{1m}}{\rho} = f_{1m}$$

one obtains

$$f_m = f_{1m}\frac{V_1}{V_1 + V_2} + \frac{V_2}{V_1 + V_2}. \tag{2.31}$$

By differentiating Equation 2.31, the increment of the ratio of mobile dislocations reads

$$\mathrm{d}f_m = \mathrm{d}f_{1m}\frac{V_1}{V_1 + V_2} + f_{1m}\mathrm{d}\left(\frac{V_1}{V_1 + V_2}\right) + \mathrm{d}\left(\frac{V_2}{V_1 + V_2}\right). \tag{2.32}$$

Because $V_1 \gg V_2$ and $f_{1m} \ll 1$, Equation 2.32 can be simplified as

$$\mathrm{d}f_m \approx \mathrm{d}f_{1m} + \mathrm{d}\left(\frac{V_2}{V_1}\right). \tag{2.33}$$

Equation 2.33 states that the increment of mean fraction of mobile dislocations consists of two terms. The first means a change in the mobile dislocation fraction within the "undamaged" volume V_1 while the second expresses the related contribution of voids.

When approaching the ultimate tensile stress, the density of dislocations rapidly increases and a relative contribution of dislocation trapping to the density of mobile dislocations starts to dominate that of their multiplication. This means that the exponential term in Equation 2.29 starts to control the density of mobile dislocations [235]. Unlike in the active volume V_2, this trend will continue further inside the inactive (undamaged) volume V_1 after reaching the ultimate stress (during the necking). Therefore, the total density of dislocations in V_1 will not substantially change and the mean relative density of mobile dislocations in V_1 can be assumed to follow the "trapping" rule

$$f_{1m} \approx \Xi \exp\left(-\frac{D + H\varepsilon_p}{\sigma_s}\right), \tag{2.34}$$

where Ξ is a dimensionless constant. Since the true mean stress σ_s remains nearly constant during the necking, the differentiation of Equation 2.34 gives

$$\mathrm{d}f_{1m} \approx -\Xi\frac{H}{\sigma_s}\exp\left(-\frac{D + H\varepsilon_p}{\sigma_s}\right)\mathrm{d}\varepsilon_p = -\Phi f_{1m}\mathrm{d}\varepsilon_p, \tag{2.35}$$

where $\Phi = H/\sigma_s$. When assuming $f_{1m} \approx f_m$ ($V_2 \ll V_1$), Equation 2.35 can be expressed in an approximate form:

$$\mathrm{d}f_{1m} \approx -\Phi f_m \mathrm{d}\varepsilon_p.$$

The second term in Equation 2.33 represents the increment of mobile dislocations due to the void growth. In order to specify this term in more detail, Equations 2.25 and 2.26 can be used while considering two cases: (1) voids are of the same initial size and their number remains constant – the ideal model and (2) the initial size and the number of voids are functions of plastic strain and the size distribution of nucleating particles – the real model. In the context of the ideal model, and by assuming the elementary volumes near voids as in Figure 2.30, Equation 2.29 takes on the following form (see Appendix C):

$$\begin{aligned}\dot{\varepsilon}_p &\approx \mu b v_c(\rho_u + M\varepsilon_p) f_m^u \exp\left\{-\frac{D_{us} + H\varepsilon_p}{\langle\sigma_s\rangle_k}\right\} + \\ &+ \mu b v_c(\rho_u + M\varepsilon_p)\frac{\Theta}{\gamma + \Phi}\exp\left\{-\frac{D_{us} + H\varepsilon_p}{\langle\sigma_s\rangle_k}\right\}\left[\mathrm{e}^{(\gamma+\Phi)\varepsilon_p} - 1\right],\end{aligned} \tag{2.36}$$

where ρ_u and D_u are respectively the density of dislocations and the friction stress at the onset of necking, $\Theta = 4\pi a_0^2\gamma n\delta/V_0$, n is the number of voids in the active volume and $\gamma = 1 + \kappa$.

The first term in Equation 2.36 is nearly equal to Equation 2.29 and expresses the process of hardening, while the second term gives the softening

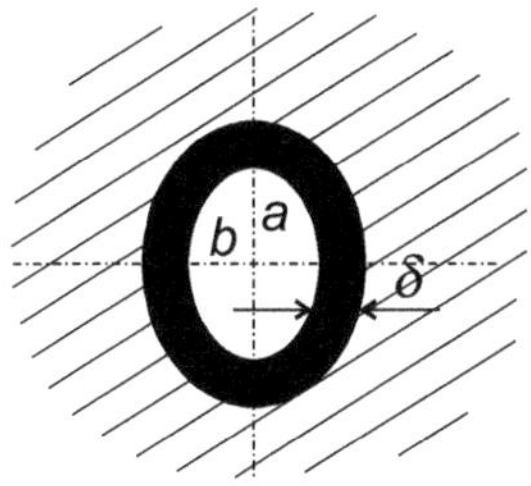

Figure 2.30 A detail of an active volume around the void

due to void growth. A more complicated relationship derived in the context of the real model (see Appendix C) is not too much different and, consequently, it does not give much better results when compared to the experiment [225].

The validity of Equation 2.36 was verified by an experimental method based on recording both the fine surface net and the neck shape by a camera during the whole tensile test performed on cylindrical specimens of a low-alloy steel [225]. This proved that the plastic strain in the thinnest part of the neck could be well approximated by a simple relation $\varepsilon_p = 2 \ln d_0/d$, where d_0 and d stand for respective initial and final diameters of the neck. The steel contained a high density of rather uniformly distributed spherical carbides that acted as void initiation sites. Measured and theoretical dependencies of $\dot{\varepsilon}_p$ on ε_p, plotted in Figure 2.31, exhibit a very good agreement.

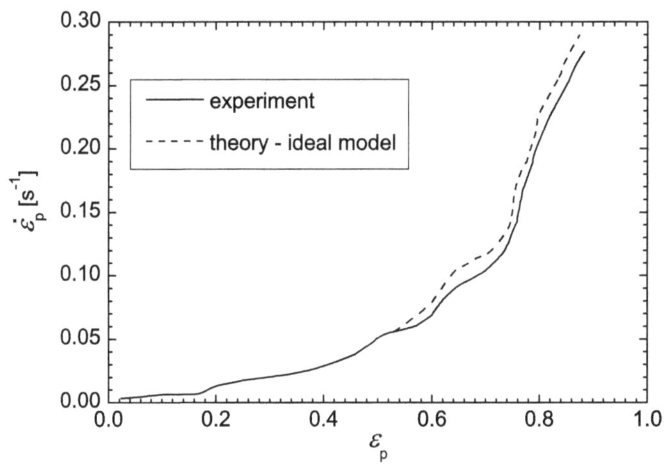

Figure 2.31 Comparison of experimental and theoretical dependences $\dot{\varepsilon}_p$ *vs* ε_p during the whole tensile test

It should be emphasized that parameters in Equation 2.36 were either taken from the literature or determined by fitting a part of the function $\dot{\varepsilon}_p$

vs ε_p during the uniform deformation until the onset of necking by Equation 2.29. The constant $\Theta = 4.5 \times 10^{-7}$ was the only fitting parameter used in the unstable region of necking. This parameter roughly reflects the volume ratio V_2/V_1 and reveals that a negligible volume of mobile dislocations activated around the voids can induce a substantial acceleration of the plastic flow inside the neck.

2.3.2 Fracture Strain

The attempts to predict theoretically the fracture strain by means of void growth and coalescence models started in the late 1960s and are still in progress. In the first part of this subsection, theoretical predictions of ductile contraction along with experimental data obtained from tensile tests on cylindrical bars made of different steels (void nucleation on carbides) are presented as published in the early 1980s [226]. In the second part, a generalized diagram of fracture strain is presented.

2.3.2.1 Prediction of Ultimate Contraction During the Tensile Test

The models of void growth and coalescence are based on three different ideas concerning the critical conditions of the coalescence process that precedes the moment of final unstable fracture: (1) continuous shrinkage of ligaments (McClintock), (2) unstable collapse of intervoid ligaments (Brown and Emburry) and (3) reaching a percolation threshold (Staněk and Pokluda).

Continuous Shrinkage of Ligaments

This model uses Equation 2.28, the relation at Equation 2.24 and the formula

$$\frac{r}{R} = 2.93(1 - \mathrm{e}^{-\varepsilon_i/3}) \tag{2.37}$$

proposed by Bridgman [238], where

$$\varepsilon_i = \left[\frac{2}{9}(\varepsilon_1 - \varepsilon_2)^2 + (\varepsilon_1 - \varepsilon_3)^2 + (\varepsilon_2 - \varepsilon_3)^2\right]^{1/2}$$

is the strain intensity in terms of true principal strains $\varepsilon_k = \ln(l_k/l_{0k})$ ($k = 1, 2, 3$). During the uniform deformation up to the ultimate strength it holds that $\sigma_2 = \sigma_3 = 0$, $\varepsilon_{pu} = n$ and the voids do not grow in the transverse direction. Consequently, the total fracture strain is

$$\varepsilon_f = n + \varepsilon'_f. \tag{2.38}$$

Taking Equation 2.30 into account, the contraction of the bar at the moment of final fracture can be expressed as

$$\psi = 1 - \exp(-\varepsilon_f). \tag{2.39}$$

By substitution of $\varepsilon_i = \varepsilon'_f/2$ or $\varepsilon_i = \varepsilon'_f$ (mean or maximal stress triaxiality) in Equation 2.37 and combining Equations 2.24, 2.28 and 2.37 we obtain an implicit equation for ε'_f. Substituting the obtained value of ε'_f into Equation 2.38 and combining it with Equation 2.39, we receive predicted ultimate contractions ψ for both considered stress triaxialities.

Unstable Collapse of Intervoid Ligaments

According to Brown and Emburry [228], the unstable collapse of ligaments between voids starts when the axial size of neighbouring voids reaches the distance between their centres, i.e., $\lambda/2a = 1$ (see Figure 2.32). By considering the void-grow law in the form of Equation 2.27, the fracture strain can be expressed as

$$\varepsilon_f = n + \frac{1}{\bar{\kappa}} \ln \left\{ \frac{1}{2} \left(\frac{\lambda}{2a_0} + 1 \right) \right\}, \tag{2.40}$$

where the mean value $\bar{\kappa}$ of the triaxiality parameter during the necking process is considered and λ is the mean distance of nucleating particles (carbides). The initial size of the voids $a_0 = 2a_{0s}$, where a_{0s} is the mean size of particles initiating voids just at the onset of necking.

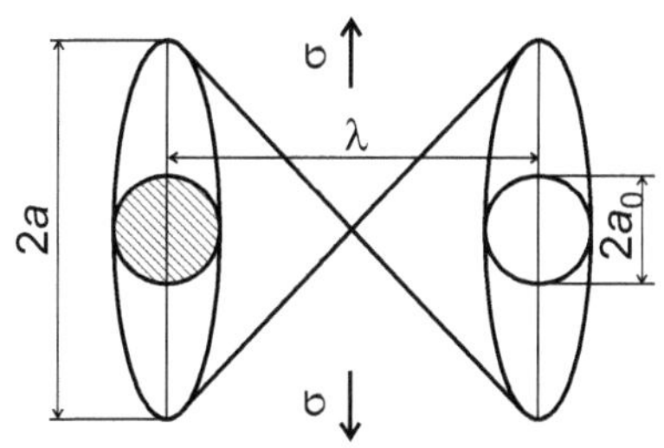

Figure 2.32 Scheme of the onset of the coalescence of axially prolonged voids according to the Brown–Emburry model

The contraction of the bar at fracture can, again, be obtained by utilizing Equation 2.39.

Percolation Threshold

This model is based on the theory of percolations, which is part of the theory of graphs [239]. While in the first stages of necking only a short-range interaction of voids takes place, the long-range interaction of void clusters in the final stages of fracture has to be taken into account. Then, the moment of unstable fracture can be related to the percolation threshold. This means that, in terms of the theory of graphs, a non-zero probability exists that a certain element of the graph (the void) belongs to an infinite cluster (the fracture surface) within an infinite graph characterizing a physical system (here an arrangement of voids in the bulk). Vertices of the characterizing graph in our model were associated with a constant number of growing voids, arranged in hexagonal or square lattices. By using stereometric rules in the context of the cluster theory and Equation 2.27, the fracture strain can be obtained as

$$\varepsilon_f = n + \frac{1}{\bar{\kappa}} \ln \left\{ \frac{1}{2} + \frac{1}{2} \left[-\frac{2S_2}{3\pi} \left(\frac{\Lambda}{2a_0} \right)^3 \ln(1 - p_c^s) \right]^{1/3} \right\}, \tag{2.41}$$

where S_2 is the limit number of in-plane void touching sides and p_c^s is the percolation threshold [226]. The theoretical values $S_2^s = 4$ and $p_c^s = 0.5$ stand for the square lattice whereas $S_2^s = 6$ and $p_c^s = 0.35$ correspond to the hexagonal one [239]. After substituting these values into Equation 2.41 and using Equation 2.39, the relevant values of contraction can be calculated.

2.3.2.2 Comparison of Theoretical and Experimental Data

In order to make a comparison between theory and experiment, the values of $\bar{\kappa}$ and $\Lambda/2a_o$ were experimentally assessed. Tensile tests on cylindrical specimens of pearlitic steel 12 010 and low-alloy steels 14 109 and 14 331 of the Czech provenance were performed by means of the Zwick machine equipped with a special camera [226]. This allowed the assessment of the mean value of triaxiality during necking by observing both the surface grid and the neck shape. In this way, the $\bar{\kappa}$-values of 0.8, 0.6 and 0.85 were established for 12 010, 14 109 and 14 331 steels, respectively. The values $\Lambda/2a_o = \Lambda/4a_{os}$ were determined from the distribution functions of carbide sizes obtained by a careful examination of carbide extraction replicas. The identification of voids on polished metallographic samples cut from the sample in the initial stage of necking led to the conclusion that the smallest size of carbides nucleating voids was about 10 µm. Therefore, only the related quantiles of the probability density functions were used to determine the mean values a_{os} for investigated steels.

The theoretical predictions and experimental values of contractions for all studied steels and models are summarized in Tables 2.4, 2.5 and 2.6, along with relevant values of n, κ, Λ/a_o, ε_f and ε'_f. The best agreement between theory and experiment was achieved by using the model based on the unstable collapse of intervoid ligaments (Brown–Emburry). Satisfactory, but slightly conservative results were also obtained by means of the percolation model. On the other hand, the fracture strains and contractions obtained by utilizing the McClintock's model highly overestimated the experimental values.

Table 2.4 Comparison of the McClintock model with experimental data

Steel	$\frac{\Lambda}{2a_0}$	n	ε'_f [a]	ε'_f [b]	$\psi'_{(MC)}$ [a] [%]	$\psi'_{(MC)}$ [b] [%]	ψ_{exp} [%]
12 010	3.9	0.25	1.88	1.41	88.1	81.0	74.6
14 109	2.0	0.22	1.45	1.06	81.2	72.2	59.6
14 331	2.8	0.22	1.80	1.33	86.7	78.8	63.8

[a] for $\varepsilon_i = \varepsilon'_f/2$
[b] for $\varepsilon_i = \varepsilon'_f$

Table 2.5 Comparison of the Brown–Emburry model with experimental data

Steel	κ	ε_f	ψ_{BE}	ψ_{exp}
12 010	0.80	1.370	74.6	74.6
14 109	0.60	0.896	59.2	59.6
14 331	0.85	0.971	62.3	63.8

Table 2.6 Comparison of the percolation model with experimental data

Steel	ε_f $(s_{(2)} = 6)$	ε_f $(s_{(2)} = 4)$	ψ_{PP} $(s_{(2)} = 4)$	ψ_{exp}
12 010	1.17	1.20	70.0	74.6
14 109	0.68	0.71	50.1	59.6
14 331	0.80	0.83	56.4	63.8

2.3.2.3 Generalized Diagram of Fracture Strain

In the early 1950s, Russian scientists proposed a generalized diagram of fracture strain [240] that was subsequently widely used in forging technology. They found that the influence of stress state on the true fracture strain (in terms of the true strain intensity) can be successfully described by the following exponential function:

$$\varepsilon_f = k_1 \exp\{-k_2\kappa\}. \tag{2.42}$$

There are only two fitting parameters k_1 and k_2 in Equation 2.42. Values of κ for selected special loading modes are as follows:

1. isotropic tension ($\sigma_1 = \sigma_2 = \sigma_3 > 0 \Rightarrow \kappa \to \infty$);
2. crack tip region ($\sigma_1 = \sigma_2 \approx 1.6\sigma_3 \Rightarrow \kappa \to 2.2$);
3. uniaxial tension ($\sigma_1 > 0, \sigma_2 = \sigma_3 = 0 \Rightarrow \kappa = 1/3$);
4. torsion ($\sigma_1 = -\sigma_2 > 0, \sigma_3 = 0 \Rightarrow \kappa = 0$);
5. uniaxial compression ($\sigma_1 < 0, \sigma_2 = \sigma_3 = 0 \Rightarrow \kappa = -1/3$);
6. isotropic compression ($\sigma_1 = \sigma_2 = \sigma_3 < 0 \Rightarrow \kappa \to -\infty$).

Thus, Equation 2.42 can, in principle, be represented by a diagram that is constructed using true fracture strain data obtained from the torsion test and the uniaxial tensile test. The compression test can be used as well. It should be noted that some stress triaxiality might be induced in the later stages of the tensile test of ductile materials. Consequently, the related mean value of κ is usually slightly higher than 1/3. In the case of the compression test, similarly, the mean value of κ might also be somewhat higher due to specimen bulging. The value of κ for the crack tip region (plane strain) is also only approximate because the stress state in volume elements inside the plastic zone changes during the loading (see the next paragraph).

For positive values of κ, a simple relation

$$\varepsilon_f = \frac{0.8}{\kappa} \ln \frac{d_0}{d} \tag{2.43}$$

can be also used [241].

Note that both empirical relations at Equations 2.42 and 2.43 are consistent with more physically justified Equations 2.28 and 2.40 (or 2.41).

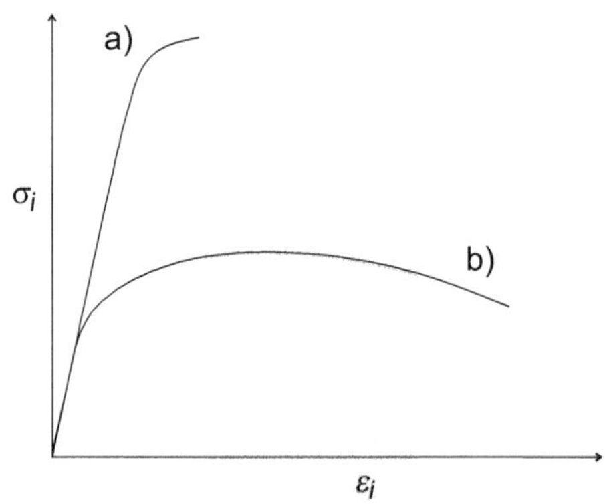

Figure 2.33 The schematic stress-strain intensity diagram of a ductile material under triaxial tension within the plastic zone (curve a) and uniaxial tension (curve b)

Important information allowing the applicability of LEFM to ductile materials can be directly obtained from the diagram of fracture strain. Even materials of a very high ductility measured in terms of the uniaxial tensile test ($\kappa \approx 0.33$) generally exhibit a low fracture strain at the crack tip under plane strain ($\kappa \approx 2.2$). According to the von Mises plasticity surface, moreover, the yield strength becomes about three times higher than that related to the uniaxial tension. Therefore, the ductile material inside the plastic zone behaves in a quasi-brittle manner. This can be clearly seen from generalized stress-strain diagrams of a ductile material under both uniaxial and triaxial stress states, as schematically plotted in Figure 2.33. Indeed, the behaviour of the ductile material inside the crack-tip plastic zone is not far away from the elastic response. This is one of the reasons why, in the case of small scale yielding, the application of LEFM reasonably predicts the crack stability.

2.3.3 Assessment of Fracture Toughness from Basic Mechanical Characteristics

During the initial phase of fracture toughness test of high-strength steels and aluminium alloys, the crack tip grows in a stable way by blunting. Since the yield stress of those materials is high enough, small scale yielding conditions can easily be realized, and valid K_{Ic} tests can be performed at room temperature by using rather small samples. This means that a localized ductile damage process always precedes the unstable fracture and, as a consequence, the so-called stretch zone near the fatigue pre-crack can usually be identified on the fracture surface. Because of the triaxial state of stress in the process (plastic) zone ahead of the crack front, the growth rate of voids is very high and can be described by means of the previously mentioned mathematical apparatus. When considering the standard fracture toughness test of relatively small deformation rates, energy dissipation in the form of elastic waves can be neglected, and Equation 2.4 can be utilized for the description of the fracture process. Practically all energy supplied by external forces and/or released by elastic relaxation is consumed in the plastic zone during the ductile damage process preceding the unstable crack advance. This means that the term 2γ can be neglected in comparison to $w_p(K, \gamma)$. Thus, the energy consumed within the plastic zone up to the moment of unstable fracture can be expressed as

$$\frac{1-\nu^2}{E} K_I^2 = w_p(K_I, \gamma) \approx 2r_p \int_0^{\varepsilon_f} \sigma_i(\varepsilon_{ip}) \mathrm{d}\varepsilon_{ip}, \tag{2.44}$$

where ε_{ip} and σ_i are the respective intensities of plastic strain and stress, ε_f is the fracture strain and r_p is the plastic zone size [242]. Let us consider the Hollomon approximation of the tensile stress-strain diagram

$$\sigma_i = A\varepsilon_{ip}^n,$$

where A is the hardening coefficient and n the hardening exponent, together with the relation $r_p \approx Bn^2$ ($B \approx 0.025\,\mathrm{m}$) first introduced by Hahn and Rosenfield [157]. Equation 2.44 can be then transformed to

$$K_{Ic} = \left[\frac{2C\varepsilon_f^{n+1}}{(1-\nu^2)(n+1)}\right]^{1/2} n, \tag{2.45}$$

where $C = ABE$. The only difficulty with a direct application of Equation 2.45 to the assessment of K_{Ic} is the fracture strain ε_f. This value is much lower than that obtained from the uniaxial tensile test due to a different (triaxial) state of stress at the crack tip. As already mentioned, the state of stress inside the plastic zone changes due to the crack tip blunting. However, this problem can be solved by a method proposed by Staněk and Pokluda [241]. During the crack tip blunting, each element of the process zone at the crack tip experiences a deformation trajectory that can be calculated numerically using the parameter κ [243,244]. In the range $\kappa \in (0.6, 2.5)$, this trajectory can be well approximated by the relation

$$\varepsilon_{ip}(\kappa) = 7.7\exp(-2.9\kappa). \tag{2.46}$$

The deformation of each element starts at the point $\varepsilon_{ip} = 0, \kappa = (\pi + 1)/\sqrt{3}]$ and, during the crack tip blunting, it proceeds along the trajectory given by Equation 2.46 under decreasing triaxial stress state defined by the parameter κ. Let us accept the hypothesis of linear damage accumulation, and define an elementary damage increment as $\mathrm{d}\varepsilon_p/\mathrm{d}\varepsilon_f$ with respect to the diagram of generalized fracture strain (Equation 2.43). Then the onset of fracture is determined by the relation

$$\int_{\kappa_c}^{(\pi+1)/\sqrt{3}} \frac{\mathrm{d}\varepsilon_{ip}(\kappa)}{\varepsilon_f(\kappa)} = 1, \tag{2.47}$$

which means that the critical damage level of 1 was reached in one of the process zone elements. A combination of Equations 2.43, 2.46 and 2.47 gives

$$(7.7\kappa_c + 2.655)\exp(-2.9\kappa_c) = 0.8\ln\frac{d_0}{d} + 0.0205. \tag{2.48}$$

Equation 2.48 enables us to do a simple numerical calculation of the critical parameter κ_c that determines the moment of unstable crack initiation within the process zone during blunting of the pre-crack tip. Substituting κ_c

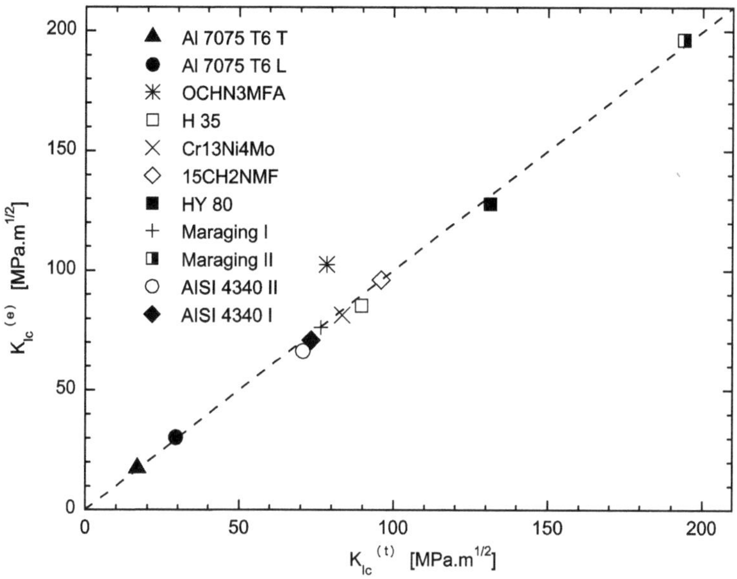

Figure 2.34 Comparison of experimental and theoretical fracture toughness data for some high-strength steels and aluminium alloys

back into Equation 2.46 one obtains the value ε_f. This value can be used in Equation 2.45 for the assessment of fracture toughness K_{Ic}.

Thus, in order to estimate the K_{Ic} value of a particular high-strength material, one needs only standard experimental characteristics of the tensile test: parameters A and n of the Hollomon approximation and the ratio d/d_0 on longitudinal profiles of fractured tensile samples. A comparison of experimental and theoretically estimated values of K_{Ic} for high-strength steels and aluminium alloys is given in Figure 2.34 [245]. All fracture toughness tests were performed at room temperature except for the steel AISI 4340 II, where the samples had to be cooled to $-120°$C. Even at such a low temperature the fracture surfaces revealed ductile patterns because the steel was tempered at 650°C. The agreement between theory and experiment is very good. It seems to be even better when taking the usual scatter of fracture toughness tests into account.

Chapter 3
Fatigue Fracture

From the practical point of view, fatigue fracture is the most important damage process. Available statistics show that, including corrosive assistance, fatigue is a leading cause of material failures registered during a long-term performance of engineering components and structures [246]. From the micromechanical point of view, the fatigue process can be understood as a sequence of the following stages: nucleation of cracks, stable propagation of short (small) cracks, stable propagation of long cracks and unstable fracture [247].

This chapter is divided into four sections. In Section 3.1, morphological patterns reflecting all crack growth stages on the fracture surface are briefly described. Moreover, the topological methods widely utilized in fatigue research and the quantitative fractography are outlined. These topics are important for all subsequent sections of the chapter.

The second section is devoted to the propagation of fatigue cracks under the remote opening mode (mode I). Mechanisms of nucleation and growth of short cracks are briefly reported, although these initial stages of the fatigue process were not a special subject of our research. Inclusion of these topics was, however, inevitable in order to provide a self-contained description of fatigue micromechanisms. The growth of short cracks is governed by shear stress components in favourably oriented crystallographic slip systems that are inclined at about 45° with respect to the maximal principal stress. This means that the short cracks grow in a local mixed-mode I+II and the picture of dislocation emission from the crack tip as well as the related growth micromechanisms are completely different from those related to long cracks.

After a certain incipient period that incorporates crystallographic and transient growths of short cracks, the fatigue cracks incline towards a direction perpendicular to the maximal principal stress, i.e., nearly towards mode I loading of the crack tip. This means that the long cracks keep propagating so that the crack tip plasticity is produced in the opening loading mode. Thus, the second section provides the reader with a micromechanical interpretation of all important phenomena accompanying this, most frequent, type of fatigue crack growth. Knowledge of these micromechanisms is essential not only for materials scientists and mechanical engineers, but also for technologists attempting to design structural materials exhibiting better resistance to fatigue crack propagation.

When the shear components of the applied stress are dominant, both short and long cracks can grow macroscopically under shear loading modes II or III. As usual, however, pure shear-mode crack propagation persists only for a limited number of loading cycles and the cracks incline or branch to get loaded in mode I. This leads to a local mixed-mode I+II, I+III or I+II+III crack propagation. The factory roof formation under torsion loading is an instructive example of such behaviour. Therefore, Section 3.3 refers to both shear-mode and mixed-mode crack growth. The first subsections introduce theoretical models and experimental results concerning crack growth under pure-shear and torsional loading. Since a combined cyclic bending-torsion is applied to many structural components, a rather extended part of the third section reports on the results of fracture tests performed under this kind of loading.

The final Section, 3.4, is devoted to the application of quantitative fractography to failure analysis. The fracture morphology can purvey a direct link between damage micromechanisms and both initiation and propagation of cracks. This section directly outlines how the knowledge of fracture micromechanisms can help to identify the reasons for failures of structural components in service. Therefore, its content can be useful for machine designers, especially for those working in the transport industry.

In general, Chapter 3 attempts to convince the reader of how useful the unified nano- micro- meso- macroscopic approach can be when trying to describe and interpret the behaviour of fatigue cracks.

3.1 Quantitative Fractography

Micromechanisms of fatigue crack propagation can be advantageously studied by means of fractographic tools. Indeed, the fracture surface morphology reflects many important stages of both stable and unstable fatigue crack propagation. Moreover, quantitative fractography is a powerful tool for failure analysis. Consequently, all sections of this fatigue chapter more or less refer to morphological patterns and fractographical results. Therefore, the quantitative fractography section was placed at the very beginning of this chapter.

The term "fractography" was first used by Carl A. Zapffe for the procedure of descriptive analysis of fracture surfaces in 1945 [248]. The output of this analysis is a set of numerical characteristics (number, shape, size, orientation, distribution) related to morphological patterns or parameters (roughness, fractality, texture) of the global surface topography. In both cases, the accuracy of these data is determined by knowledge of space coordinates of points on the fracture surface investigated. A very intensive development of quantitative fractography is obviously directly associated with increasing accuracy of measuring methods as well as with a rapidly growing capacity and computing rate of computers. During the second half of the last century, two-dimensional fractography in the scanning electron microscope was widely developed. In the last 20 years, however, an extended utilization of computer-aided topography techniques has enabled enormous progress in three-dimensional methods. Similarly, a number of descriptive concepts, distinguished by both the extent and the quality of utilized parameters, have been developed in the area of quantitative fractography.

Nevertheless, two various approaches can be distinguished here. The first utilizes various parameters of roughness, fractality or texture of fracture surfaces in order to find relationships between the fracture topology on one side, and the loading mode or the crack growth rate on the other side. In the first subsection, therefore, definitions of basic topological parameters are outlined. A rather different problem, usually demanding extensive research experience, constitutes the correct identification of morphological patterns as fracture facets, ridges, beach marks, tire tracks or fatigue striations. Thus, the second approach deals with the quantification of morphological patterns and their relationships to the loading parameters.

3.1.1 Topological Analysis

A general description of fracture surfaces by means of topographical parameters should be able to provide topological characteristics as well as morphological patterns as special cases. This might, in principle, be achieved by an analysis of a global set and relevant subsets of topological data. However, the basic problem here is represented by a high variability of fracture surface topology measured at different resolutions as well as an extreme complexity of the related microreliefs. In spite of a rather long history, no general definition of the surface topology and, consequently, no universal methodology of its quantification has been commonly accepted up to now [249–254].

In order to acquire a sufficiently wide and relevant set of topological parameters, advanced three-dimensional topological methods are to be employed. A great majority of results presented in this book was obtained by application of two methods that are based on different physical principles. Stereophotogrammetry is a method that makes use of the stereoscopic principles in order to obtain topological data of the fracture surface under investigation. Inputs to the method are two images of the analyzed region taken from different angles of view (so-called stereoimages or the stereopair) and some additional parameters that characterize a projection used during their acquisition. Usually, a scanning electron microscope (SEM) equipped with a eucentric holder is employed and the stereopair is obtained by tilting the specimen in the

SEM chamber by an angle that depends on the local roughness of the surface. The stereopair is processed via a matching algorithm in order to find corresponding points on both images (homologous points) and the relative z-coordinates of these points are calculated. The 3D model of the depicted surface area usually consists of ten to twenty thousand non-equidistant points and so the Delaunay triangulation must be performed [255].

Optical chromatography represents another method useful for a 3D reconstitution of the fracture surface micromorphology. The profilometer MicroProf FRT, Fries Research & Technology GmbH, makes use of the chromatic aberration of the optical lens. Different light monochromatic components are focused at different heights from a reference plane at the output of the optical fibre. The light intensity exhibits a maximum at the wavelength exactly focused on the surface and the height of the surface irregularities is deduced by using a calibration table. This optical method was usually employed only for verification of selected results obtained by stereophotogrammetry.

According to their mathematical basis, recently used topological parameters can be divided into five main categories. This classification is based on published works [250–253] and, in particular, on the work of Petropoulos *et al.* [256]. First two categories represent vertical (altitudinal) and length roughness parameters which characterize vertical and horizontal distributions of surface points, respectively. The third group involves hybrid parameters simultaneously describing more than one of the above-mentioned aspects. The fourth and fifth groups respectively consist of spectral and fractal characteristics of the fracture surface.

3.1.1.1 Roughness Parameters

In this brief overview only vertical, length, hybrid, spectral and fractal parameters are mentioned in more detail. The description of other parameters can be found elsewhere [214,249–251]. For the sake of simplicity, the assumption of non-overlapping surface elements is accepted hereafter. This means that each pair of coordinates (x_i, y_i) that determine the location of a point on the reference plane perpendicular to the macroscopic fracture surface is uniquely related to one altitudinal coordinate z_i. Because of a discrete data set of points utilized here to quantify surface topography, only a discrete form of definition of individual parameters is presented. Integral definition of many topological characteristics can be found, e.g., in [251,252].

Vertical Parameters

Most important altitudinal parameters are characteristics associated with the probability of realization of the value z_i in terms of central moments, defined as

$$\mu_k = \frac{1}{n}\sum_{i=0}^{n-1}(z_i - \langle z\rangle)^k,$$

where k is the order of the central moment, n is the range of the analyzed set of data and $\langle z\rangle$ is the arithmetic average height:

$$\langle z\rangle = \frac{1}{n}\sum_{i=0}^{n-1} z_i.$$

With respect to properties of $\langle z\rangle$, the first central moment is zero. The second moment (variance) yields information on the width of the distribution. This parameter is connected with the standard deviation R_q by the relation

$$R_q = \sqrt{\mu_2}.$$

The characteristic of the third central moment, the skewness, describes a symmetry of the function $p(z)$. It is defined as

$$\gamma_1 = \frac{\mu_3}{\mu_2^{3/2}} = \frac{\mu_3}{R_q^3}.$$

A normalized form of the fourth central moment is the kurtosis:

$$\gamma_2 = \frac{\mu_4}{\mu_2^2} - 3 = \frac{\mu_4}{R_q^4} - 3.$$

This parameter becomes zero for the Gaussian distribution and a negative value indicates a more flat distribution. Another parameter often used in the literature is the arithmetic roughness (the centre line average):

$$R_a = \frac{1}{n}\sum_{i=0}^{n-1}|z_i - \langle z\rangle|.$$

This parameter, however, is usually nearly proportional to the standard deviation R_q [250].

The second group of vertical parameters involves characteristics based on extreme values of the set Z [257]. The most widely used parameters are the highest height R_p ($R_p = z_{\max} - \langle z\rangle$) and the maximum depth R_v ($R_v = \langle z\rangle - z_{\min}$) which are the maximal and minimal values of the relative coordinates associated with a selected altitudinal level $\langle z\rangle$, respectively. Another parameter is the vertical range, R_z, representing their sum $R_z = z_{\max} - z_{\min}$ (Figure 3.1).

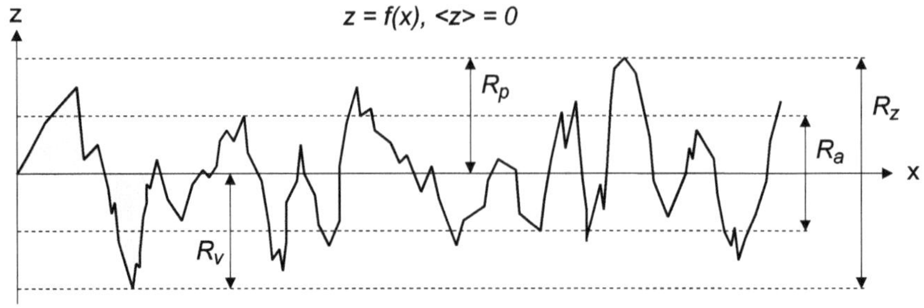

Figure 3.1 Graphical definition of basic extreme vertical parameters: the maximum hight R_p, the maximum depth R_v and the vertical range R_z. For comparison, the arithmetic roughness R_a is also depicted. The *grey area* marks the first element of the profile

Length Parameters

Length parameters describe the distribution of specific altitudinal levels of the surface in the horizontal plane x–y. These parameters are particularly useful for quality control in manufacturing processes. Their application in the fractography is rather rare. The most commonly used length parameters are, for example, the average spacing of profile elements on the mean line, S_m, the number of profile intersections with the mean line n_0 or the number of peaks per the length unit m_0.

By evaluation of the parameter m_0, for example, the peak is counted only when its horizontal distance from the previously counted peak is higher than $\frac{1}{10}$ of the vertical range R_z [251]. Consequently, the relation $x_i^p - x_{i-1}^p > \frac{1}{10} R_z$ must be fulfilled, where x_i^p is the x-coordinate of the i-th peak and x_{i-1}^p is that of the previously counted peak.

Hybrid Parameters

Hybrid parameters can be understood as a combination of altitudinal and length characteristics [250, 252]. In quantitative fractography, the linear roughness R_L and the area roughness R_A have been used for some time. These dimensionless characteristics, sometimes respectively called the relative profile length and the relative surface area, are defined as

$$R_L = \frac{L}{L'},$$

$$R_A = \frac{S}{S'},$$

where L is the fracture profile length, S is the area of the fracture surface, L' is the profile projection length and S' is the surface area projection into the macroscopic fracture plane. In the case of the profile composed of z randomly oriented linear segments $R_L = \frac{\pi}{2}$, whereas for the fracture surface composed of randomly oriented (nonoverlapping) facets $R_A = 2$ [214]. The area roughness R_A can be roughly assessed by means of the linear roughness R_L as

$$R_A = \left(\frac{4}{\pi}\right)(\langle R_L \rangle - 1) + 1.$$

The average slope Δ_a and its standard deviation Δ_q are defined in the following manner:

$$\Delta_a = \frac{1}{n-1} \sum_{i=0}^{n-2} \frac{|z_{i+1} - z_i|}{(x_{i+1} - x_i)},$$

$$\Delta_q = \left[\frac{1}{n-1} \sum_{i=0}^{n-2} \left(\frac{|z_{i+1} - z_i|}{(x_{i+1} - x_i)} - \Delta_a\right)\right]^{\frac{1}{2}}.$$

It should be noted that the parameters R_L and R_A also provide information about the angular distribution of surface elements [258].

Spectral Parameters

Spectral character of the profile can be described by means of the autocorrelation function which is a quantitative measure of similarity of the "original" surface to its laterally shifted version [251, 253]. Thus, the autocorrelation function expresses the level of interrelations of surface points to neighbouring ones. In the case of the fracture profile described by a set of n equidistant points (or $m \times n$ points obtained by sampling using constant steps in the directions of coordinate axes x, y), the autocorrelation function is defined by the following relations:

$$R(p) = \frac{1}{(n-p)} \sum_{i=0}^{n-p-1} (z_i - \langle z \rangle)(z_{i+p} - \langle z \rangle),$$

$$R(p,q) = \frac{1}{(m-p)(n-q)} \sum_{i=0}^{m-p-1} \sum_{j=0}^{n-q-1} (z_{i,j} - \langle z \rangle)(z_{i+p,j+q} - \langle z \rangle),$$

where p and q are shifts in the directions of x and y. The autocorrelation function has the following properties:

1. $R(0) = \mu_2$ or $R(0,0) = \mu_2$;
2. $R(p) = R(-p)$ or $R(p,q) = R(-p,-q)$;
3. $R(0) \geq |R(p)|$ or $R(0,0) \geq |R(p,q)|$ which means that the autocorrelation function attains a maximum for zero shifts.

With respect to the first attribute, the autocorrelation function is often normalized so that $R(0) = 1$ or $R(0,0) = 1$. The normalized autocorrelation function is usually denoted as $r(p)$ or $r(p,q)$:

$$-1 \leq r(p) = \frac{R(p)}{\mu_2} \leq 1, \qquad -1 \leq r(p,q) = \frac{R(p,q)}{\mu_2} \leq 1.$$

Autocorrelation lengths β_p and β_q are defined as shifts p, q corresponding to a drop of the autocorrelation function to a given fraction of its initial value. The fractions $\frac{1}{10}$ and $\frac{1}{e}$ are most frequently utilized [251,253]. Consequently, the surface points more distant than β_p, β_q can be assumed to be uncorrelated. This means that the related part of the fracture surface was created by another, rather independent, process of surface generation.

The character of the spectral surface can also be described in the Fourier space. The most important characteristic is the power spectral density

$$G(\omega_p) = |F(\omega_p)|^2,$$

$$G(\omega_p, \omega_q) = |F(\omega_p, \omega_q)|^2,$$

where ω_p and ω_q are space frequencies in the directions of coordinate axes x and y [251, 254]. Functions $F(\omega_p)$ and $F(\omega_p, \omega_q)$ represent relevant Fourier transforms of the fracture surface:

$$F(\omega_p) = \frac{1}{n} \sum_{k=0}^{n-1} z_k \exp\left\{-\mathrm{i}2\pi\left(\frac{k\omega_p}{n}\right)\right\}, \tag{3.1}$$

$$F(\omega_p, \omega_q) = \frac{1}{mn} \sum_{k=0}^{m-1} \sum_{l=0}^{n-1} z_{k,l} \exp\left\{-\mathrm{i}2\pi\left(\frac{k\omega_p}{m} + \frac{l\omega_q}{n}\right)\right\}, \tag{3.2}$$

where i is the imaginary unit [254]. Equations 3.1 and 3.2 define the so-called discrete Fourier transform (DFT). The conventional factors $\frac{1}{n}$ and $\frac{1}{mn}$ might differ for various applications. Instead of the highly computationally demanding DFT the fast Fourier transform is often utilized.

Fractal Parameters

Fractal geometry is a mathematical discipline introduced by Mandelbrot [259] in the early 1980s. It is widely utilized as a suitable tool for the description of jagged natural objects of complicated geometrical structure. Fundamental properties of fractal objects are so-called self-similarity or self-affinity which mean an invariance with respect to scale changes. As a measure of the fractality the Hausdorff (fractal) dimension D_H is often used. The metrics of D_H can be determined by means of the Hausdorff measure

$$\Gamma_H^d = \lim_{\varepsilon \to 0} \inf_{U_i} \sum_i (\mathrm{diam}\ U_i)^d . \tag{3.3}$$

When calculating the Hausdorff measure, the object is covered by cells U_i. The diameter of each cell meets the following condition: $\mathrm{diam}\ U_i = \sup\{|x-y| : x, y \in U_i\} \leq \varepsilon$. Consequently, one searches the cell network minimizing the sum in Equation 3.3 for an infinitely small diameter of covering cells ($\varepsilon \to 0$). There is only a single value of D_H fulfilling the conditions $\Gamma_H^d = 0$ for each $d > D_H$ and $\Gamma_H^d = \infty$ for each $d < D_H$. This value is called the Hausdorff (fractal) dimension of the object. In the case of a smooth (Euclidean) object $D_H = d_T$, where d_T is the topological dimension, whereas $d_T < D_H \leq (d_T + 1)$ holds for the fractal object. In general, D_H is a rational number exceeding the topological dimension. A higher D_H-value means a higher segmentation of the object. As an example of the fractal object, Von Koch's curve is depicted in Figure 3.2 along with several first steps of its construction.

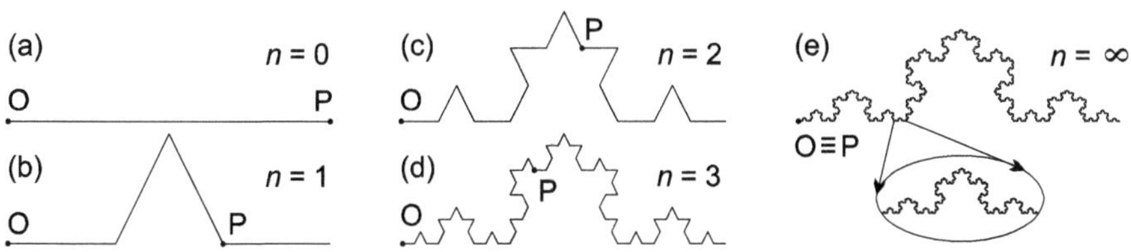

Figure 3.2 Von Koch's curve: (a) fractal initiator, (b) first iteration, (c) second iteration, (d) third iteration, and (e) final fractal ($D_H \approx 1.262$)

As can be seen from Figure 3.2, the length of the curve increases with increasing number of iterations and, for the final fractal, it becomes infinite. On the other hand, the area under the curve remains finite and practically unchanged. The infinite length of the fractal curve means that the marked points O and P retain the same distance during all iterations. Paradoxically, however, they coincide in the case of the final fractal ($\lim_{n\to\infty} |OP| = 0$). Real natural objects exhibit a statistical self-similarity rather than a perfect deterministic one. This means that the self-similarity does not hold for the object itself but only for its statistical parameters (average, variation, etc.) [260].

With respect to difficulties connected with the calculation of the Hausdorff dimension Γ_H^d directly according to Equation 3.3, various other simpler solutions were derived. The most widely used methods are depicted in Figure 3.3. By application of those methods to real objects, however, deviations from the theoretical fractal dependencies are usually observed [249,261,262]. A sigmoidal trend, obtained when calculating the parameter D_H, can serve as a typical example.

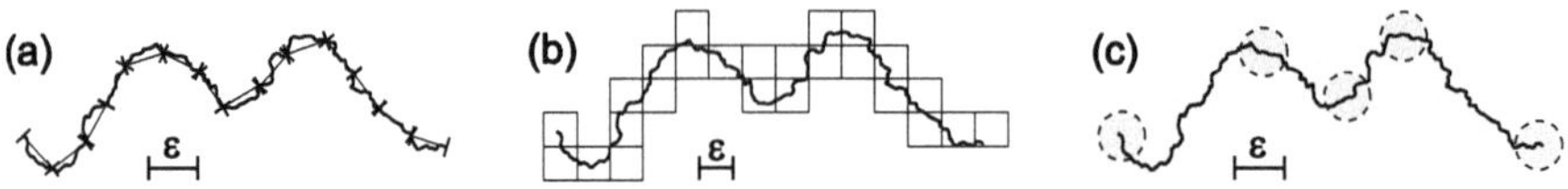

Figure 3.3 Some computation methods of the of fractal dimension: (a) perimeter method, (b) computation of squares, and (c) Minkowski method

Calculation of the of areas fractal dimension ($d_T = 2$) is more complicated and, as usual, it is performed either by means of space versions of curve methods [263, 264] or using the area-perimeter method. The latter method analyzes the fractal dimension of boundary curves of "islands" created by intersections of the horizontal plane with the object surface [264, 265].

Self-affinity is a more general form of self-similarity. A regular object exhibiting self-affinity is invariant to the transformation

$$x' \to \lambda_x x, \qquad y' \to \lambda_y y, \qquad z' \to \lambda_z z,$$

where $\lambda_y \propto \lambda_x^{\nu_y}$ and $\lambda_z \propto \lambda_x^{\nu_z}$ so that $\lambda_z \propto \lambda_y^{\nu_z/\nu_y}$. The ratio $H = \nu_z/\nu_y$ is called the Hurst exponent (the exponent of self-affinity), $H \in \langle 0; 1 \rangle$. When all contraction coefficients are equal ($\lambda_x = \lambda_y = \lambda_z$), the same transformation describes the self-similarity (see Figure 3.2(e)). In the case of isotropic surfaces, $\lambda_x = \lambda_y$ and $H = \nu_z$. This relation also refers to the arbitrary self-affine plane curve [260]. Again, the natural objects exhibit a statistical self-affinity rather than a deterministic one. Many experiments reveal that the fracture surfaces of most materials exhibit such a property [260,266,267]. Indeed, the self-similarity is usually preserved in the horizontal plane x–y (the area-perimeter method is based on that assumption), whereas the self-affinity is associated with the z-coordinate.

The Hurst exponent also yields information on a degree of internal randomness. When the object can be described by the Hurst exponent $H > H^*$, where $H^* = \frac{d_T}{d_T+1}$, the trend in the local site x (e.g., low or high z-values) is most probably followed by a similar trend in every other site $x + \Delta x$ (the persistence or the long-term memory). On the other hand, $H < H^*$ means an opposite tendency (antipersistence or short-term memory). The former type is typical for brittle fractures whereas the latter is typical for ductile ones [260].

As a rule, the Hurst exponent is calculated by means of the so-called variable bandwidth method [267, 268]. First, the profile is divided into k

movable windows (size ε) and, for each window, the local value of R_q is determined. Then the dependence of the average (global) characteristic

$$W = \frac{1}{k} \sum_{i=0}^{k-1} R_{qi}$$

on ε is established. Finally, the Hurst exponent is calculated according to the relation $W(\varepsilon) \propto \varepsilon^H$ as a slope of the exponential function plotted in bi-logarithmic coordinates (similarly to the fractal dimension).

3.1.2 Morphological Patterns in Fatigue

The morphological features in fatigue are different when looking at the fracture surface under low and high magnifications. Therefore, they can simply be separated into macroscopic and microscopic patterns [149, 187, 269, 270].

Macroscopic Patterns

The fractographic investigation of fatigue fracture surfaces usually starts visually or by low-magnification optical microscopy. Because of a change from a stage I slip plane that is inclined to the fracture plane macroscopically perpendicular to the principal loading axis, one can often distinguish a boundary between stage I and stage II crack propagations. It should be noted that the stage I fracture never extends beyond about three grains around the crack initiation site at the surface. The transition from stage I to stage II is usually accompanied by ridges parallel to the crack propagation direction. These ridges are formed by local shears that merge many plateaus corresponding to different fracture plains in individual grains.

One of the most important features usually found on fatigue fracture surfaces is beach marks (arrest marks), which are centred around the point of fatigue crack origin. These patterns correspond to stage II of crack propagation, and occur as a result of changes in loading or frequency or by oxidation of fracture surfaces during periods of crack arrest from intermittent service of the component.

The final-fracture zone can usually be identified by a fibrous morphology, which is different to that of the stage II crack growth. The size of this zone depends on the magnitude of loading, and its shape depends on the mode of external loading. Therefore, the transition boundary (line) between the stage II and final fracture zones can be related to the value of critical stress intensity factor K_c, associated with the fast (unstable) fracture. In ductile materials, shear lips (plane-stress) at approximately 45° to the fracture surface appear at the end of the final-fracture zone related to the free surface of the component. These lips enable us to differentiate clearly the final fracture zone from that of the initial crack growth stage.

Microscopic Patterns

During stage II of fatigue crack propagation, striations are formed in the way shown in Section 3.2.3. According to the model relevant for the Paris–Erdogan region, the striation spacing should be equal to the local crack growth rate, which is particularly true for ductile and homogeneous materials. Fatigue striations often bow out in the direction of crack propagation and generally tend to align perpendicular to the principal (macroscopic) crack growth direction. However, the crack locally propagates along multiple plateaus (facets) so that the local crack propagation directions in individual facets are usually different. The SEM picture of a striation field which documents this phenomenon is presented in Figure 3.4. Moreover, the plateaus can lie at different elevations with respect to each other and join by tear ridges or walls that also contain striations.

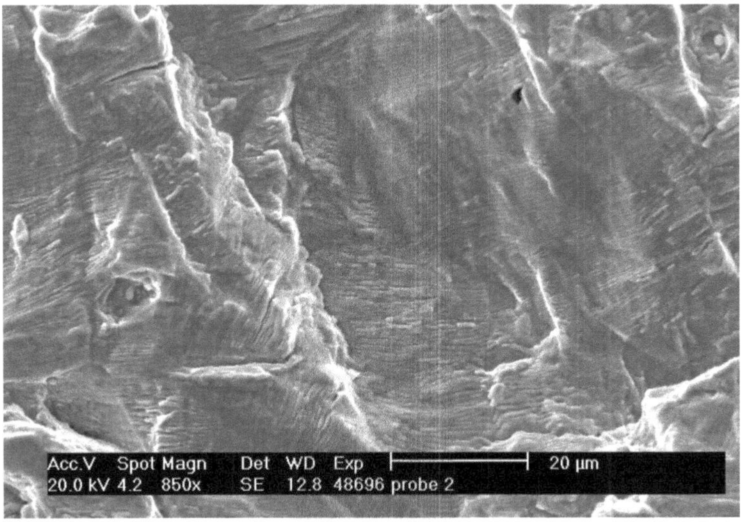

Figure 3.4 The striation field in the austenitic steel

Therefore, when comparing the macroscopic crack growth rate with the striation spacing for a given crack length, one has to consider an average of local crack growth rates indicated within the related striation field. This can be done by assuming the average of projections of local crack growth rates, perpendicular to the striations on individual facets, to the macroscopic propagation direction by using the formula

$$\bar{s} = \frac{1}{n}\sum_{i=1}^{n} s_i \cos\alpha_i, \tag{3.4}$$

where n is the number of facets, s_i is the striation spacing on the i-th facet and α_i are the related projection angles (see Figure 3.5)

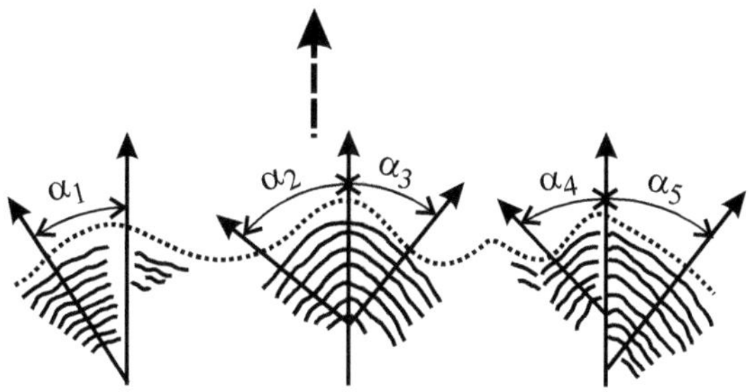

Figure 3.5 The scheme of local inclination angles of striation fields. The macroscopic crack growth direction is marked by the *arrow*

On striation facets which are at an angle to the average plane of crack growth, so-called fissures can be formed [271, 272]. They appear as short (secondary) cracks related only to one surface of the local crack, are regularly spaced, and penetrate a distance below the fracture surface which is much larger than the variations in fracture surface topography. The fissures form at striations most likely due to their stress concentration effect. Because the formation of fissures causes local crack branching, the further striations created on the inclined facet start to be immediately shielded especially from the local tensile stress parallel to the fracture surface. As the crack front moves away from the fissure, the local tensile stress builds up to the point where a new fissure is formed, and the process is repeated. Therefore, the fissure spacing does not directly correspond to the local crack growth rate, similarly to the spacing of false striations created in the near-threshold region (Section 3.2).

There are also other periodic patterns, sometimes observed on fatigue fracture surfaces generated by a shear loading, known as fibrous patterns and tire tracks. Both these markings are produced by a contact wear of crack-flank asperities. The fibrous patterns are, unlike striations, parallel to the crack growth direction and can be clearly seen in Figure 3.6. The tire tracks, resembling the tracks left by the tread pattern of a tire, are the result of particles or protrusions on the fracture surface being successively impressed into the mating surface part during the closing portion of the loading cycle. They are nicely depicted especially in Figure 3.52 (Section 3.3). An appearance of both fibrous patterns and tire tracks always indicates a presence of either remote or local shear mode II at the crack front. The local shear mode can also be induced in the case of a pure remote mode I by inclinations of crack front elements from the main crack plane perpendicular to the remote loading direction (see also Section 3.3). The direction of the tire tracks and the change in the spacing of indentations can also indicate both the magnitude and the type of displacements that occurred during the fracture process, such as lateral movement from shear or torsional loading. There is, again, no simple correspondence between the spacing of tire tracks (or fibrous patterns) and the crack growth rate.

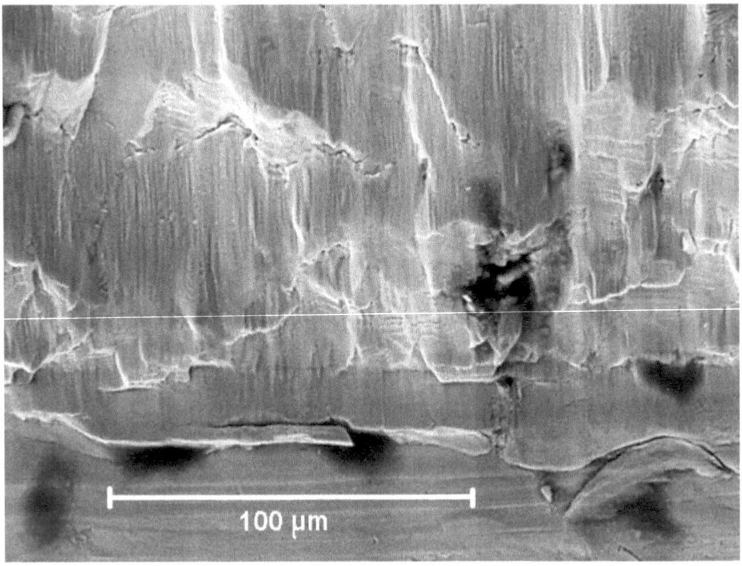

Figure 3.6 Fibrous patterns and tire tracks formed under the shear (mode II) crack propagation in austenitic steel. The crack growth direction is from the bottom to the top

Sometimes, periodic microstructural phases (pearlite, martensite laths, etc.) or slip traces can also be observed on fracture surfaces. Such patterns, obviously, have nothing to do with the crack growth rate. Thus, the fields of fatigue striations are the only relevant patterns that can be directly correlated with the rate of crack front propagation. A clear distinction between the striations and other periodical features is not trivial. Therefore, when performing a reconstitution of the fatigue process based on the fracture surface micromorphology, one should possess a sufficient level of fractographic experience.

3.2 Opening Loading Mode

There is common agreement within the international scientific community concerning a principal physical difference between the driving force of the fatigue (stable) crack growth and that of the brittle (unstable) fracture. While the latter is directly associated with a drop in the elastic energy (elastic strain) and the critical value of the stress intensity factor K_c, the driving force in fatigue is directly related to the range of the cyclic plastic strain at the crack tip. Because the maximum K-value during the stable crack growth lies below K_c, the crack growth can proceed only when supported by the work of external cyclic forces. Similarly, the stable growth in the case of stress corrosion cracking presumes the assistance of the chemical driving force. In 1963, Paris and Erdogan [273] proved that the diagram $\mathrm{d}a/\mathrm{d}N$ *vs* ΔK for so-called long cracks in the small-scale yielding range (the high-cycle fatigue) retains the advantage of LEFM, namely a satisfactory invariance in the shape and size of cracked solids. It might seem to be surprising that the linear elastic parameter also allows us to describe successfully the rate of plastic processes at the crack tip. Several years later, however, Rice [274] brought to light a theoretical reason justifying the present opinion: the small-scale cyclic plasticity (the cyclic plastic zone) at the crack tip is, indeed, controlled by the value of ΔK.

The local value of ΔK at the crack tip is determined by both external and internal stresses resulting from external forces and local plastic deformations (or generally from microstructural defects), respectively. The tensile and compressive elastic energies associated with internal stresses are mutually compensated all over the bulk and, therefore, they do not contribute to a global tensile elastic energy of stressed solids. Consequently, they cannot be released to support the unstable fracture process. At the same time, limited amounts of elastic energy that can be released by relaxation of local tensile internal stresses in small volumes adjacent to the crack tip can significantly influence neither the onset of brittle fracture nor its crack growth rate. On the other hand, the level of internal stresses can substantially influence the stable fatigue growth rate in each step of the crack advance, since the emission of dislocations from the crack tip occurs at very low stress intensity factors. This means that even very small changes of local K-values at the crack tip can considerably modify the stable crack growth rate. Thus, unlike in the case of unstable fracture, the internal stresses created by dislocation configurations and secondary phases are to be considered as an important additional factor affecting the fatigue crack propagation rate. It is particularly this difference that elucidates a much higher complexity of shielding (or anti-shielding) effects accompanying the fatigue crack growth when compared to brittle fracture. Unlike in brittle fracture, for example, the contact shielding in fatigue also occurs under the opening loading mode, which causes so-called crack closure phenomena. Moreover, many important phenomena associated especially with the small scale yielding, e.g., the existence of fatigue thresholds or a delay in ratcheting, can be explained only when the behaviour of discrete dislocations is taken into account. This underlines the necessity to utilize the multiscale approaches to the fatigue crack growth phenomenon which is documented in Chapter 3. Unfortunately, those fundamental diversities of stable and unstable crack growths still do not seem to be widely understood, particularly among "classical" mechanical engineers dealing solely with continuum mechanics.

There is, however, another important factor strongly influencing the stable crack growth rate in contrast to the unstable one. This is the effect of environment, even that of the air. Simply, the stronger the chemical interaction between the environment and the material, the higher the fatigue crack growth rate. For a majority of metallic materials, the crack growth rates in air are two orders in magnitude higher than those in vacuum [275]. Although the detailed mechanisms of chemical processes are not examined here, the micromechanical reasons for their strong influence are clearly outlined hereafter. On the other hand, the growth rate of unstable (brittle) cracks remains unaffected by the environment because there is not enough time available to produce a chemical damage during the fast fracture.

Let us emphasize that the mechanism of cyclic plasticity immediately elucidates why the fatigue cracks avoid propagating in hard (or brittle) microstructural phases. In such phases, the movement of dislocations is strongly limited and, therefore, the fatigue crack is always repelled to a softer material [276, 277]. The same qualitative explanation holds for an increase in the fatigue limit with an increasing hardness (or ultimate strength). Indeed, the fatigue limit is related to a critical stress under which the microstructurally short cracks still remain arrested [149, 247]. These cracks are smaller in high-strength materials (finer microstructure) so that the stress necessary for their further growth is higher. The fundamental difference in micromechanisms of brittle and fatigue crack propagation also has a consequence in the following well–known phenomenon: whilst a continuous increase in the materials ductility along the path of propagating brittle cracks (induced, e.g., by a temperature gradient) causes their arrest, this is generally not the case of fatigue cracks.

Large deformations inside the plastic zone in both the Paris–Erdogan and the near fracture regions of long–crack propagation can be described by classical continuum theories (Figure 3.7). However, these regions are of less engineering importance than those of the short crack growth and the long–crack threshold. Here, on the other hand, the plastic zone is relatively small and the dislocation activity is confined to one or two favourable slip systems. Under such conditions, the experimentally observed phenomena can be sufficiently elucidated only when taking the discrete nature of plasticity into account. Therefore, Sections 3.2.1, 3.2.2 and 3.2.3 are devoted to discrete dislocation models of cyclic plasticity and fatigue crack growth.

The results presented here are essential for a sufficient grasp of both the crack closure effects and the unified model of the crack-tip shielding, as de-

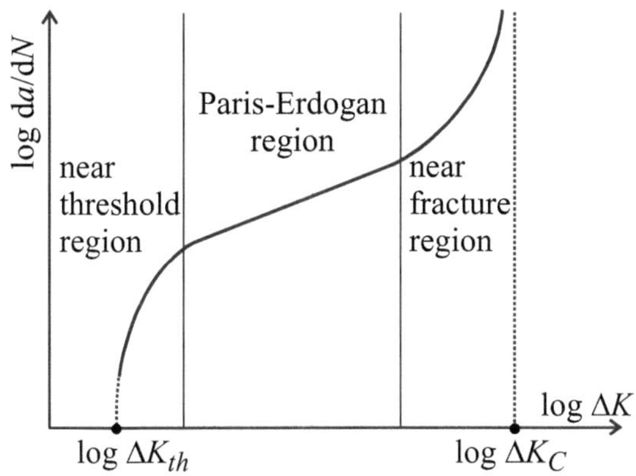

Figure 3.7 Scheme of the crack-growth rate *vs* ΔK dependence for long fatigue cracks

scribed in Sections 3.2.4 and 3.2.5. The unified model was developed as a multiscale concept involving three basic levels; micro – crystal defects, meso – grain (phase) microstructure and macro – continuum. The discrete dislocation theory and the size ratio effect constitute links between these three levels. It is important to emphasize, however, that a full comprehension of the crack growth threshold phenomenon requires an insight into the atomistics of dislocation emissions from the crack tip. Section 3.2.6 presents results of an extended application of the unified model to identify the shielding components as well as the intrinsic resistance to the near-threshold fatigue crack growth in various metallic materials. Finally, Section 3.2.7 is devoted to the influence of shielding effects on the crack growth rate in the Paris–Erdogan region.

3.2.1 Discrete Dislocation Models of Mechanical Hysteresis

3.2.1.1 Hysteresis Loop

The process of cyclic plastic deformation controls both the crack initiation and the rate of fatigue crack propagation. The Nabarro–Cottrell analysis [156] represents a rather simplified model of mechanical hysteresis but it is very useful for understanding the physical background of that process. This analysis was applied to provide an insight into the early stage of cyclic softening as well as to the micromechanism of ratcheting (the cyclic creep) in polycrystalline materials at room temperature [149, 278, 279].

A polycrystalline material at the onset of cyclic softening can be considered to be a nearly elastic aggregate containing a small number of perfectly plastic grains. The macroscopic stress-strain response of such a system of n grains corresponds to a composite of n_{el} elastic grains with Young's modulus E (a major phase) and n_{pl} plastic grains with Young's modulus $E_2 \ll E$ (a minor phase). Denoting the relative number of plastic grains as $\alpha = n_{pl}/n$ ($n \approx n_{el}$), the composite modulus E_c can then be expressed as

$$E_c = E\left(1 - 1.9\frac{\alpha}{1+\alpha} + 0.9\frac{\alpha^2}{(1+\alpha)^2}\right) \approx E(1-2\alpha),$$

in analogy to a porosity influence in ceramic materials [149, 280].
The aggregate strain can be expressed as

$$\begin{aligned} \varepsilon &= \sigma/E && \text{for } \sigma < \sigma_0, \\ \varepsilon &= \varepsilon_e + \varepsilon_p = \sigma/E + 2\alpha(\sigma - \sigma_0)/E && \text{for } \sigma > \sigma_0, \end{aligned} \tag{3.5}$$

where σ_0 is the yield stress of the composite. The stress-strain response of such aggregate is depicted in Figure 3.8 in $\tau - \gamma$ shear coordinates ($E \to G, \sigma \to \tau, \varepsilon \to \gamma$).

By raising the applied tensile stress τ from zero, the strain remains elastic up to the point A in Figure 3.8(a), where $\tau = \tau_0$. At this moment the first dislocations are emitted from Frank–Read (F-R) sources and create pile-ups at grain boundaries. These pile-ups hinder further dislocation emissions by producing an increasing back stress to F-R sources. Note that, in engineering applications, this stress is usually declared as a residual stress that remains in the material after unloading. Thus, the dislocations are emitted from the F-R source and produce plastic strain as long as the increasing effective shear stress $\tau - \tau_0$ compensates the back stress. At the point B the applied stress stops increasing ($\tau = \tau_{\max}$), and the following equilibrium equation holds:

$$\tau_{\max} - \tau_0 - \tau_B = 0, \tag{3.6}$$

where τ_B is the back stress.

Now we start to reduce the tensile stress from $\tau_{\max} \to \tau_{\max} - \Delta\tau$. Before a sufficiently high value of $\Delta\tau$ is reached, the dislocations cannot move back and the reverse deformation proceeds in an elastic manner (the segment BC in Figure 3.8(a)). The backward motion of dislocations begins only after a stress reversal at the point C, where the sum of the applied (reversed) stress and the back stress becomes equal to the compressive yield stress:

$$\tau_{\max} - \Delta\tau - \tau_B + \tau_0 = 0. \tag{3.7}$$

Combining Equations 3.6 and 3.7 one obtains

$$\Delta\tau = 2\tau_0.$$

This simple relation reveals that, for the unloading strain path, a doubled yield stress is to be taken into account. The value of the compressive yield stress $\tau_{\max} - 2\tau_0$ corresponding to the point C is lower than that of the tensile

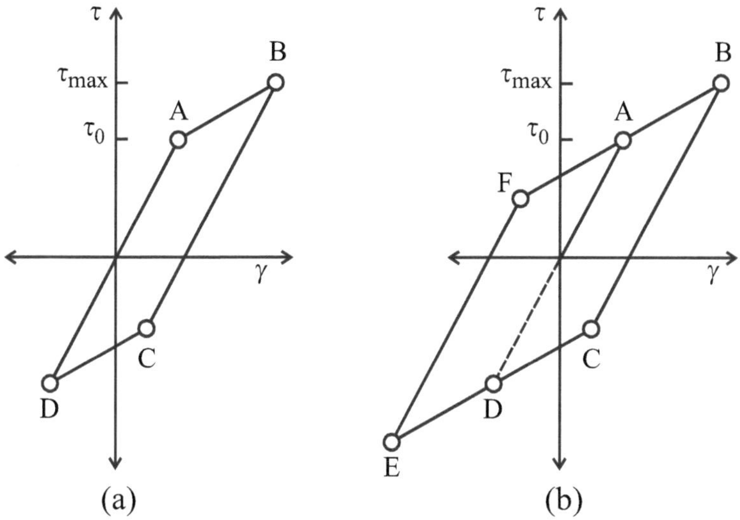

Figure 3.8 Scheme of the mechanical hysteresis behaviour in the frame of the Nabarro–Cottrell analysis: (a) the asymmetric loading cycle, and (b) the symmetric loading cycle

yield stress. This is the well-known Bauschinger effect. Beyond the point C, the dislocations move backwards while gradually reducing the back stress up to the point D, where $\tau_B = 0$. If the applied compressive stress starts to be reduced now, the whole process is repeated in the frame of a closed hysteresis loop. Such a loop is also created when the reverse deformation starts before reaching the point B or proceeds beyond the point D. An example of the latter case for a symmetrical loading is shown in Figure 3.8(b). Indeed, an inverse back stress is created within the segment DE, which causes an appropriate reduction of the tensile yield stress at the point F, similarly to that of the compressive yield stress at the point C. Let us finally emphasize that such a behaviour presumes a totally reversible dislocation slip.

As a rule, the elastic parts of real hysteresis loops in engineering materials are somewhat shorter, while the plastic ones are longer and curved. A main reason for that fact constitutes a statistical distribution of values of the yield stress in various grains [281]. It should also be noted that all the discrete dislocation models presented in this book utilize only dislocations inevitable for simulation of cyclic plasticity phenomena, the so-called geometrically necessary dislocations. There are also other dislocations forming the dislocation structure in real polycrystals, the so-called statistically stored dislocations. For more details about the geometrically necessary dislocations, see [8].

3.2.1.2 Ratcheting

The ratcheting (cyclic creep) is a process of gradual plastic elongation or contraction of a sample during cycling loading of a constant nominal stress amplitude. The first experimental observation of the ratcheting at room temperature was reported by Kennedy [282]. This process can lead to a substantial reduction of fatigue life in comparison with the strain-controlled loading of the same stress range [283, 284]. For example, failures of helicopter airscrews can be induced by ratcheting [285]. When loadings of a high asymmetry near the fatigue limit are applied, the cyclic microcreep can cause undesirable shape changes of precisely manufactured components as spiral springs or turbine blades.

In most practical cases, the prescribed load (work) level is achieved only after a certain period of time (the so-called ramp-loading). During that starting period, the plastic strain is increasing and the ratcheting process can only appear when the plastic strain range exceeds a certain critical value $\Delta\varepsilon_{pc}$ which is characteristic for a particular material. This means that, during the loading, no ratcheting occurs when either the highest achieved value of the plastic strain range $\Delta\varepsilon_p$ is less than $\Delta\varepsilon_{pc}$ or a cyclic hardening starts before reaching $\Delta\varepsilon_{pc}$. In the case of cyclically softening materials such behaviour is observed even when the work load is reached just in the first loading cycle. The scheme of a typical hysteresis behaviour during the asymmetric loading of cyclically softening materials before and after the onset of ratcheting is shown in Figure 3.9. However, remarkable elongations may occur even when a symmetrical loading cycle is applied (the cyclic ratio $R = \sigma_{\min}/\sigma_{\max} = -1$) [286, 287].

The value of $\Delta\varepsilon_{pc}$ is a material characteristic and depends on the cyclic ratio (decreases with increasing R) – see e.g., [278, 287]. The elongation or contraction per one loading cycle (the ratcheting rate) increases with increasing both the cyclic softening rate and the cycle asymmetry [149, 288].

As first described in [279], the analysis based on the discrete dislocation model of the hysteresis loop can be utilized to elucidate all the experimentally observed phenomena. Obviously, a closed hysteresis loop means that no ratcheting can take place. According to the above-mentioned Nabarro–Cottrell analysis, the incomplete closure of the hysteresis loop must be caused by micromechanisms producing an irreversibility of the cyclic plastic strain. During initial stages of a global cyclic softening, the plastic deformation is gradually transferred from pile-ups in most favourably oriented grains into the adjacent grains [289]. Simultaneously, a cyclic hardening starts to take place in some of the already plasticized grains. At the end of the global cyclic softening stage, all grains become plastic and the global response changes to the cyclic hardening. During the global cyclic softening stage the density of dislocations increases by a cooperative operation of Frank–Read sources and, in this way, primary dislocation networks and loop patches are successively created. In their vicinity, a gradual increase of internal stresses results in an activation of secondary slip systems. Schemes of both the hysteresis loop and the primary slip system in Figure 3.10(a) correspond to a moment just before the activation of the secondary slip. The interaction of primary and secondary dislocations leads to a formation of sessile dislocations, i.e., either to the Lommer–Cottrell barriers (in fcc metals) or to the [001] dislocations (in bcc metals). This means a start of the slip irreversibility, since the primary dislocations become locked in between the secondary (newly created) barrier and the primary (microstructural) barrier. Consequently, they cannot return to the Frank–Read source during the reverse half-cycle. Moreover, the secondary slip always reduces the back stress of the pile-up (regardless to a creation of the secondary barrier). This corresponds to an increase in the compressive yield stress. These micromechanisms are schematically depicted in Figure 3.10(b) ($\Delta\tau_B$ is the increment of the compressive yield stress, GB is the grain boundary, SD is the sessile dislocation and SS is the secondary slip system). Both processes substantially reduce the reverse plasticity, and leave a residual tensile plastic strain γ_{rb}, i.e., initiate the ratcheting [278, 290]. The first experimental observation of a direct connection between the secondary slip activity and the ratcheting process was reported by Lorenzo and Laird [291]. Due to superimposed internal and external stresses, the creation of barriers is related to a peak of the applied true stress. If the peak is tensile (compressive), the ratcheting causes the elongation (contraction) of the specimen.

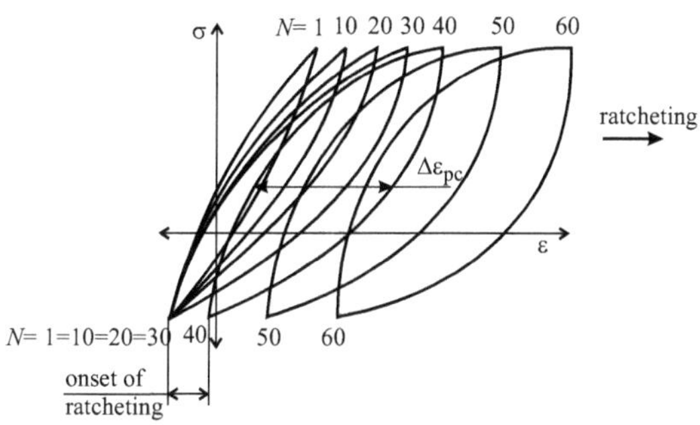

Figure 3.9 The scheme of the hysteresis behaviour of cyclic softening materials before and after the onset of ratcheting. The ratcheting starts after reaching the critical range of plastic strain $\Delta\varepsilon_{pc}$

In order to quantify the residual plastic deformation associated with the loop disclosure, let us consider the reverse slip of the pile-up in more detail. The Burgers vector density $B(x)$ within the pile-up can be expressed as

$$B(x) = \frac{\tau}{\pi c_0} \frac{x}{\sqrt{a^2 - x^2}},$$

where $c_0 = \mu/[2\pi(1-\nu)]$, ν is the Poisson's ratio, x is the coordinate along the pile-up and a is the length of the pile-up. In the first half-cycle the slip

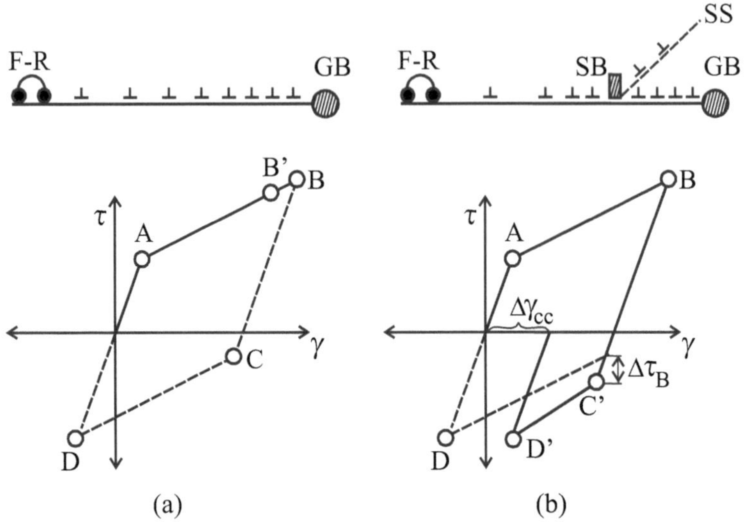

Figure 3.10 Incomplete closure of hysteresis loops produced by the secondary slip mechanism: (a) the moment just before a secondary slip activation (point B'), and (b) disclosure of the loop caused by the secondary barrier

of pile-up dislocations emitted from the F-R source is restricted only by the microstructural barrier. Consequently, the related shear strain

$$\gamma_r \sim \int_0^a B(x)x\mathrm{d}x = \frac{\tau}{\pi c_0}\int_0^a \frac{x^2}{\sqrt{a^2-x^2}}\mathrm{d}x. \tag{3.8}$$

When the maximum applied stress $\tau_{\max}$ is reached, the secondary slip creates the sessile dislocation. One can assume that the probability $P(x)$ of its location at the point x inside the pile-up is proportional to the density $B(x)$, i.e., $P(x) = DB(x)$. The probability of finding the dislocation barrier anywhere in between the F-R source and the primary obstacle (grain boundary) must be equal to 1, and, therefore, the following relation holds:

$$1 = \int_0^a P(x)\mathrm{d}x \Rightarrow 1 = \frac{\tau D}{\pi c_0}\int_0^a \frac{x}{\sqrt{a^2-x^2}}\mathrm{d}x \Rightarrow P(x) = \frac{1}{a}B(x). \tag{3.9}$$

After the creation of the secondary barrier, only dislocations located between the F-R source and that barrier can move back to the F-R source. These dislocations are positioned in the range $(0, y)$, where y is the coordinate of the secondary obstacle occurring with the probability $P(y)$. This means that, according to Equations 3.9 and 3.8, the total reverse shear displacement can be expressed as

$$\gamma_{rb} \sim \int_0^a \int_0^y P(y)B(x)x\mathrm{d}x\mathrm{d}y = \frac{\tau}{\pi c_0 a}\int_0^a \frac{y}{\sqrt{a^2-y^2}}\int_0^y \frac{x^2}{\sqrt{a^2-x^2}}\mathrm{d}x\mathrm{d}y. \quad (3.10)$$

After integrating Equations 3.8 and 3.10 one obtains

$$\gamma_{rb}/\gamma_r = 0.41.$$

This result shows that the secondary barrier prohibits at least a half of dislocations from returning back to the F-R source in the unloading half-cycle, which leads to a disclosure of the hysteresis loop. Obviously, this demands a sufficient density of primary dislocations to be generated during the cyclic softening in order to allow the creation of secondary barriers in the preferentially oriented large grains. From a macroscopic point of view it corresponds to a critical amount of a global cyclic plastic strain, i.e., the critical plastic strain range $\Delta\varepsilon_{pc}$ associated with the onset of ratcheting during the ramp-loading (generally during the global cyclic softening stage). This is a micromechanical interpretation of initial delays in the cyclic creep process observed for many metallic materials and various cyclic ratios [286,288]. The existence of ratcheting in case of symmetrical loading can be elucidated in a similar way. Due to a transverse contraction of the specimen (component), the peaks of the true stress come up in the tensile half-cycles. Because the creation of the first barriers is related to these peaks, the cyclic creep follows the tensile direction and elongates the specimen.

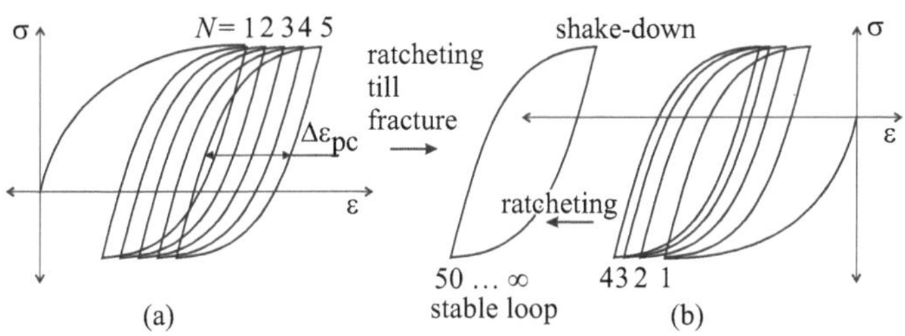

Figure 3.11 The scheme of the hysteresis behaviour when the critical plastic strain range is exceeded in all loading cycles: (a) the loading starts in the tensile direction, and (b) the loading starts in the compressive direction (the shake-down behaviour)

When the work load is reached just in the first loading cycle and, simultaneously, the critical value $\Delta\varepsilon_{pc}$ is exceeded, the kinetics of ratcheting strongly depends on the starting loading direction and the plastic strain range in the following cycles [292]. Such multislip plastic behaviour can already be described by continuum plasticity theories involving kinematic hardening [293,294]. The starting tensile (compressive) half-cycle produces the elongation (contraction) after a completion of the first cycle – see Figure 3.11. The ratcheting process proceeds only when the plastic strain range also remains sufficiently high in the next cycles (Figure 3.11(a)). This case corresponds to exceeding the so-called shakedown limit, usually expressed in terms of a critical applied stress value. When the ratcheting is oriented into the compressive direction, however, it eventually stops after a certain number of cycles due to a decreasing true stress range (an increasing diameter of the specimen) – see Figure 3.11(b). Such behaviour is known as a plastic shakedown. When the material hardens and the plastic strain range is relatively small, a rather slow ratcheting, reversed to that of the first half-cycle, is usually observed (Figure 3.12). As a rule, however, the loop stabilizes and the ratcheting shakes down after a certain number of cycles. This backward ratcheting is caused by high residual stresses which are created in the first half-cycle. During a certain period of a cyclic plastic deformation, these stresses gradually become relaxed by multislip.

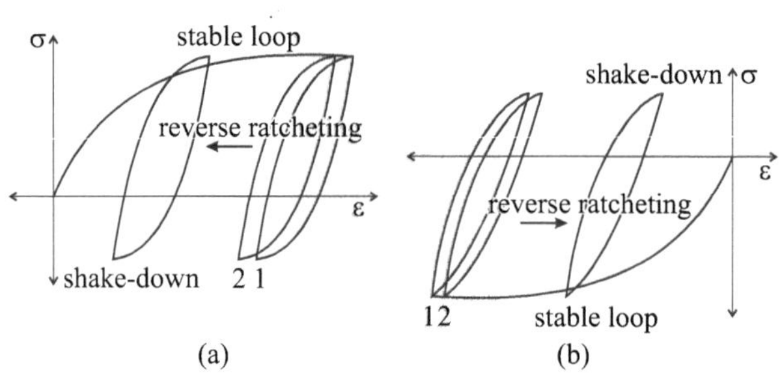

Figure 3.12 The scheme of the shake-down behaviour when the critical plastic strain range is exceeded only in the first loading cycle: (a) starting in the tensile direction, and (b) starting in the compressive direction

In order to assess the ratcheting rate in the initial softening stage, the difference between the plastic strains produced in the loading and unloading parts of the cycle is to be determined. Let us consider that during each loading cycle the reverse slip is restricted to one half due to the secondary slip in $\Delta\alpha$ grains. This means that the reversed plasticity will be reduced proportionally to $\Delta\alpha/2$. By using Equation 3.5, and respecting the reduction of the reversed plasticity, it leads to the following result:

$$\begin{aligned}\Delta\gamma_{cc} = \gamma_p - \gamma_{pR} &\approx \pm\left(\frac{2\alpha}{\mu}(\tau_{\max}-\tau_0) - \frac{2\alpha-\Delta\alpha}{\mu}(\tau_{\max}-\tau_0)\right) = \\ &= \pm\frac{\Delta\alpha}{\mu}(\tau_{\max}-\tau_0) = \pm\frac{\Delta\alpha}{\mu}\left(\frac{2}{1-R}\tau_a - \tau_0\right).\end{aligned} \tag{3.11}$$

The positive sign holds for $R > -1$ (elongation) and the negative sign for $R < -1$ (contraction). According to Equation 3.11 the ratcheting rate increases with both the increasing cycle asymmetry and the plastic strain range. Indeed, one can assume $\Delta\alpha \propto \Delta\varepsilon_p$, where $\Delta\varepsilon_p$ is the push-pull plastic strain range. This is in general agreement with experimental observations as well as with continuum plasticity models [295]. It should be emphasized, however, that the continuum mechanics is unable to interpret the two, already discussed, micromechanically induced phenomena: the critical plastic strain range related to the onset of ratcheting at the end of a delay and the tensile cyclic creep in the case of symmetrical loading.

Equation 3.11 was quantitatively verified by a simulation of initial ratcheting stages in ultra-high-strength steels [296]. Because $\Delta\alpha \approx k\Delta\varepsilon_p$ and the push-pull stresses σ and strains ε are proportional to the shear quantities τ and γ, Equation 3.11 can be rewritten in terms of σ and ε as

$$\frac{\Delta\varepsilon_{cc}}{\Delta\varepsilon_p} \approx \frac{k}{E}(\sigma_{\max} - \sigma_0). \tag{3.12}$$

According to Equation 3.12, the dependence $\Delta\varepsilon_{cc}/\Delta\varepsilon_p$ *vs* $\sigma_{\max}$ should be linear. Thus, the linear interpolation of experimental points is plotted in Figure 3.13 along with associated scatter ranges.

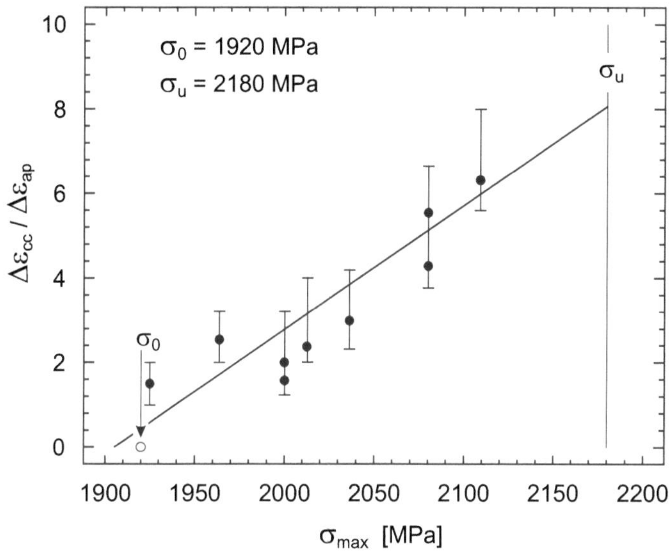

Figure 3.13 The normalized ratcheting rate as a function of the maximum loading stress. The *straight line* corresponds to the theoretical model

The plastic strain range $\Delta\varepsilon_p$ in the experiment was of the order of 10^{-5}. The extrapolated value $\sigma_0 = 1905\,\mathrm{MPa\,m^{1/2}}$ for the zero ratcheting rate agrees well with the experimental one ($\sigma_0 = 1920\,\mathrm{MPa\,m^{1/2}}$). The constant k is of the order of 10^4 and $\Delta\alpha \approx 10^{-1}$. This means that, in every loading cycle, the reversed dislocation slip was suppressed approximately in one of each ten grains. This seems to be a plausible result as well.

Let us finally note that similar irreversible micromechanisms also operate in the cyclic plastic zone ahead of an advancing fatigue crack front (see the next subsection), and contribute to the roughness-induced shielding (see Section 3.2.4).

3.2.2 Nucleation and Growth of Short Cracks

3.2.2.1 Mechanisms of Crack Nucleation

Fatigue cracks in metallic materials are nucleated by local interactions of dislocation slip bands or pile-ups with microstructural heterogeneities and defects as grain boundaries, phase boundaries, large secondary phase particles or free surfaces [149, 192, 297]. These interactions are induced by elastic mismatch strains evolving at boundaries of grains and microstructural phases with different stress-strain characteristics during the external loading. Although the related high elastic incompatibility stresses can be relaxed by plastic deformation (dislocation movements) in a nearly entire volume of a deformed solid, relatively high long-range stresses of dislocation arrangements in slip bands and pile-ups remain locally at microstructural boundaries. Since the cohesive strength (fracture energy) of incoherent boundaries can be very low, these stresses often lead to microcrack initiation at the weakest sites. Discrete dislocation models [298–300] revealed that the peak stresses could be higher than the cohesion stress, which depends on the surface and grain boundary energies. A high hydrostatic component of the long-range stresses intensifies a diffusion of interstitials into the stressed regions, thereby reducing the cohesive strength of the boundary. Following the molecular dynamics computations of Van der Ven and Ceder [301], a fraction of 40% of oxygen (hydrogen) atoms in a $\{111\}$ aluminium plane induces a relative reduction in the cohesive strength by a factor of two (three). In order to propagate further the nucleated microcrack, however, a sufficient amount of irreversible plasticity must be produced at its front. This is conditioned by a permanent flux of oxygen to avoid a repeating recovering of newly created fracture surfaces (see hereafter). Therefore, an intergranular propagation from the bulk interior along the grain or phase boundaries by repeating decohesion mechanism is usually observed only in strong corrosive environments. On the other hand, a transgranular crack propagation from the internal nucleus at an inclusion was often observed in specimens with a reinforced surface layers or in an ultra-high cycle regime (see Section 3.3.4).

It should be emphasized, however, that the fatigue cracks preferentially initiate at the surface [149, 192]. Indeed, the free surface is a boundary that separates media with an extremely high difference in properties (lack of interatomic bonds in air). In this particular case, the movement of dislocations towards the free surface is caused by attractive mirror forces (see also Section 2.3). During the first loading cycles progressive changes in the dislocation structure start to proceed. The damage processes begins at the sites of cyclic strain localization, usually called persistent slip bands (PSBs) and results in the formation of sharp surface slip patterns called persistent slip markings (PSMs). PSMs consist of extrusions and intrusions which develop on the initially flat surface at emerging PSBs (see Figure 3.14). While the amplitude of reversed plastic slip in the interior of PSBs or pile-ups is highly constrained by surrounding elastic matrix, this is not the case for the parts near the PSMs on the free surface where high local plastic deformations appear. Indeed, atomic force microscopy revealed that these strains can be two orders of magnitude higher than the applied macroscopic ones (e.g., [302]).

The slip localization in surface thin bands can be explained by several mechanisms. For example, cycling of precipitation hardened alloys induces formation of thin slip bands in zones with easily shareable (coherent) precipitates [303]. In ductile fcc and bcc polycrystals subjected to cyclic loading the PSBs are often formed within the surface grains. The PSBs consist of hard walls (high density of edge dislocation dipoles) and soft channels (low density of screw dislocations which glide and cross-slip) creating the well-known ladder-like structure (e.g., [304, 305]). This rather regular dislocation arrangement lowers the internal energy and enables high local shear deformations [306, 307]. Although the ladder-like structure is usually not observed in metals and alloys possessing a low stacking-fault energy, the PSMs occur in these materials [302].

The amount of plastic slip inside the slip bands and PSBs can be modelled by considering an elongated bulk inclusion embedded in the matrix which mimics the whole polycrystal [308]. By considering analytical solutions derived by Eshelby for bulk inclusions [309] one can show that the primary slip in the vicinity of the free surface is higher than that in the bulk, where the deformation is constrained by the grain boundary surrounded by a hard elastic material [310]. The slip within the band can be defined as the ratio between the displacement along the most active Burgers vector and the slip band thickness, h. Then the shear deformation within the surface slip bands can be expressed as

$$\gamma_p = r(1-\nu)\frac{L}{h}\left(1+\frac{h}{L}\right)^2 \frac{0.5\Sigma - \tau_{crs}}{G}, \tag{3.13}$$

where L is the length of the slip band (usually the grain size), $r = 1.9$, ν is the Poisson's ratio, G is the shear modulus, Σ is the applied stress and τ_{crs} is the critical resolved shear stress in the channel [310]. The factor 0.5 corresponds to a highest possible Schmid factor of the slip band inclined 45° from the loading axis. Equation 3.13 was recently verified by finite element calculations [311]. Since in large grains the values of L are tens of microns and the values of h are units or tens of nanometers, the ratio $h/L \approx 0.005 \ll 1$. With regard to Equation 3.13 this means that the surface plastic deformation in the slip band is proportional to L/h, i.e., it increases with increasing aspect ratio of the band. Clearly, the Eshelby inclusion elongated in the direction of the applied stress experiences lower back stress. Thus, the ratio of the strain ε_L localized in slip bands at the surface and the applied (nearly elastic) strain ε can reach values as high as $\varepsilon_L/\varepsilon \approx 200$. Note that the localization ratio related to the inner end of the slip band (impinged by a grain boundary) is about ten times lower [311].

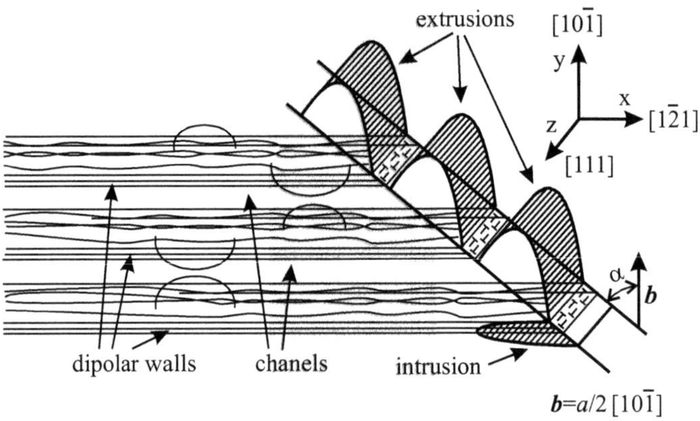

Figure 3.14 A scheme of extrusion and intrusion patterns at the intersection of the persistent slip band with a free surface

The high local reversed plasticity in slip bands and channels of PSBs produces a surface microroughness in the form of extrusions and intrusions (e.g., [312]). The movement of screw dislocations with jogs generates a surplus of vacancies in the channels. The counterbalancing flux of atoms inside the channels causes extrusions at the free surface, the volume of which is much higher than that of intrusions, as shown in Figure 3.14. On the other hand, volumes near the surface and close to the outer boundary of the channels become depleted by atoms (or enriched by vacancies). This usually leads to formation of thin intrusions next to the extrusions [313]. Such a model of combined movements of dislocations and point defects well reflects geometrical proportions of extrusions and intrusions [314]. However, there are also many other older models of the surface relief evolution (e.g., [315, 316]).

High stress concentrations around extrusions and, particularly, at the tip of intrusions leads to a formation of short surface cracks which start to propagate along the slip bands or PSBs inside the bulk. This growth is controlled by irreversible emission and absorption of dislocations at the fronts of these cracks. Owing to the localization ratio and Equation 3.13, such a damage process is most probable in the largest surface grains along slip planes with the highest Schmid factors. One should also note that the average level of cyclic plasticity is raising when transferring from a high-cycle regime to a low-cycle one. Consequently, the concentration of nucleated surface cracks follows the same trend.

3.2.2.2 Propagation of Short Cracks

There is no clear notion about the moment of transition from the vacancy-assisted intrusion growth into the dislocation based propagation of the related short crack. Nevertheless, there are some plausible models explaining possible mechanisms of initial growth stages (e.g., [149, 312, 317]). Since the crack nuclei are strictly aligned with the slip planes of PSBs, their further advance proceeds along these planes. This means that the cracks propagate under the shear stress coupled with the tensile normal stress (the mixed-mode I+II) that can somewhat facilitate the emission of dislocations from the crack tip by reducing the ideal shear strength (see Section 1.1). During cyclic loading in air or other corrosive environments, a passive oxide layer always forms on metallic surfaces. If the oxide layer is fractured as a consequence of deformation in the underlying polycrystal, the passivation is lost and a re-oxidation occurs. The process of repeated fracture and re-oxidation is a central principle of slip-oxidation models that are widely used to elucidate the mechanism of propagation of both short and long cracks [149, 317].

The simplest model assuming the propagation along a single slip plane is depicted in Figure 3.15. In the tensile half-cycle, the edge dislocations are emitted from (or absorbed at) the crack tip which generates a new fracture surface on one of the crack flanks. The size of the new surface is equal to the number of dislocations times the Burgers vector and the length of the crack front. When assuming an immediate surface oxidization, the dislocations returning during the unloading cycle cannot remove the new surface. Instead of that, they will form a new fracture surface on the other crack flank, the size of which is, again, equal to the number of returning dislocations times the Burgers vector and the crack front length. Under a constant loading amplitude the number of returning dislocations to the crack tip is nearly equal to the number of dislocations generated during loading. Hence, the crack extension per cycle is equal to the number of dislocations generated at the crack tip times the Burgers vector. In other words it is equal to the cyclic crack tip opening displacement. In this way, the crack advances during each loading cycle.

In general, the crack growth rate $\mathrm{d}a/\mathrm{d}N$ can be assumed to be nearly proportional to the frequency of the oxide-layer fractures which is, again, proportional to the plastic strain γ_p in the slip band. Consequently, the crack growth rate can be assumed to be proportional to $(L/h)^m$, where $m \in (0.3, 0.8)$ is commonly accepted [311, 317]. This also means that the growth rate of short cracks is higher in the long PSBs embedded in large surface grains.

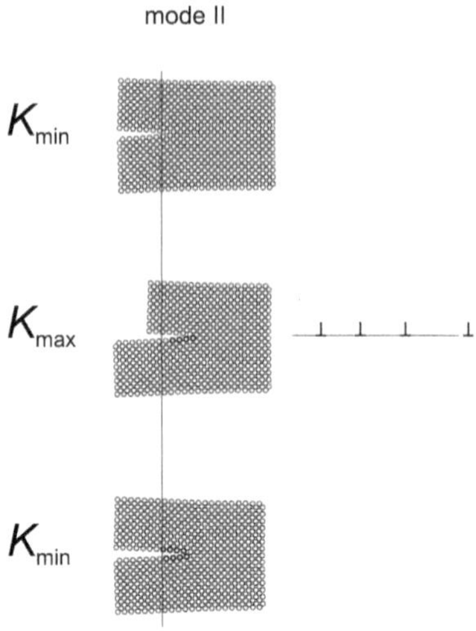

Figure 3.15 The single-slip model of short crack propagation

When the growing short crack approaches a grain boundary, the emission of crack tip dislocations starts to be restricted by the back stress from a creating pile-up. As a result, the growth rate rapidly decreases and, eventually, the crack could be arrested at the grain boundary. In general, the crack growth of these so-called microstructurally short cracks (MSCs) becomes retarded by various microstructural barriers of different strength. The maximal length of MSCs is determined by the distance b_s of the strongest barriers such as grain or phase boundaries. As was sufficiently verified particularly by the group of K. J. Miller in Sheffield [318], the fatigue limit corresponds to a maximal stress still not high enough to propagate the longest short cracks, i.e., to transfer it through the boundary of the surface grain to adjacent bulk grains. The distance b_s also corresponds well to the maximal size of non-damaging cracks in the well known Kitagawa–Takahashi diagram [319]. In specimens fractured close to the fatigue limit, therefore, many small surface cracks arrested at grain or phase boundaries can be found.

Unlike in the case of long cracks, a description of MSC propagation in terms of ΔK does not make too much sense. Indeed, the relative size of the plastic zone r_p/a determined by the zone of emitted dislocations at the tip is not small enough to fulfil the conditions of small-scale yielding and plane strain. Moreover, there is a rapid change in the T-stress during the crack propagation from the surface towards the grain boundary. Therefore, the growth rate of MSCs is often described in the form

$$\mathrm{d}a/\mathrm{d}N = A\Delta\varepsilon_p^l(b_s - a)^k,$$

where $\Delta\varepsilon_p$ is the applied plastic strain range, and A, l and k are material parameters ($k \leq 1$) [247]. This relationship implies a proportionality between the applied and the local plastic strains and reflects well the retardation and the subsequent arrest of MSCs at the barriers ($\mathrm{d}a/\mathrm{d}N = 0$ for $a = b_s$). It should be emphasized that the MSCs can grow at substantially lower applied stresses than the long cracks, i.e., well below the fatigue limit. This is mainly caused by an absence of crack closure effects (see Section 3.2.5 for more details). Indeed, the crack flanks of MSCs are ideally flat so that there is no roughness-induced closure. A highly elongated shear-mode plastic zone lies within the confines of one or two slip planes and no plastic blunting of the crack tip occurs. Therefore, there is also a lack of plasticity-induced crack closure. In the very first growth stages, the closure level might even be negative due to the fact that the Peierls–Nabarro stress prevents individual dislocations from the backward motion within the plastic zone.

For a further advancement of MSCs, the applied stress must be raised to reinitiate the crack in the adjacent grains. Indeed, the superposition of the remote and local stresses can also create sufficiently high local plastic strains in the slip planes with lower Schmid factors. When the crack overcomes the grain boundaries, it starts to be spatially tortuous. This means that the crack flanks become rougher and the growing friction forces decelerate the crack growth under the shear mode II. To avoid friction, the crack attempts to incline to the plane perpendicular to the direction of the applied stress in order to get a more open crack tip by a higher mode I loading component. The cracks occurring in such a transient stage are called physically short cracks (PSCs). As a rule, MSCs start to be PSCs after subsequent passing through one or two grain boundaries. The plastic zone size of a growing PSC steadily increase to embrace more than a single grain. While the crack closure effect increases, the influence of microstructure on the crack growth rate decreases. The PSC completely converts into a long crack (LC) after passing more than ten grain boundaries. At this growth stage, the crack tip plastic zone already embraces several grains and the microstructural influence becomes negligible.

The dependence of the crack growth rate on the range of the stress intensity factor ΔK during all the stages of MSC, PSC and LC is schematically depicted in Figure 3.16. An enormous variation of crack growth rate due to interactions with individual grain boundaries is typical for the MSC stage. The threshold range ΔK_{th} indicates the limit below which the long cracks do not propagate.

3.2.3 Discrete Dislocation Models of Mode I Growth of Long Cracks

3.2.3.1 Near-threshold Crack Tip Plasticity

In the near threshold region of long fatigue cracks, the range of the stress intensity factor (SIF) is very small (units of $\mathrm{MPa\,m^{1/2}}$), and the plastic zone is

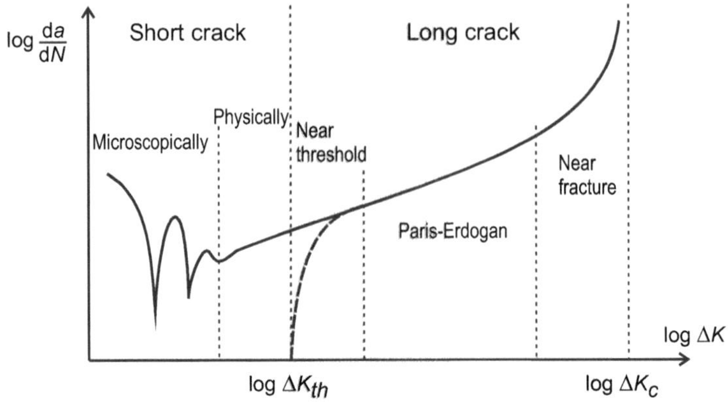

Figure 3.16 The scheme of the crack growth rate of short and long cracks showing the characteristic ranges of stress intensity factors. The near-threshold behaviour of long cracks is indicated by the *bold dashed line*

bounded to a close vicinity of the crack tip. Therefore, the internal dislocation sources are likely not present, and the dislocations can be assumed to be generated at the crack tip. Atomistic models developed by Rice and Thomson [134] and Rice [139] for atomically sharp cracks show that the spontaneous emission of dislocations occurs when the SIF reaches a critical value k_e. This value can be estimated as

$$k_e^2 = \frac{(1 + (1 - \nu)\tan \Phi)}{(1 + \cos\theta)\sin^2\theta} \frac{16G\gamma_{us}}{(1 - \nu)}.$$

Here γ_{us} is the unstable stacking fault energy and θ, Φ are angles characterizing the activated slip system. According to [139] and [320], the predicted values of k_e are $1.3\,\mathrm{MPa\,m^{1/2}}$ for Fe, $0.3\,\mathrm{MPa\,m^{1/2}}$ for Al and $0.5\,\mathrm{MPa\,m^{1/2}}$ for Cu. However, the stress fields produced by emitted dislocations shield the crack tip, so that the value of the local SIF k_{loc} becomes significantly lower than that of K_{rem} transmitted from the remote loading. One can write

$$k_{\mathrm{loc}} = K_{\mathrm{rem}} + \sum_i k_{di}, \tag{3.14}$$

where $k_{di} < 0$ is the SIF originated from i-th single dislocation [320]. In the crack tip plasticity models, the movement of an already emitted, individual, dislocation is determined by the Peach–Koehler force

$$\mathrm{d}\mathbf{f} = (\sigma \cdot \mathbf{b}) \times \mathrm{d}\mathbf{l},$$

where $\mathbf{b}$ is the Burgers vector, σ is the stress tensor and $\mathrm{d}\mathbf{f}$ is the force on the line segment $\mathrm{d}\mathbf{l}$. The stress tensor is defined by interaction with other dislocations, free surfaces (e.g., a crack) and remote stresses. The dislocation remains at rest whenever the slip component of the Peach–Koehler force is less

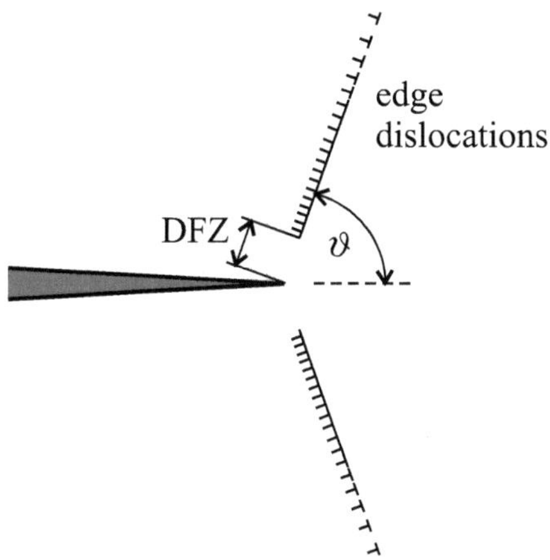

Figure 3.17 Configuration of geometrically necessary dislocations emitted from the crack tip during a loading sequence (DFZ – the dislocation free zone)

than the lattice resistance τ_0 ($\tau_0 \approx G/2000$ for bcc metals). The dislocations may return to the crack, where they disappear by decreasing the crack tip opening displacement (CTOD).

The mode I crack tip plasticity can be numerically simulated as motion of edge dislocations, which are parallel to the crack front (plane strain) [9]. Dislocations are symmetrically emitted in pairs and the angle between the slip planes and the crack propagation plane $\vartheta = 70°$ – see Figure 3.17. This value meets the results of both theoretical and experimental investigations [22]. When K_{rem} is increased from its zero value to $K_{\max}$ and then unloaded back to zero, the model shows the following main features of dislocation behaviour:

1. $K_{\max} < k_e$: no dislocations are generated and the material behaves elastically.
2. $k_e \leq K_{\max} < 1.3k_e$: several pairs of positive (shielding) dislocations are generated and pushed into the bulk so that the k_{loc} becomes lower than K_{rem}. Upon unloading the back stress never reaches the negative friction stress $-\tau_0$. Consequently, the material behaves elastically in all further cycles.
3. $1.3k_e \leq K_{\max} < 3.5k_e$: many pairs of positive dislocations are generated and, during the unloading sequence, some of them return to the crack again. This means that the crack-tip cyclic plasticity appears first within this range of $K_{\max}$ values. No negative dislocations (with opposite Burgers vectors) are emitted during unloading.
4. $3.5k_e \leq K_{\max} < 4k_e$: during the unloading some pairs of negative (anti-shielding) dislocations are emitted but they immediately annihilate with their positive counterparts (their mutual distance is less than $10\mathbf{b}$).

5. $4k_e \le K_{\max}$: some pairs of negative dislocations, generated during the unloading half-cycle, remain stable.

The fully developed dislocation configurations look like inverse pile-ups (Figure 3.17). A dislocation free zone (DFZ) is found in the immediate vicinity of the crack tip. Such zones were often observed in thin foils (e.g., [321]) and their existence is an exclusive consequence of the discrete nature of plasticity. Similarly, the cyclic plasticity starts to operate only when $K_{\max} \approx 2k_e$ is reached (the point 3). This condition can be interpreted as an intrinsic threshold $K_{in,th}$ of the stationary mode I fatigue crack. The coefficient 2 was chosen here instead of 1.3 (the lower limit), because the emitted dislocations blunt the crack tip. As a consequence, the real critical values for spontaneous dislocation emission ($K_{in,th} \approx 2.6\,\mathrm{MPa\,m^{1/2}}$ for Fe, $0.6\,\mathrm{MPa\,m^{1/2}}$ for Al and $1\,\mathrm{MPa\,m^{1/2}}$ for Cu) are expected to be somewhat higher than those calculated for the atomically sharp cracks. Thus, the experimentally verified existence of the intrinsic threshold can be nicely interpreted by the discrete dislocation model.

3.2.3.2 Crack Growth in Near-threshold Region

In order to study the development of dislocation configurations and related details of the crack front geometry during the loading history, the discrete dislocation model of growing fatigue crack was developed by Riemelmoser *et al.* [320, 322]. It is a blunting model very similar to the original concept of Pelloux [323], but the plastic deformation is modelled as motion of edge dislocations.

Blunting and growing of the modelled crack is schematically shown in Figure 3.18. The crack growth increment per emission of one dislocation pair is $\Delta a = b \cos \theta$, where b is the magnitude of the Burgers vector.

The crack resharpens again when the emitted dislocations return to the crack tip (or crack flanks). As usual, an assumption of full irreversibility of the fracture process is accepted when modelling the fatigue crack growth in the air. It means that, due to the environmental assistance (in particular that of the oxygen), once created free surfaces do not re-weld. Consequently, after the unloading the length of the crack remains the same as it was at the moment of the preceding maximum load. Thus, in principle, only one dislocation that was emitted, and returned back to the crack tip causes an elementary fatigue crack advance. This is the microstructural interpretation of the previously mentioned fact that the range of local cyclic plasticity can be considered to be the fatigue crack driving force.

A result of the numerical analysis of a growing crack is demonstrated in Figure 3.19. In the first loading half-cycle many dislocations are emitted as a slip band. During unloading m of them return to the crack, where they annihilate on the free surface (Figure 3.19(b)). In the second cycle the same m dislocations are generated and annihilated again, while the crack propa-

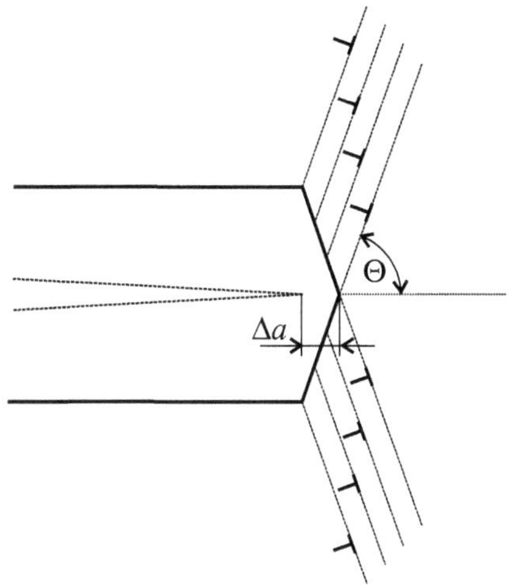

Figure 3.18 Scheme of the crack tip advancement produced by its blunting due to dislocation emission. Δa is the crack advancement caused by emission of four dislocation pairs

gates. The same occurs in many further cycles, until the crack advancement becomes about 200 Burgers vectors. Then one further pair of dislocations remains in the bulk, since the repulsive stress from the slip band, acting at the advanced crack tip, has become somewhat lower (Figure 3.19(c)). This happens every 200 Burgers vector distances after the crack advancement is so that a dislocation wall parallel with the crack is formed, see Figure 3.19(d). When the influence of the first slip band on the moving-away crack tip becomes small enough, the newly generated dislocations can pass the wall to form a second slip band, as also shown in Figure 3.19(d). In the following unloading half-cycle the dislocations in the wall experience the back stress from the dislocations in the secondary band. Many of them return to the crack tip and disappear. This situation is depicted in Figure 3.19(e). From that moment on, the whole process starts to repeat periodically which leads to a dislocation arrangement as shown in Figure 3.19(f). The width of each periodicg999 segment is about 2500 Burgers vectors (0.6 mm). The segments are accompanied by surface steps (adjacent to the slip bands) that can be visible on fracture surfaces as so-called subcritical (near-threshold) striations. Their spacing of about 0.6 mm is, unlike that of supercritical striations (see hereafter), fully independent of the crack growth rate.

Thus, the growing near-threshold fatigue crack leaves a plastically deformed band, the static plastic zone, along its flanks behind the crack tip. In real crystals, this zone consists of both the geometrically necessary dislocations and the statistically stored ones. Although only geometrically necessary dislocations are considered here, the dislocation arrangements observed in the electron microscopy look very similar, e.g., [324]. Such a periodic configura-

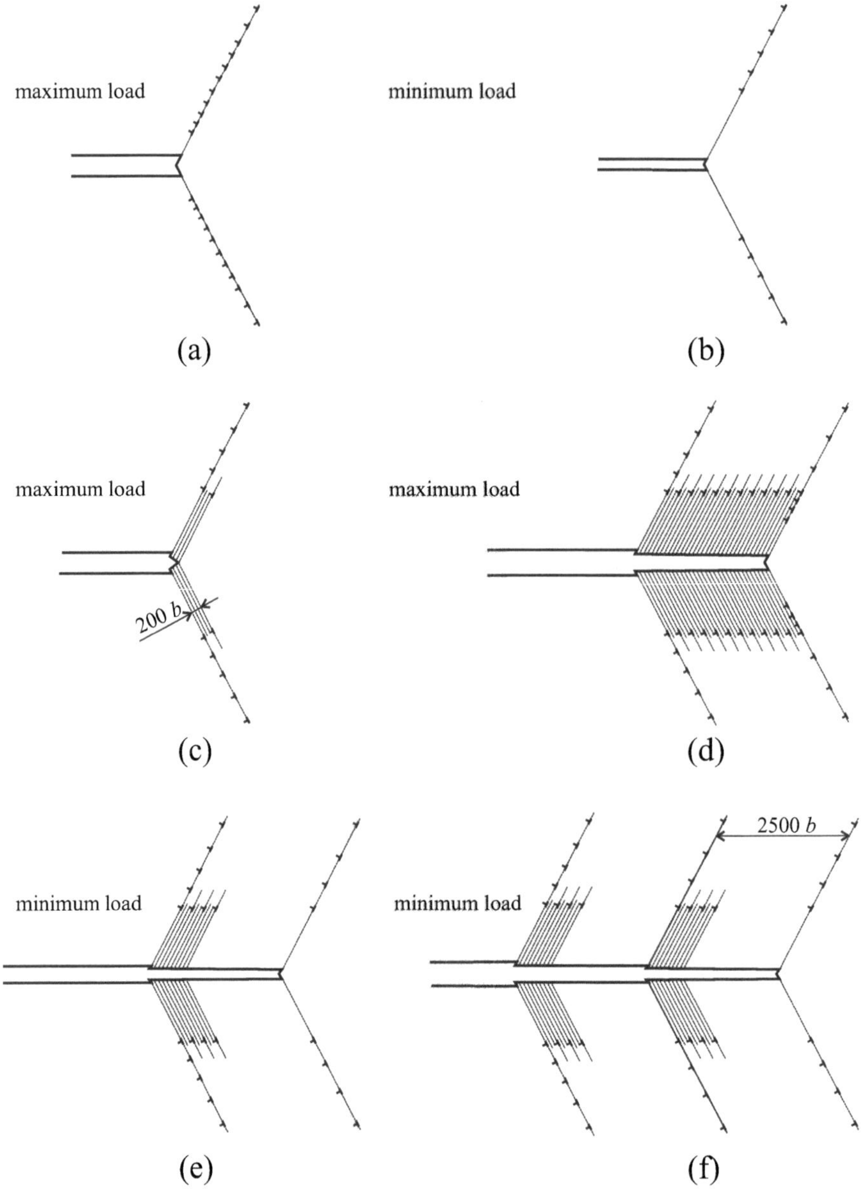

Figure 3.19 Prediction of geometrically necessary dislocation arrangements in the wake of a growing near-threshold crack according to the discrete dislocation model: (a) the first growing stage, (b) after releasing the load some dislocations disappear to free surfaces, (c) generation of the first wall dislocation, (d) generation of the second slip band, (e) vanishing of a part of wall dislocations in the unloading phase, and (f) creation of a periodic dislocation structure in the crack wake

tion of slip bands, remaining in the wake of a propagating crack, causes a tilting of the crystal lattice near crack flanks. This can produce a significant contact shielding of the crack tip (see Section 3.2.4).

The model can also well reproduce the intrinsic crack growth curve [9] near the threshold. On the other hand, the continuum plasticity models (e.g., [274]) are not able to reproduce correctly any of above-mentioned phenomena. The difference in prediction of the near-threshold crack growth behaviour by the continuum approach and the discrete dislocation model is schematically depicted in Figure 3.20.

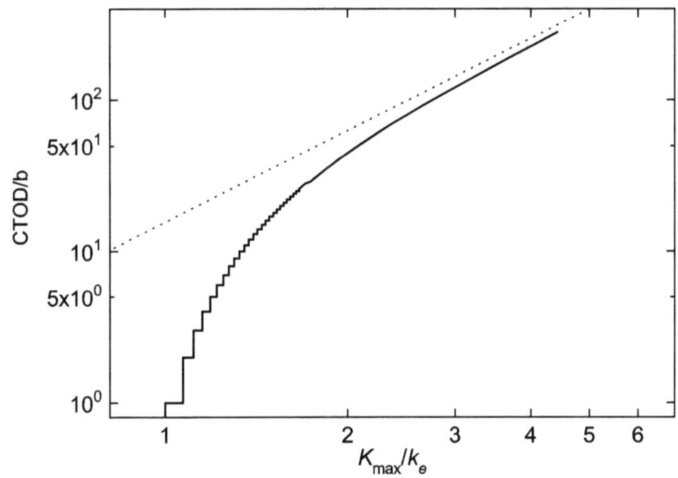

Figure 3.20 The scheme of the near-threshold crack growth rate according to the continuum approach (*dotted line*) and the discrete dislocation model (*solid line*)

3.2.3.3 Crack Growth in Paris–Erdogan Region

When the crack proceeds to reach the Paris–Erdogan region, the SIF at its tip keeps raising steadily. The ability of dislocation emission increases and, consequently, the width of periodic segments decreases. Due to a much higher number of emitted dislocations, simultaneously, the crack advance per cycle becomes much larger. After reaching the Paris–Erdogan region, both these phenomena lead to a vanishing of the periodic crack-wake dislocation structure shown in Figure 3.19(f). Many dislocation slip bands are emitted during each loading cycle, which results in a significant crack blunting. The related qualitative change in the micromechanism of crack propagation is schematically depicted in Figure 3.21. In order to make this demonstration clear, only four slip bands are assumed to be emitted per each cycle.

During the loading phase, large slip bands are alternatively emitted on both sides of the crack tip. This is shown in Figure 3.21(a,b). Owing to their repulsive stress, the subsequently emitted bands appear shorter than the previous ones (Figure 3.21(c,d)). During unloading m dislocations of the second band return to the crack, where they annihilate on the free surface. This leads to resharpening of the near-tip part of crack flanks (Figure 3.21(e)). The same sequence is carried on in further loading cycles while leaving periodic patterns on crack flanks which are known as (supercritical) striations. Their morphology on both fracture surfaces is repeating in each cycle, which is clearly seen in Figure 3.21(f), where the crack flanks are pressed together during a compressive part of the loading cycle. Here the microscopic tortuosity of the crack path is suppressed in order to highlight important periodic patterns. Indeed, such created striations successfully mark locations of the crack front and their spacing corresponds to a local rate of the crack growth (microscopic growth rate). Therefore, they play a significant role in the failure analysis (see Section 3.4). Naturally, the schematically described "Paris" crack also leaves the geometrically necessary dislocations in individual bands on both sides of the crack flanks.

Another simple model for crack advance was proposed by Neumann [325]. It assumes an alternating shear only on two slip planes, primary and secondary, which intersect the crack front as shown in Figure 3.22. The work hardening on primary slip planes leads to alternating shear on secondary planes. In the reverse half-cycles a majority of dislocations returns back to the crack tip and disappears. Some of dislocations remain in the crack wake to create the residual plastic zone and saw-tooth supercritical striations. This model is also applicable to ductile metal single crystals as well as to the near threshold growth in polycrystals. In the latter case, however, the repulsive forces of remaining shear-band dislocations do not allow emission of the shear bands again in many subsequent loading cycles. During these cycles the crack proceeds along a finer zig-zag path (without leaving observable saw-tooth striations) in a way similar to that described in connection with Figures 3.19(c,d). It should be noted that the original Neumann's model did not assume any dislocations left behind the advancing crack tip.

In engineering materials, the crack path exhibits significant tortuosity on many scales owing to different microstructural barriers to dislocation motion. The largest fluctuations are of the order of the grain size. Especially in the near threshold region much smaller deflections can occur which can be induced by dislocation-dislocation interaction or by microstructural features much smaller than the grain size [10]. The smallest deviations are of the order of lattice spacing. They are caused by remaining ledges formed by dislocations generated at the tip or entered on the fracture surface. Therefore, the crack path geometry is usually simply drawn in a zig-zag manner as in Figure 3.22(c). Let us emphasize that the extent of created dislocation bands (the plastic zone size) in relation to the length of largest deflected crack segments constitutes a very important parameter. It determines a symmetry

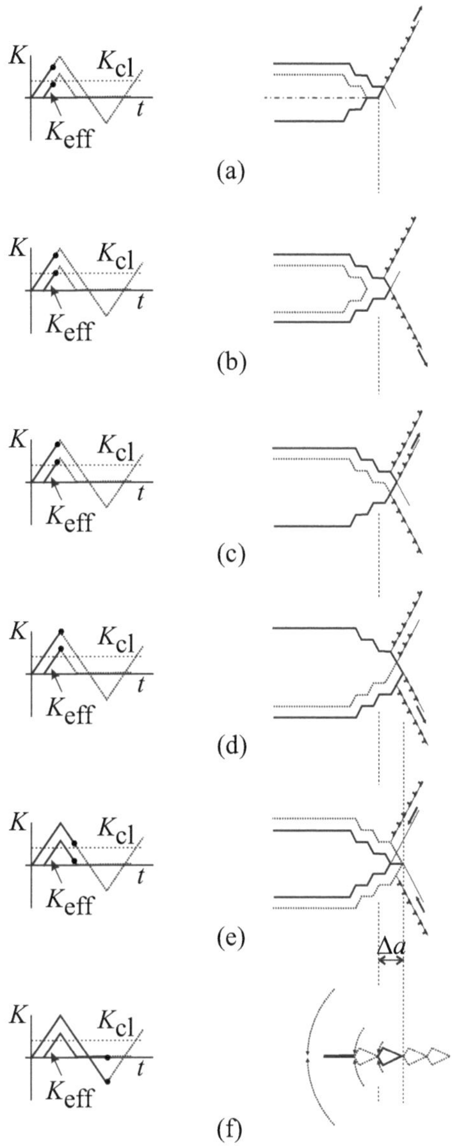

Figure 3.21 A scheme of the crack tip advance by dislocation emission during one loading cycle in the Paris–Erdogan region. The points on the K *vs* t diagram show subsequent loading stages. In the last compressive stage the crack flanks are pressed together which highlights ductile striations

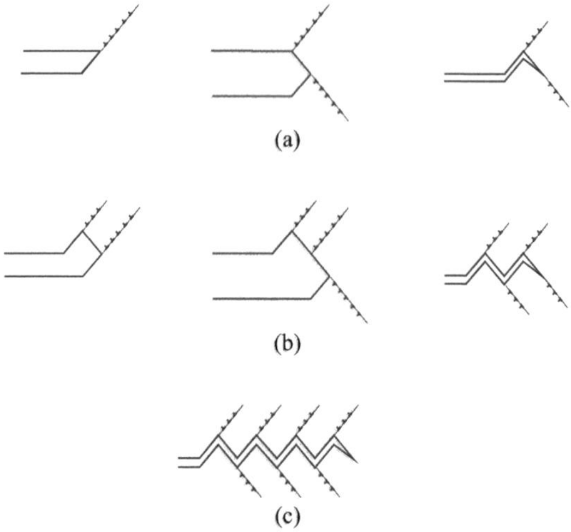

Figure 3.22 Alternating slip model after Neumann: (a) the first loading cycle, (b) the second loading cycle, and (c) the resulting crack path leaving saw-tooth striations on fracture surfaces

of the crack-wake dislocation arrangements and decides about the level of roughness-induced crack closure (see the next subsection).

3.2.4 Crack Closure Mechanisms

The fatigue crack growth behaviour, particularly near the threshold of the applied SIF, is affected by both the chemical composition and the microstructure. Various shielding mechanisms that are used to explain the diversity in the threshold values can be, according to Ritchie [165], categorized as intrinsic and extrinsic (see also Chapter 2). A general scheme of extrinsic mechanisms is depicted in Figure 3.23. In this figure the applied (external) ΔK consists of two basic components: the closure range ΔK_{cl} and the effective range ΔK_{eff}. The range ΔK_{cl} is divided into three parts (from the left to the right): the oxide-induced closure, the roughness-induced closure and the plasticity-induced closure. The range ΔK_{eff} consists of two parts: the geometrically-induced shielding ΔK_{br} (crack kinking and branching) and the intrinsic range ΔK_{in}. While all the closure mechanisms operate in the crack wake, the crack kinking and branching produce shielding in front of the crack tip.

Thus, besides the geometrical shielding that has already been described in the brittle-fracture section, several closure (or contact shielding) mechanisms become important in fatigue.

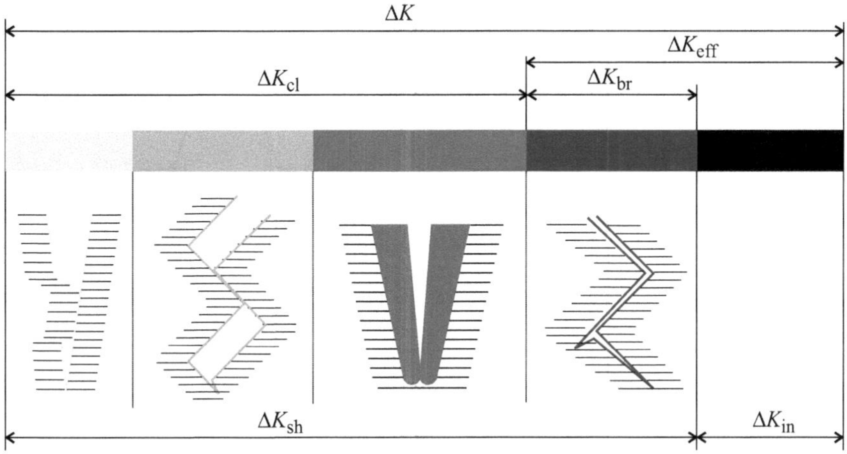

Figure 3.23 A general scheme of extrinsic mechanisms occurring during the fatigue crack propagation; ΔK – the applied SIF range, ΔK_{cl} – the closure SIF range, ΔK_{eff} – the effective SIF range, ΔK_{in} – the intrinsic SIF range

As recognized by Elber [326], the crack remains to be closed not only in the compressive part of the loading cycle but, to some extent, also during its tensile part. This means that only the $\Delta K_{eff} = \Delta K - K_{cl}$ part of the applied ΔK is effective as a crack driving force of the fatigue crack growth – see Figure 3.24. When the minimum applied SIF is higher than K_{cl}, the crack remains steadily open (Figure 3.25). However, this happens only for very high cyclic ratios ($R > 0.6$).

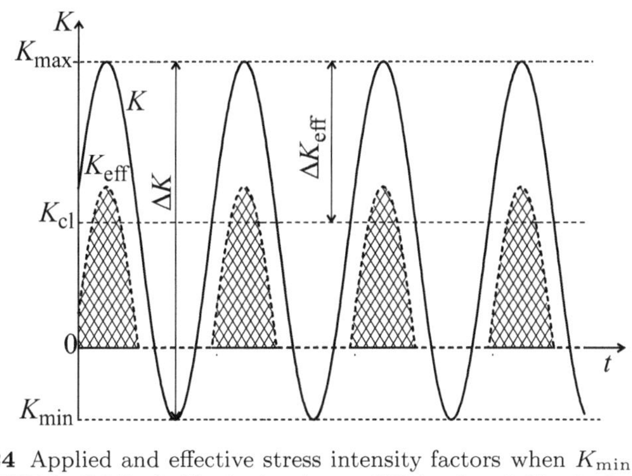

Figure 3.24 Applied and effective stress intensity factors when $K_{\min} < K_{cl}$. K is the applied SIF (*solid line*), K_{cl} is the closure SIF level and K_{eff} is the effective SIF (*dotted line*). The *dashed areas* show ranges, where the crack tip is open

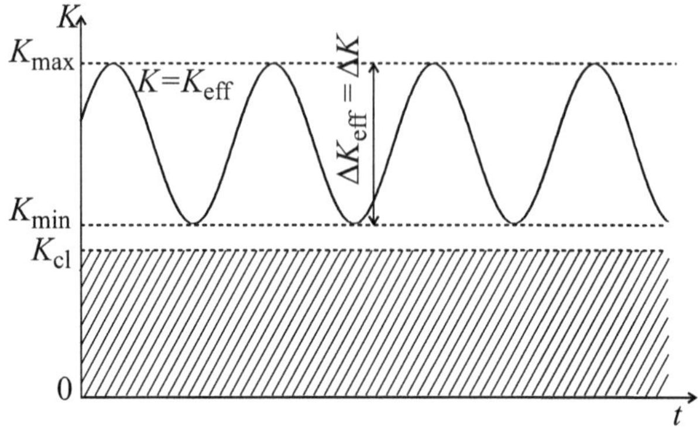

Figure 3.25 Applied and effective stress intensity factors when $K_{\min} > K_{cl}$ ($\Delta K_{eff} = \Delta K$), and the crack remains steadily open during the loading. The *dashed area* shows the K-range of a potential crack closure

In metallic materials fatigued in air or vacuum, both the roughness-induced crack closure (RICC) and the plasticity-induced crack closure (PICC) are the most important closure mechanisms. There can also be other shielding contributions [165] such as, for example, the oxide-induced closure (OICC) and the crack-wake bridging that occur in corrosive environments and composite materials, respectively. It should be emphasized that there is no experimental tool for measuring individual closure components; only the total crack closure level can be measured using the methods based on extension or electrical resistance (e.g., [327]).

When the crack front branches (splits) during the crack propagation, the effective driving force becomes further reduced to ΔK_{in} as marked in Figure 3.23. An additional reduction can be caused by the crack deflection and meandering. However, this fully holds only for brittle materials as high-strength alloys or ceramics (see Chapter 2 for details). Another shielding (or anti-shielding) component can be associated with large particles or microvoids embedded in the matrix. These effects are important only in nodular ductile irons (see Section 3.2.7) or in particle reinforced composites (see Chapter 2). The value of $\Delta K_{in,th}$ represents the intrinsic threshold or the intrinsic resistance to the fatigue crack growth. Because a partial recovery of newly created surfaces usually takes place in engineering materials during the unloading cycle, the real values of $\Delta K_{in,th}$ are somewhat higher than those of 0.5–2.6 MPa m$^{1/2}$ resulting from the discrete dislocation model. However, they can be almost an order of magnitude lower than the measured fatigue thresholds ΔK_{th}, as shown in Section 3.2.6.

Note that there is no experimental method available for measuring the geometrical shielding component and, consequently, the values of the intrinsic threshold $\Delta K_{in,th}$. Nevertheless, they can be assessed theoretically by employing rather standard materials data in the frame of the unified analytical model of extrinsic shielding, as presented hereafter.

3.2.4.1 Plasticity-induced Crack Closure

The well known reason for the PICC in thin solids (the plane stress condition) is the plastic wedge in the wake of the crack front; see Figure 3.26. This wedge is created by tensile necking within the crack-tip plastic zone. Experimental data lead mostly to the conclusion that the ratio of the closure stress level to the maximum stress (the so-called closure ratio) $\sigma_{cl}/\sigma_{\max} \approx 0.6$ is rather independent of the material. There is, however, a moderate dependence on the cyclic ratio R. Theoretical models published in the last 30 years confirm these results [192]. In the case of thin specimens, however, the LEFM concept loses its validity, which constitutes a considerable disadvantage.

In a rather more important case of plane strain (thick solids), the PICC can also be detected, albeit to a much lower extent. Indeed, the closure ratio $K_{cl}/K_{\max} \approx 0.25$. However, the mechanism of that phenomenon remained unclear for rather a long time. Under the plane strain, indeed, the necking associated with a transverse contraction is not allowed by definition. The problem was solved by Pippan *et al.* in 2004 [10]. A static plastic zone of a constant height is always formed by both the geometrically necessary and the statistically stored crack-wake dislocations; see Figure 3.27 (compare also Section 3.2.3). Unlike those statistically stored, the geometrically necessary dislocations have a long-range tilting effect as shown in Figure 3.28. Only two narrow bands of dislocations, forming low-angle tilt boundaries, are considered here since the effect of the real distribution can simply be obtained by integrating over many such dislocation bands. The tilting of the volume elements in the plastic wake is associated with shear in the direction of the crack propagation. This shear causes a transfer of the material to the crack tip – a reason for the PICC under plane strain conditions.

The dislocation bands shield the crack tip from the remote SIF range ΔK, as generally expressed by Equation 3.14. In the case of an unloaded crack, one dislocation band produces the local shielding stress intensity factors k_1^s and k_2^s at the crack tip as

$$\begin{pmatrix} k_1^s \\ k_2^s \end{pmatrix} = Gb\sqrt{r}\frac{1}{D}\begin{pmatrix} c_1 \\ c_2 \end{pmatrix}, \tag{3.15}$$

where r is the distance of a single band of dislocations from the crack flanks, D is the spacing between individual dislocations in the array and c_I and c_{II} are constants of about -1 [10]. In the near tip regime, the displacements at the crack flanks are given by the well known LEFM equations

$$\begin{pmatrix} u_1 \\ u_2 \end{pmatrix} = \frac{2(1-\nu)}{G}\sqrt{\frac{|x_c|}{2\pi}}\begin{pmatrix} k_2^s \\ k_1^s \end{pmatrix}, \tag{3.16}$$

where x_c is the distance of the crack-wake contact of crack flanks from the crack tip. Indices 1 and 2 correspond to the shear mode II and the opening mode I, respectively. Note that an overlap of the crack flanks (PICC) takes

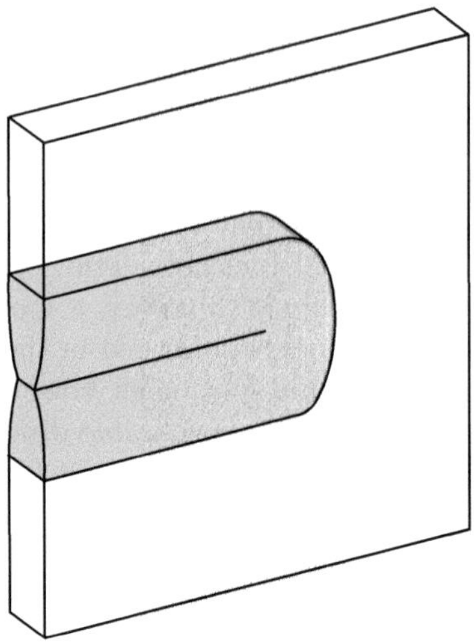

Figure 3.26 The scheme of the plastic wedge in the crack-wake causing the plasticity-induced crack closure in thin specimens (the plane stress condition)

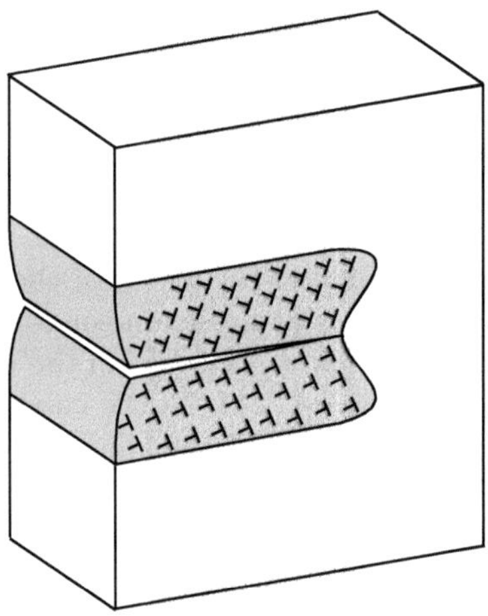

Figure 3.27 The scheme of geometrically necessary crack-wake dislocations causing the plasticity-induced crack closure in thick solids (the plane strain condition)

place since k_1^s and u_2 are negative. Dislocation arrays on both sides of the crack flanks contribute to PICC. Consequently, the closure distance for PICC can be assessed by combining Equations 3.15 and 3.16 to get

$$\delta_{cl,p} \approx \frac{4(1-\nu)b}{D}\sqrt{\frac{|x_c|}{2\pi}}. \tag{3.17}$$

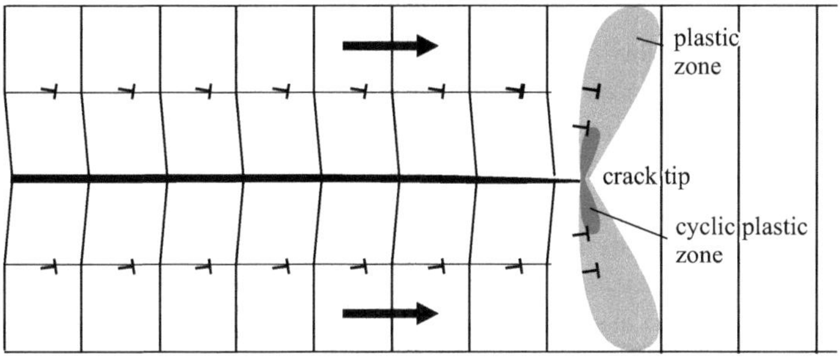

Figure 3.28 Two narrow dislocation bands approximating the real arrangement of geometrically necessary dislocations

The spacing D is inversely proportional to the number n of geometrically necessary dislocations per unit extension of the crack. That number must be proportional to the mean strain at the distance r of the dislocation arrays from the crack flanks. The mean strain is proportional to the yield stress σ_y divided by the Young's modulus $E = 2G(1+\nu)$. That reasoning leads to

$$D \approx 2BG\,(1+\nu)/\sigma_y, \tag{3.18}$$

where $B \approx 10^{-10}$ m can be assumed to be rather material independent [11].

The maximal applied SIF can be expressed as

$$K_{I\max} \approx \sigma_y (2\pi r_p)^{1/2} \tag{3.19}$$

and the maximal opening displacement as

$$\delta_{\max} = 2u_2(|x_c|, K_{I,\max}). \tag{3.20}$$

A half-size of the static plastic zone r_p constitutes a plausible estimate of the distance of dislocation strips from the crack flanks, i.e., $r \approx r_p/2$. By combining Equations 3.17, 3.18, 3.20 and 3.19, the closure ratio reads

$$\frac{\delta_{cl,p}}{\delta_{\max}} = 2C, \tag{3.21}$$

where $C = b/(10\sqrt{\pi}B) \approx 0.1$ is a dimensionless constant [204]. Equation 3.21 is in agreement with the experimentally determined value of the closure ratio of about 0.25, nearly independent of the material.

3.2.4.2 Roughness-induced Crack Closure

The existence of RICC is connected with asperities on fracture surfaces in the wake of the tortuous crack front. Suresh and Ritchie [328] proposed a simple two-dimensional LEFM model describing clearly the essential mechanism of that phenomenon. Irreversible operation of the local mode II at the front of the tortuous crack causes a mutual horizontal shifting of fracture surfaces and their premature contact. Although this geometrical scheme is principally correct, the model does not yield an explicit relation to the microstructure, since the averaged crack path inclination angle used in that analysis does not change with the microstructure coarseness. Thus, in fact, this model is not very useful for a quantitative assessment of the role of the microstructure in the RICC. In order to assess the dependence of the mode II displacement on the mean grain size, Wang and Mueller [258,329] proposed a more recent concept that deals with an irreversible dislocation pile-up adjacent to the crack tip and restricted by a grain boundary. This one-parameter approach requires a complicated analysis of the fracture surface roughness and, moreover, the assumption of a planar slip mode is necessary for its validity.

It should be emphasized that all the previous analytical models could not take the so-called long-range component of the RICC into account because, as mentioned above, that important component was described only recently [10].

Maximal Long-range Component

When the arrangement of crack-wake dislocations becomes asymmetric as in the extreme case in Figure 3.29, the shear displacements of both crack flanks are different. This shear misfit causes the long-range RICC on rough fracture surfaces. Indeed, the asymmetric arrangements of crack wake dislocations can produce the RICC far behind the crack tip, in contrast to shear displacements induced by the irreversible slip (see the next paragraph). Thus, both the asymmetry of crack-wake dislocations and the roughness of fracture surfaces can be assumed to be necessary conditions for the appearance of long-range RICC.

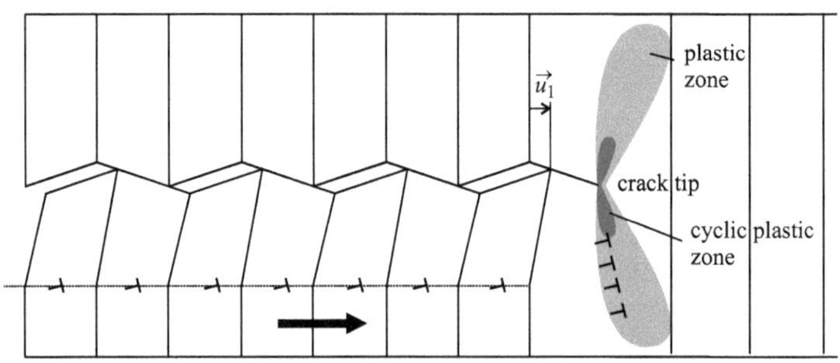

Figure 3.29 The scheme of a shear misfit and the roughness-induced crack closure produced by an asymmetric arrangement of crack-wake dislocations

As follows from theoretical and experimental results [11,168,204], the necessary conditions for RICC are well fulfilled when $S_R > 1$, where $S_R = d/r_p$ is the ratio of the characteristic microstructural distance d (the mean grain size or the interparticle spacing) and the plastic zone size r_p (see also Chapter 2). On the other hand, when the plastic zone embraces several grains or particles, i.e., $S_R \ll 1$, the phenomenon of RICC can be neglected. The reason for these opposite effects is schematically elucidated in Figures 3.30 and 3.31.

In the case of $S_R > 1$, an alternating (zig-zag) single-slip accompanied by dislocation pile-ups represents a strongly prevalent crack propagation mechanism in all metallic materials [192,194]. The static plastic zone is constrained within individual grains containing the crack front. The grain boundaries or secondary phase particles constitute obstacles for crack growth since the atomistic structure of these defects is quite different from that of the matrix, where the dislocations possess a minimum energy and can slide in an optimal manner. Additionally, there is no possibility for their further gliding on the same slip planes in the adjacent grain or inside the secondary particle. Therefore, the crack often grows along grain boundaries and secondary phases or, eventually, even intergranular or interface propagation appears. This obviously leads to a microscopically tortuous crack path associated with a high surface roughness (Figure 3.30). The deviations from the straight growth direction are comparable to the grain size, particle size or their spacing. Such kind of crack advance must also be associated with highly asymmetric crack-wake plasticity varying from grain to grain. The distance of the characteristic dislocation strip from the crack flanks is lower than the grain size. It means that, even when the strips regularly alternate their positions towards the crack flanks, the interaction of dislocation configurations in adjacent grains (reducing the shear asymmetry) can be nearly neglected along the whole length of the crack flanks. This is schematically depicted in Figure 3.30), where only the short segments of the cracks flanks (marked bold) experience the same shear. All the above-mentioned effects result in the maximal level of the RICC.

In the case of $S_R \ll 1$, on the other hand, grain boundaries and secondary phases do not constitute obstacles for the crack growth owing to a large-scale crack-tip plastic deformation embracing many grains or particles (Figure 3.31). The plastic zone is not restricted (constrained) by barriers on either side of the crack flanks and, obviously, there is no reason for the asymmetry of the crack-wake plasticity. The symmetry of the large-scale multislip deformation results in rather straight crack propagation, where crack advance during one cycle (or the CTOD) is higher than the microtortuosity of the crack path (the width and the height of asperities on the fracture surface). The distance of the characteristic dislocation strips from the crack flanks is higher than the average grain size. All those effects result in a negligible level of the RICC.

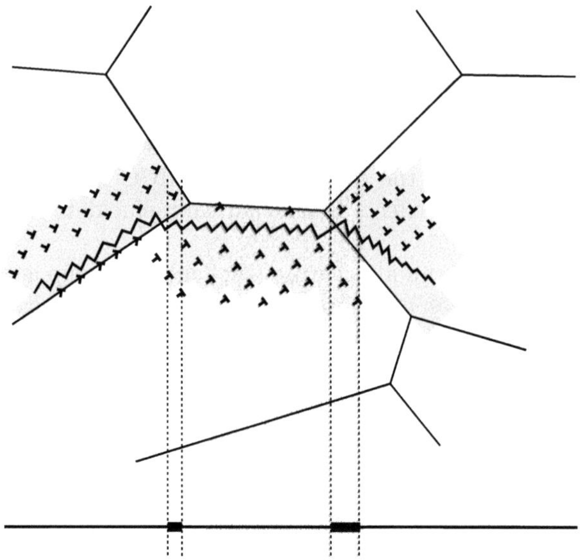

Figure 3.30 The scheme of asymmetric dislocation configurations produced in the case of high size ratios when the plastic zone size (*the grey region*) is smaller than the mean grain size. The highlighted short segments on the projected crack path correspond to nearly symmetric dislocation arrangements

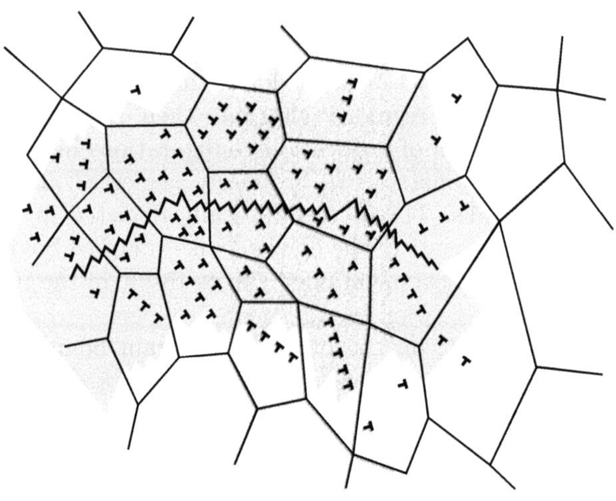

Figure 3.31 The scheme of symmetric dislocation configurations produced in the case of low size ratios (the plastic zone is larger than the mean grain size). The dislocation arrangement is nearly symmetric along the whole crack path

Thus, the statistical concept of the size ratio effect, as already described in Chapter 2, also holds valid for the RICC phenomenon. On the other hand, the size ratio effect is not so significant for the PICC since the dislocation bands on both sides of the crack flanks contribute to that effect. Nevertheless, a small influence might still be expected when $S_R > 1$ since, due to the asymmetry of dislocation strips, the total closure effect might be somewhat less than that described by Equation 3.21. In our further considerations, however, this small inaccuracy will be neglected.

The maximum possible level of RICC can be determined by assuming only one characteristic dislocation band that produces the local mode II SIF k_2^s and the crack-flank shear displacement u_1 according to Equations 3.15 and 3.16, respectively. The simple scheme of RICC in Figure 3.32 relates the closure distance $\delta_{cl,rl}$ to both the shear displacement u_1 and the surface roughness $R_A = 1/\cos\theta$ (in this 2D model $R_L = R_A$) as follows:

$$\delta_{cl,rl} = |u_1| \sqrt{R_A^2 - 1}. \tag{3.22}$$

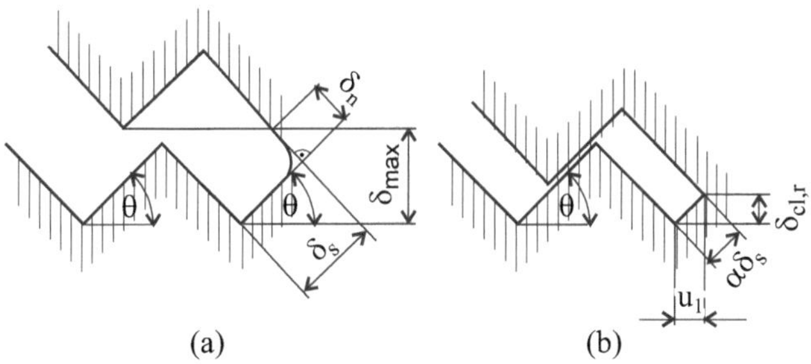

Figure 3.32 The scheme of the roughness-induced crack closure mechanism: (a) the crack tip opening at a maximal tensile load, and (b) the shear shift after unloading

Then a combination of Equations 3.16, 3.18, 3.19, 3.20 and 3.22 yields

$$\frac{\delta_{cl,rl}}{\delta_{max}} = C\sqrt{(R_A^2 - 1)}. \tag{3.23}$$

This is the final expression for a maximum possible long-range closure ratio in metallic materials.

Maximal Short-range Component

Besides the long-range component of RICC, a short-range component has to be considered as well. As already mentioned, this component is associated with the local mode II induced by the zig-zag crack path, and it is created by an irreversible slip at the crack tip – see Figure 3.32. It was the mechanism that was solely considered in the previous RICC models [168,328,329]. Unlike the long-range component, the overlap of crack flanks caused by the short-range mechanism quickly decays at larger distances from the crack tip. It completely disappears at distances comparable to the mean grain size.

During the loading part of the cycle, the crack tip is opened in a local mixed-mode reaching the maximum CTOD denoted as $\delta_{\max}$ at the moment of the peak stress (Figure 3.32(a)). This displacement is considered to be composed of both the reversible normal component δ_n and the irreversible shear component δ_s defining the local mixed-mode 1+2. Since the shear displacement created in the loading phase is not totally compensated during the unloading phase, the crack flanks remain shifted (Figure 3.32(b)). The main reason for this irreversibility, namely the interaction of dislocations of primary and secondary slip systems, has already been described in Section 3.2.1. Thus, the maximum value of the ratio of backward/forward shear displacements (the irreversibility level) is about 0.5. Obviously, this value is associated with sufficiently long pile-ups (a high probability of activating the secondary slip), i.e., with the size ratio $S_R \geq 1$. On the other hand, $S_R \ll 1$ means no residual shift of crack flanks, i.e., zero irreversibility level. Indeed, the related multislip mechanism cannot produce the residual shear shift, which is apparent from Figure 3.32. Moreover, the crack advance per cycle is comparable to (or higher than) the width of the asperities. Thus, by following the scheme in Figure 3.32(b), and assuming the irreversibility parameter $\alpha \approx 0.5\eta$, one can easily derive the expression at Equation 3.24 for the maximal short-range closure ratio:

$$\frac{\delta_{cl,rs}}{\delta_{\max}} = \frac{\eta \delta_s \sin\vartheta}{2(\delta_n \cos\vartheta + \delta_s \sin\vartheta)}. \tag{3.24}$$

In the framework of the saw-tooth (zig-zag) approximation of the crack flanks [330] it holds that

$$\frac{\delta_n}{\delta_s} = \frac{k_1}{k_2} = \sqrt{\frac{2}{3}} \cot\frac{\vartheta}{2}, \tag{3.25}$$

where k_1 and k_2 are the local SIFs induced by the remote K_I at the deflected crack tip. Note that the linear relationship $\delta \propto k$ is used here (as in Equation 3.16), instead of a quadratic one in the previous models. This is because the real contact point, determined by a sum of the closure effects, lies in the crack-wake. Because $R_A = 1/\cos\theta$, the final relation for the maximal short-range RICC ratio can be obtained by combining Equations 3.24 and 3.25 as follows:

$$\frac{\delta_{cl,rs}}{\delta_{\max}} = \frac{3\eta\,(R_A - 1)}{2\sqrt{6} + 6\,(R_A - 1)}. \tag{3.26}$$

The real short-range closure effect can, obviously, be composed of both the mode II and the mode III contributions (twisted crack segments). The corresponding shear displacements seem to influence the crack closure more or less independently. Hence, the RICC is determined predominantly by those sides of the fracture surface that exhibit the highest local values of either mode II or mode III. These values are associated with the highest and the steepest asperities and, therefore, the standard deviation D_R of the surface roughness seems to be also an important roughness parameter [329]. Although the quantities D_R and R_A are usually well correlated, the level of RICC according to Equation 3.26 can be underestimated in the case of very irregular fracture profiles of extremely high D_R (highly anisotropic and heterogeneous materials).

It should be noted that the roughness-induced crack closure is caused not only by the lateral displacement of two geometrically identical fracture surfaces. Due to local plasticity the shapes of the two rough surfaces are not identical, which leads to a premature contact at the top of the asperities. Such a mechanism, however, can dominate the RICC only when the plastic zone size is very small in comparison with the deflection length (see [10] for more details).

Let us finally mention that stress levels, corresponding to the moments of crack closure and crack opening, are practically equal. Most probably, however, there is a subtle difference related to the fact that the crack propagation is slightly postponed beyond the opening stress level (see [327] for details). However, this inconsistency practically does not affect the effective range ΔK_{eff}.

3.2.5 Unified Model of Crack-tip Shielding

The total crack closure effect represents a sum of the PICC and the RICC components ($\delta_{cl} = \delta_{cl,p} + \delta_{cl,rl} + \delta_{cl,rs}$). Because of the statistics of the size ratio in engineering materials, the parameter η determines the weight of the RICC components similarly to that of the geometrical shielding term in the brittle fracture case (see Chapter 2). Regarding that fact and Equations 3.21, 3.23 and 3.26, the total contact shielding ratio reads

$$\frac{\delta_{cl}}{\delta_{\max}} = \frac{K_{cl}}{K_{\max}} = C\eta\sqrt{R_S^2 - 1} + \frac{3\eta\,(R_S - 1)}{2\sqrt{6} + 6\,(R_S - 1)} + 2C. \tag{3.27}$$

Equation 3.27 reflects many interesting phenomena that were observed in connection with the fatigue crack behaviour:

1. Both terms of the RICC must approach zero for microscopically straight (planar) cracks, i.e., the microstructurally short fatigue cracks in polycrystals or shear cracks in single crystals. This obviously corresponds to $R_S = 0$ in Equation 3.27.

2. The total RICC level must also approach zero in the near fracture region of fatigue crack growth. In that region, clearly, the mean size ratio $S_{Rm} \to 0$ and $\eta \to 0$. Thus, the crack closure ratio is independent of the material and fully determined by the PICC.
3. The maximal level of RICC is detected in the near-threshold region, where, obviously, $S_{Rm} \to 1$ and $\eta \to 1$ is to be expected.

Equation 3.27 also reflects all the transient crack growth phenomena observed after overloads or during the random loading. For example, a sudden increase in $\delta_{\max}$ following an overload cycle increases the total level of the closure displacement δ_{cl}, and reduces the effective crack driving force ΔK_{eff} in the next cycles. The crack growth rate decelerates and remains lower as long as the crack front propagates through the plastic zone created by the overload cycle. After passing that zone, the crack growth rate approaches the previous "steady-state" value. Note that this transient effect is mainly due to the sudden change in the PICC because the overload causes a simultaneous increase in the size ratio (the parameter η). This keeps the level of the RICC close to that before the overload.

The effective crack driving force ΔK_{eff} can be expressed by means of Equation 3.27 and relations $K_{\mathit{eff}} = K_{\max} - K_{cl}$, $K_{\max} = \Delta K/(1-R)$. After some rearrangements one obtains

$$\begin{aligned}\Delta K_{\mathit{eff}} &= \left(1 - C\eta\sqrt{R_S^2 - 1} - \frac{3\eta\,(R_S - 1)}{2\sqrt{6} + 6\,(R_S - 1)} - 2C\right)\frac{\Delta K}{1-R} = \\ &= \Omega_1 \frac{\Delta K}{1-R}.\end{aligned} \tag{3.28}$$

The main advantage of the relation at Equation 3.28 is a possibility of its direct comparison with experimental ΔK_{eff} data, as shown in Section 3.2.6. Although a very moderate decrease in C with decreasing R is to be expected due to a slight drop in the crack-wake dislocation density, the constant C must lie within the range of 0.2 – 0.3. Therefore, in fact, the parameter $S_{Rc} \in (0.1, 1)$ is the only one real fitting parameter in this analysis. Anyway, both dimensionless parameters C and S_{Rc} possess a clear physical meaning and their values must lie within narrow ranges for all metallic materials. The relation at Equation 3.28 predicts that, for a particular material and a fixed applied ΔK, the effective driving force decreases with:
1. decreasing cyclic ratio;
2. increasing surface roughness;
3. decreasing plastic zone size (decreasing η).

This is in agreement with experimental observations [192]. Equation 3.28 also predicts that, in nanomaterials, the RICC level must be negligible in the whole fatigue crack growth region (not only near fracture). Indeed, the value of the intrinsic (effective) fatigue threshold range is lower than $3\,\mathrm{MPa\,m}^{1/2}$

(see Section 3.2.3). Therefore, assuming $K_{\max,th} \approx 5\,\mathrm{MPa\,m}^{1/2}$ as a minimum value for $R < 0.5$, and considering $\sigma_y \approx 1000\,\mathrm{MPa}$, one obtains $r_p \approx 2500\,\mathrm{nm}$ in the near-threshold region. For nanomaterials with $d_m < 100\,\mathrm{nm}$ it means that $S_{Rm} < 0.04 \ll S_{Rc} \approx 0.5 \rightarrow \eta \rightarrow 0$ and $(\delta_{cl}/\delta_{\max})_{RICC} \rightarrow 0$. In nanomaterials, consequently, a negligible level of the RICC ratio applies even to the near-threshold region. Recent experimental results [331] confirm this conclusion.

The values of ΔK_{eff} (or the effective resistance to the crack growth) can be measured by indicating the contact of crack flanks using extensometrical or resistometrical methods. Some authors have used a much simpler procedure based on the fact that, for cycle asymmetries $R > 0.7$, the crack remains steadily open. However, this method was recently challenged by Kondo [332], who indicated a decrease in ΔK_{eff} with increasing $K_{\max}$ for $R > 0.7$ in high-strength steels. This is, most probably, caused by an increasing environmental effect. Related extreme crack tip opening displacements (and associated lattice dilatations) enable a massive penetration of hydrogen into the plastic zone, which results in a reduction of the intrinsic matrix resistance.

When the crack tip opening displacement (CTOD) is much less than the deflections (kinks) or branches (splits) of the crack front, the geometrical shielding can take place [168,192,328]. These geometrical irregularities induce a reduction of the crack driving force (both the k_I and the k_{eff}) as has already been discussed in Chapter 2 in detail. The contact- and geometrical-shielding components can be unified on the basis of the size ratio statistics. With regard to Equations 2.23 and 3.28, the full effect of the geometrical shielding can be included in the model as follows:

$$\begin{aligned}\Delta K_{in} &= \left(1 - C\eta\sqrt{R_S^2 - 1} - \frac{3\eta\,(R_S - 1)}{2\sqrt{6} + 6\,(R_S - 1)} - 2C\right) \times \\ &\times \left[\left(1 - \eta + \eta\sqrt{\frac{\bar{g}_{eff,r}}{R_S}}\right)(1 - A_B) + 0.5A_B\right]\frac{\Delta K}{1 - R} = \\ &= \Omega_1\Omega_2\frac{\Delta K}{1 - R}.\end{aligned} \tag{3.29}$$

Equation 3.29 for calculation of the intrinsic crack driving force ΔK_{in} represents the most general form of the unified model. This relation holds in both the near-threshold and the Paris–Erdogan regions, but its application is limited when special shielding and anti-shielding effects are induced by large secondary phase particles (see Section 3.2.7). In cases of very high cycle asymmetry ($R > 0.6$), the crack closure level lies below the minimum SIF value ($K_{cl} < K_{\min}$) – see Figure 3.25. Under such circumstances no closure occurs and, obviously, Equation 3.29 remains no longer valid.

A typical diagram of relative contributions of individual mechanisms to crack tip shielding ($\Delta\bar{K}_{br} = \Delta K_{br}/\Delta K, \Delta\bar{K}_{in} = \Delta K_{in}/\Delta K, \Delta\bar{K}_{cl} = \Delta K_{cl}/\Delta K$) in dependence on both the applied ΔK (constant d_m) and the mean grain size d_m (constant ΔK) according to Equation 3.29 is schematically depicted in Figures 3.33 and 3.34 for a material of particular chemical composition. Note that the decreasing (increasing) influence of the RICC mechanism with increasing ΔK (d_m) almost completely determines the related dependencies of global shielding or intrinsic driving force.

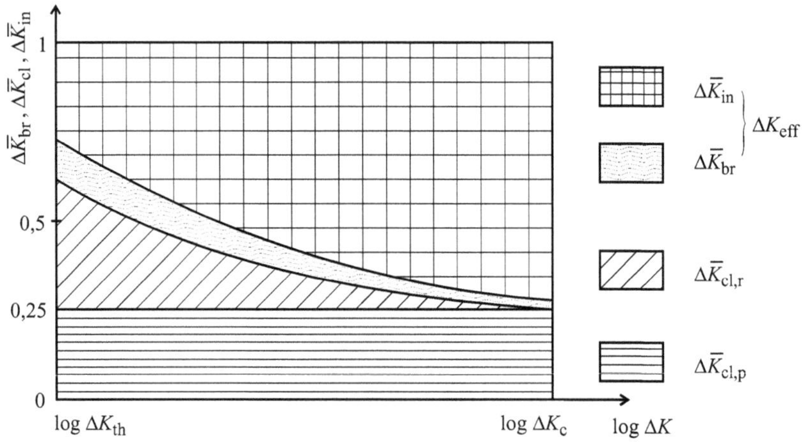

Figure 3.33 Diagrams of relative participation of individual mechanisms in the crack tip shielding as functions of the applied ΔK (constant d_m)

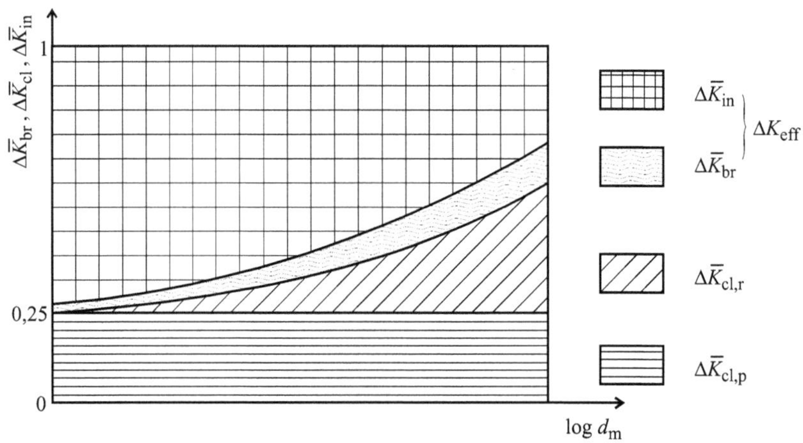

Figure 3.34 Diagrams of relative participation of individual mechanisms in the crack tip shielding as functions of the mean grain size d_m (constant ΔK)

By applications of Equation 3.29 to metallic materials, however, one has to pay attention to the following problem. Unlike in brittle fracture, the fatigue crack growth mechanism in metallic materials is closely related to a cyclic movement of dislocations that is controlled by the shear stress. As shown by Pippan [333], the deflections like kinks or double-kinks do not cause any reduction in the maximal shear stress at the crack tip. Thus, that kind of shielding can be neglected in a majority of metallic materials, perhaps except for those exhibiting a quasi-brittle fracture mechanism. On the other hand, the crack branching causes both the reduction of the local shear stress and the increase in the resistance to the crack growth (the doubled fracture energy) [168,169]. Although the branching process is not typical of the fatigue cracks in metallic materials, its occurrence must always be explicitly reflected when applying the unified model to a determination of both the shielding components and the intrinsic thresholds in metallic materials.

For a material of a particular chemical and phase composition, the values of the intrinsic thresholds should be practically independent of either the microstructure coarseness or the cyclic ratio. A slight dependence on the former might result from a difference in the resistance of grain or phase boundaries to fatigue crack penetration. As already mentioned, a small change in the crack-wake dislocation density with variation in R is also expected. These effects are, most probably, the main reasons why there is some scatter in the calculated values of $\Delta K_{eff,th}$ or $\Delta K_{in,th}$ (see the following subsection).

3.2.6 Applications of the Unified Model

Equations 3.28 and 3.29 of the unified model can be simply applied to the assessment of the intrinsic thresholds and the shielding components in engineering metallic materials. Indeed, only a few rather standard material characteristics and loading parameters are necessary: σ_y, d_m, R_S, ΔK and R. A question important for practical measurements of the surface roughness R_S arises in connection with a semi-fractal character of the fracture surface morphology (e.g., [334]). As already mentioned in Section 3.2.3, almost all cracks in metallic materials can be assumed to be tortuous within a considerably wide range of scaling. Consequently, one should be able to decide about the fracture surface roughness R_S of a particular material under specific loading conditions. Fortunately, the size ratio effect gives a rather clear answer to that question: the asperities smaller than about one half of the static plastic zone size do not contribute to the RICC. This statement is justified by the fact that the size of asperities is closely related to the characteristic microstructural parameter. Thus, for example, $K_{\max,th} \approx 5\,\mathrm{MPa\,m}^{1/2}$ is generally appropriate for the near-threshold region. When considering the yield stress $\sigma_y \approx 500\,\mathrm{MPa}$, the facet sizes of the order of units of μm might still be significant. In the Paris–Erdogan region ($K_{\max} \approx 20 - 30\,\mathrm{MPa\,m^{1/2}}$), however, the finest plausible scale constitutes only tens of microns.

The unified concept of crack tip shielding was applied in several studies of the intrinsic thresholds in aluminium alloys, α-titanium, titanium alloys and steels [168,173,335]. The geometrical shielding effects were neglected for almost all investigated metals ($\Delta K_{in,th} \approx \Delta K_{eff,th}$), which means that, in most cases, Equation 3.28 was used to determine the intrinsic thresholds. An exception to the rule was the duplex steel, where a pronounced interphase cracking and crack branching took place. The analyses are briefly reported hereafter. Although some results were slightly corrected according to the current state of the theoretical concept, the changes do not have any impact on conclusions stated in the works published earlier.

3.2.6.1 Aluminium Alloys

The crack closure contribution in the near-threshold region has been experimentally studied using underaged and overaged compact tension samples of a 7475 aluminium alloy in air and vacuum [336]. Different thermomechanical treatments have produced microstructures with grain sizes of 18 µm and 80 µm. The experimental and theoretical data are displayed in Table 3.1.

Table 3.1 Experimental and calculated fatigue threshold data for 7475 aluminium alloy in air and vacuum (Λ is the percentage of extrinsic shielding)

Envir.	ΔK_{th}^{exp} [MPa m$^{1/2}$]	d_m [µm]	R_S	σ_y [MPa]	η	$\Delta K_{eff,th}$ [MPa m$^{1/2}$]	Λ [%]
Air	2.60	18	1.30	505	1.00	1.2	54
	1.70	18	1.21	455	1.00	0.89	48
	2.70	80	1.90	451	1.00	1.14	58
	2.20	80	1.25	445	1.00	1.08	51
Vacuum	4.00	18	1.30	505	0.97	1.87	53
	2.90	18	1.21	455	0.99	1.52	48
	8.80	80	1.90	451	0.94	1.85	79
	4.10	80	1.25	445	1.00	2.02	51

The values of $\Delta K_{eff,th}$ calculated according to Equation 3.28 for the tests in air are practically identical and independent of the grain size. They lie only slightly below the measured data of about $1.4\,\mathrm{MPa\,m^{1/2}}$ [336] and are in a good agreement with the range of $0.75 - 1.0\,\mathrm{MPa\,m^{1/2}}$ reported by Pippan [337] for aluminium alloys. The calculated values $\Delta K_{eff,th}$ for the vacuum tests are also in good agreement with the experiment. They are distinctly higher than those in air, which can be reasonably elucidated by more extended re-welding of newly created surfaces in the vacuum.

3.2.6.2 Titanium Alloys

Experimental near-threshold crack growth data for various α-titanium grades [338] are collected in Table 3.2. The large scatter of measured values of ΔK_{th} is mainly a consequence of a wide distribution of grain sizes. The average measured value $\Delta K_{eff,th} = 2.1\,\mathrm{MPa\,m^{1/2}}$ [339] lies within the computed range of intrinsic thresholds $\Delta K_{eff,th} \in (2.0, 3.8)\,\mathrm{MPa\,m^{1/2}}$.

Table 3.2 Experimental and calculated fatigue threshold data for α titanium grades

ΔK_{th}^{exp} [MPa m$^{1/2}$]	R	d_m [µm]	σ_y [MPa]	η	$\Delta K_{eff,th}$ [MPa m$^{1/2}$]	Λ [%]
6.00	0.07	40	430	0.95	2.03	66
5.00	0.35	40	430	0.89	2.56	49
5.30	0.07	35	260	0.68	2.25	58
4.30	0.35	35	260	0.48	3.02	30
7.00	0.07	230	220	0.96	2.35	66
5.80	0.35	230	220	0.91	2.91	50
6.00	0.07	20	630	0.96	2.02	66
4.30	0.35	20	630	0.95	2.08	52
10.00	0.07	210	580	1	3.23	68
8.00	0.35	210	580	0.99	3.71	54

Experimental and theoretical threshold data for two grades of the Ti-2.5%Cu alloy are displayed in Table 3.3. The microstructures consisted of coarse lamellar colonies ($d_m = 580\,\mu\mathrm{m}$) and a fine Widmanstätten microstructure ($d_m = 10\mu\mathrm{m}$), respectively [329]. The calculated intrinsic threshold values of $3.3\,\mathrm{MPa\,m^{1/2}}$ are very close to the average measured value $\Delta K_{eff,th} = 3.2\,\mathrm{MPa\,m^{1/2}}$ for Ti-6Al-4V alloys [340].

Table 3.3 Experimental and calculated fatigue threshold data for Ti-2.5%Cu alloy

ΔK_{th}^{exp} [MPa m$^{1/2}$]	R	d_m [µm]	σ_y [MPa]	η	$\Delta K_{eff,th}$ [MPa m$^{1/2}$]	Λ [%]
9.00	0.1	580	420	1.00	3.32	63
7.00	0.1	10	499	0.65	3.15	55

3.2.6.3 ARMCO Iron

Experimental and calculated data for different grades of the ARMCO iron [341, 342] are summarized in Table 3.4. Besides an enormous variance in the grain size there is also a wide range of R ratios. The measured effective thresholds of 2.75 MPa m$^{1/2}$ do not depend on the mean grain size. The theoretical data are presented only for loading cases, where a non-zero closure was predicted. The calculated $\Delta K_{eff,th}$ values in Figure 3.35 lie close to the experimental dotted and dashed line.

The dominant contribution of RICC explains well the strong dependence $\Delta K_{th}(d_m, R)$ observed in the experiment. Note that $\Delta K_{th} \approx \Delta K_{in,th} = \Delta K_{eff,th}$ for $R > 0.55$, which means no closure during cycling.

Table 3.4 Experimental and calculated fatigue threshold data for ARMCO iron

ΔK_{th}^{exp} [MPa m$^{1/2}$]	R	d_m [μm]	σ_y [MPa]	η	$\Delta K_{eff,th}$ [MPa m$^{1/2}$]	Λ [%]
4.5	0.1	2	530	0.19	2.78	38
2.9	0.7	2	530			
5.3	0.1	20	240	0.5	2.49	53
2.9	0.7	20	240			
6.8	0.1	90	150	0.55	3.03	55
3.65	0.55	90	150	0.44	3.64	0
2.8	0.7	90	150			
2.9	0.8	90	150			
8.7	0.1	410	108	0.76	2.96	66
3.2	0.7	410	108			
10.3	0.1	3550	96	0.99	2.37	77
3.6	0.7	3550	96	0.99	2.49	31

3.2.6.4 Duplex Steel

The fatigue threshold values of duplex ferritic-martensitic steels can reach 20 MPa m$^{1/2}$ [343], which is much higher than the threshold range of single phase alloys [344]. The high thresholds of multi phase alloys are typically attributed to extreme crack deflection and branching of the crack front [343, 345].

Hot rolled bars made of the austenitic-ferritic duplex stainless steel SAF 2507 (equivalent to UNS S32750) with a diameter of 80 mm (AD) were also used in our experiments [171]. The tests were performed at three different temperatures 150°C, −50°C and at room temperature (R = 0.1 and the frequency f = 10 Hz). The microstructure was evaluated using the optical microscope Olympus PMG3 after etching and the values of R_S were measured by the profilometer MicroProf FRT. The path of propagating cracks was found to be very tortuous and branched. The propagation directions change in the austenite or ferrite phase and, in particular, at the austenite/ferrite interface. Crack branching could be identified by secondary cracks

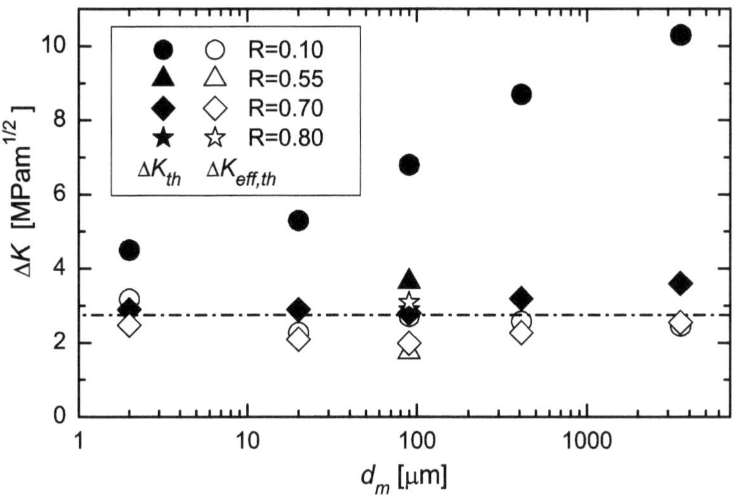

Figure 3.35 The dependence of both measured thresholds ΔK_{th} and calculated effective thresholds $\Delta K_{eff,th}$ on the mean grain size of the ARMCO iron for various cyclic ratios. Reprinted with permission from ASTM International. (see page 265)

in all specimens (see Figure 3.36). The values of the area fraction A_B were assessed by a careful counting of both the number and the size of secondary cracks. Because of a quasi-brittle character of the crack path, the shielding effect induced by crack deflections was taken into account as well.

Measured values of R_S and A_B for all investigated temperatures are listed in Table 3.5.

Table 3.5 Experimental fatigue threshold data for the duplex steel

t [°C]	ΔK_{th}^{exp} [MPa m$^{1/2}$]	A_B	R_S	σ_y [MPa]	$\Delta K_{eff,th}^{exp}$ [MPa m$^{1/2}$]
150	10.83	10	1.55	484	4.92
20	8.64	20	1.43	625	3.90
−50	14.85	30	1.56	738	5.84

Calculated values of both the statistical factor η and $\Delta K_{eff,th}$ are displayed in Table 3.6. There is good agreement between the calculated and the measured effective threshold values (compare with Table 3.5).

The effective threshold values for −50°C and 150°C are higher than the effective value of 2.75 MPa m$^{1/2}$ measured and calculated for the ferritic single-phase steel. This disagreement can be almost eliminated by calculating $\Delta K_{in,th}$ using Equation 3.29 (see Table 3.6). The remaining slight deviations can be attributed to: (1) a different intrinsic resistance of single-phase and

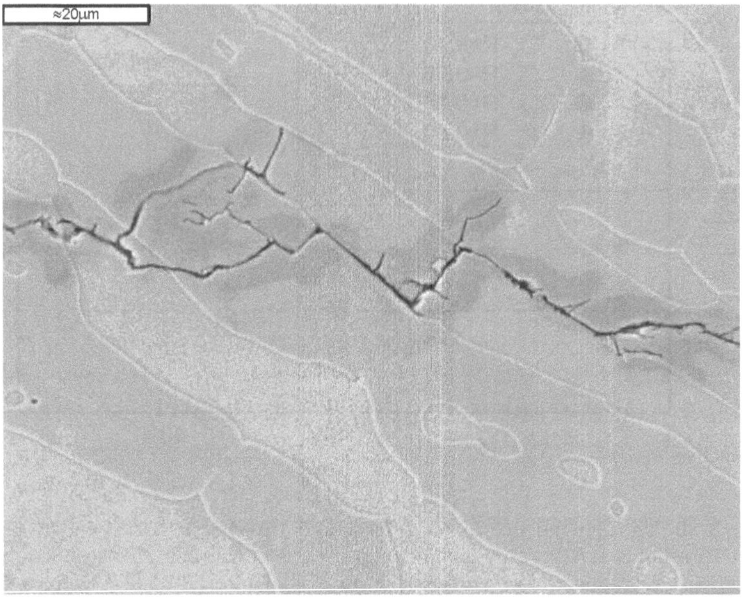

Figure 3.36 Tortuous and branched geometry of fatigue cracks in the duplex steel

Table 3.6 Calculated fatigue threshold data for the duplex steel

t [°C]	η	$\Delta K_{eff,th}$ [MPa m$^{1/2}$]	$\Delta K_{in,th}$ [MPa m$^{1/2}$]	Λ [%]
150	0.58	4.6	3.6	67
20	0.92	2.9	1.9	78
−50	0.72	5.3	3.6	76

duplex matrices; (2) imperfections in experimental methods; (3) an approximate character of the theoretical concept and (4) the temperature influence. The last column in Table 3.6 shows a contribution of the total extrinsic component (closure + shielding) of almost 80%, which is much higher than that in all other investigated materials, including austenite steel [346]. This confirms the decisive role of the extrinsic toughening induced by the duplex microstructure.

The difference in intrinsic thresholds of materials with various chemical composition is, in principle, determined by differences in their values of the critical SIF for dislocation emission (see Section 3.2.3). However, the environmental effects can also play an important role because, during experimental procedures, elementary crack advances cannot be avoided.

3.2.7 Influence of Shielding on Crack Growth Rate

When a long crack propagates under the constant applied stress range, both the plastic zone size and the crack opening displacement linearly increase with an increasing crack length. Consequently, the size ratio S_R decreases and the relative influence of surface asperities becomes smaller. Particularly in the near-fracture region the crack advancement per cycle usually becomes higher than the size of crack-wake asperities. As a consequence, the super-critical fatigue striations, created cycle-by-cycle by partially reversible large-scale blunting beyond the near-threshold region, usually remain as the most striking morphological patterns on fracture surfaces (see Section 3.4 for more details). It obviously means that the role of the roughness-induced shielding diminishes. Nevertheless, the length fraction of deviated segments along the crack path can be reasonably estimated by the mean value $\bar{\eta}$ during the crack growth.

When assuming an ideally straight crack propagating in a quasi-brittle matrix ($R = 0$), the simplest "intrinsic" form of the well known Paris–Erdogan relation

$$\left(\frac{\mathrm{d}a}{\mathrm{d}N}\right)_s = A(\Delta K_{in})^n \tag{3.30}$$

applies, where A and n are materials constants. Obviously, no shielding appears because $\eta \to 0$, $g \to 1$, $C \to 0$ $A_B \to 0$ and $R_S = 1$ in Equation 3.29, i.e., $\Delta K = \Delta K_{in}$. This intrinsic behaviour is assumed to be reproduced for all further crack types which means, at least, the same chemical composition and crystallography of the matrix. On the other hand, the presence of shielding reduces the crack driving force and, consequently, the substitution $\Delta K \to \Omega_1\Omega_2\Delta K$ is to be made in Equation 3.30. Moreover, an extended crack path due to the crack tortuosity is to be considered. The ratio of the mean crack growth rates along the straight and the tortuous paths must be inverse to that of their lengths. The length l_t of the tortuous crack path is determined by the mean value of its linear (profile) roughness $\bar{R}_L$. In a general case, the length l_m of a partially tortuous crack path can be determined by the mean values of $\bar{R}_L$ and $\bar{\eta}$ as follows:

$$\left(\frac{\mathrm{d}a}{\mathrm{d}N}\right)_s \bigg/ \left(\frac{\mathrm{d}a}{\mathrm{d}N}\right)_m = l_m/l_s \approx 1 - \bar{\eta} + \bar{\eta}\bar{R}_L .$$

This leads to

$$\left(\frac{\mathrm{d}a}{\mathrm{d}N}\right)_m \approx \frac{A}{1-\bar{\eta}+\bar{\eta}\bar{R}_L}\left(\Omega_1\Omega_2\right)^n \Delta K^n = \frac{\Omega_1^n\Omega_2^n}{1-\bar{\eta}+\bar{\eta}\bar{R}_L}\left(\frac{\mathrm{d}a}{\mathrm{d}N}\right)_s . \tag{3.31}$$

When introducing typical values $\bar{\eta} = 0.8$, $g = 0.9$, $\bar{R}_L = 1.2$, $\bar{R}_S = 1.3$, $A_B = 0.1$, $C = 0.1$ and $n = 3$ into Equation 3.31, the growth rate for a partially tortuous crack is an order of magnitude lower than that of the ideally straight one, which nearly covers the scatter observed in experiments [192].

Additional types of shielding and anti-shielding effects can appear in some special materials [177]. When analyzing the fatigue crack growth rate in ductile irons, for example, such effects are induced by graphite noduli that may occupy 25% of the fracture surface, as documented in Figure 3.37. Because this number is about twice as high as that of the volume fraction (13%), the crack path is strongly affected by the presence of nodules. The difference is clearly visible in Figure 3.38, where the left part of the picture corresponds to fatigue fracture whereas the right part corresponds to the ductile final fracture (closer to the volume fraction of noduli). This means that the interparticle distance of graphite noduli, and not the grain or the phase size, is the characteristic microstructural parameter in ductile irons. Both the stress concentration around the noduli and the crack initiation at nodule/matrix interface are to be particularly considered, as expressed in the following summary of shielding and anti-shielding interactions.

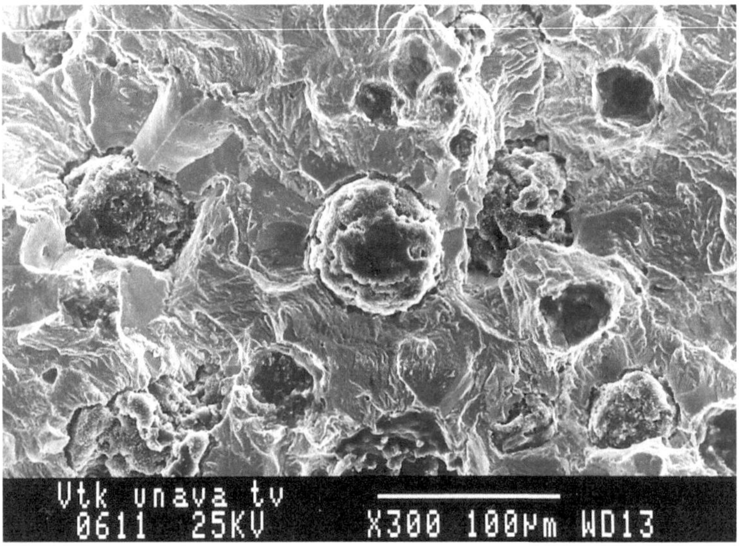

Figure 3.37 The fatigue fracture morphology of the ferritic ductile iron showing a high participation of graphite noduli in the fracture process

Shielding effects:

1. A decrease in ΔK due to a higher level of the roughness-induced crack closure.
2. An extension of the main crack path due to the crack front deflections.
3. A reduction in ΔK due to the crack front deflection and branching. Because of a quasi-brittle fracture mechanism controlled by noduli, the shielding effect caused by crack deflections is to be also taken into account.

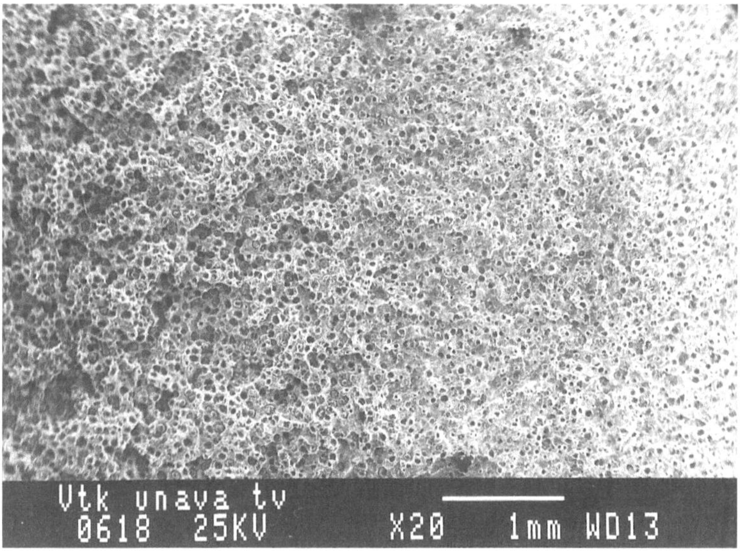

Figure 3.38 The overview fractograph documenting a more frequent presence of graphite noduli on fatigue fracture surfaces (*the left half*) than it would correspond to their bulk concentration (*the right half* – the ductile final fracture)

4. A reduction in ΔK owing to a network of microcracks connected with noduli that are located within the plastic (process) zone at the crack tip.
5. An increase in the material resistance to the crack growth caused by an additional work needed for a creation of the microcrack network;

Anti-shielding effects:

6. An increase in ΔK due to the difference between elastic moduli E_G (graphite) and E_F (ferrite): $E_G \ll E_F \rightarrow$ the stress concentration around the graphite nodule (the local effect).
7. An increase in ΔK caused by the interaction between the main crack and the microcracks at noduli in front of the main crack.
8. An increase in the crack driving force ΔG caused by the difference between the elastic moduli E_{FDI} (ferritic ductile iron) and E_F: $E_{FDI} < E_F$ (the bulk effect).
9. A reduction of the crack path due to a coalescence of the main crack front with the microcracks.

Note that the effects described by points 4–9 are not considered in Equation 3.31. Therefore, Equation 3.31 can be simply generalized as

$$\left(\frac{\mathrm{d}a}{\mathrm{d}N}\right)_t = \frac{\prod\limits_{i=1}^{n} \Omega_i}{1 - \bar{\eta} + \bar{\eta} R_L} \left(\frac{\mathrm{d}a}{\mathrm{d}N}\right)_s, \tag{3.32}$$

where n is the number of shielding effects involved ($n = 8$ for FDI, point No. 4 irrelevant). Obviously, the shielding effects correspond to $\Omega_i < 1$, whereas $\Omega_i > 1$ applies to the anti-shielding ones.

Equation 3.32 shows a general structure of the Paris–Erdogan law in terms of shielding effects. By using that equation, in particular, a variation of fatigue crack growth rates in engineering materials of the same matrix but with different microstructures can be understood in a quantitative manner.

The difference in intrinsic crack growth rates (given by difference in parameters A and n) is determined mainly by a diverse environmental influence of air on rewelding of newly created fracture surfaces in various materials. In the ultra-high vacuum, the intrinsic crack growth rates of various steels, aluminium and copper alloys are practically identical when normalized to the Young's modulus [275].

3.3 Shear and Mixed-mode Loading

When either a torsion or a pure shear loading is applied to a cracked solid, the crack front can propagate under local shear modes II and III. Therefore, the first subsection deals with basic theoretical concepts of mode II and mode III propagations of long fatigue cracks. A special emphasis is devoted to the micromechanism of mode III crack growth which seem to be much more complicated than that of the mode II. Indeed, the theoretical analysis and recent experiments show that the macroscopic mode III crack propagation can be produced by local mode II or mixed-mode II+III micromechanisms. As a consequence, mode III crack growth rate is often found to be lower than that of the mode II which is documented in the Section 3.3.3.

Pure shear-mode crack propagation usually persists only for a limited number of loading cycles and the cracks incline (or branch) to get a support of the mode I loading. Such a behaviour is predominantly caused by attempting to avoid friction stresses at shear crack flanks that reduce the driving force. This leads to a local mixed-mode I+II, I+III or I+II+III crack propagation as is documented in Sections 3.3.2 and 3.3.3 dedicated to crack propagation under cyclic torsion and pure shear. Typical products of this complicated crack path are so-called factory roofs that represent one of the most remarkable morphological patterns in fatigue. The related model, based on LEFM mixed-mode analysis and 3D fractography, reveals that the opening loading mode dominates the process of factory roof formation. However, the most important result of this analysis seems to be a definition of generalized conditions of mode I branching (kinking) from the shear-mode crack propagation.

The last two parts are devoted to fatigue life of steel specimens subjected to a combined bending-torsion loading. Many engineering components such as shafts, piston rods or gear wheels operate under this type of loading. An application of stress-based multiaxial criteria to predictions of fatigue life and comparison with experimental data obtained on virgin and nitrided specimens are presented. In addition, an extended fractographical analysis of nitrided specimens reveals changes in the fracture surface roughness and the geometry of so-called fish-eye cracks with varying proportion of bending and torsion loading components.

3.3.1 Models of Shear-mode Crack Growth

While the principal micromechanisms of fatigue crack growth in modes I and II can be clearly demonstrated by simple dislocation models, this is not the case in a pure mode III crack propagation. Nevertheless, besides a local twisting of the mode III crack front to get a mode I support [347], a micromechanical interpretation of the crack growth under the out-of-plane (mode III) cyclic shear is possible by means of alternating in-plane (mode II) shears acting either at a microscopically tortuous crack front [348] or between particles cracked near the crack front [348, 349]. Models based on such concepts are described hereafter.

3.3.1.1 Basic Assumptions

Plastic deformation in metals, alloys and intermetallics is usually caused by generation and movement of dislocations. Only in special cases do twinning, grain boundary sliding and diffusion of vacancies give an additional contribution to plastic deformation. In fact, a large amount of dislocations, including those statistically stored, on different slip systems is active during plastic deformation. However, only geometrically necessary ones are sufficient to demonstrate the local fracture processes. Deformations at the crack tip under modes I, II and III produced by geometrically necessary dislocations are schematically depicted in Figures 3.39 and 3.15, where the formation of the new fracture surface is illustrated on the atomistic scale.

As has already been described in Section 3.2, mode I deformation at the crack tip is caused by a symmetric generation of edge dislocations at the crack tip. A new fracture surface is created that is proportional to the Burgers vector times the number of generated dislocations and the crack front length. If one assumes that the new fracture surface will be immediately oxidized so that the dislocations cannot annihilate the newly generated fracture surface during unloading, the crack will propagate with a rate proportional to the cyclic crack tip opening displacement.

Mode II deformation is caused by edge dislocations in the plane of the crack. This process has already been described in the previous section (see Figure 3.15). The crack extension per cycle is, again, proportional to the

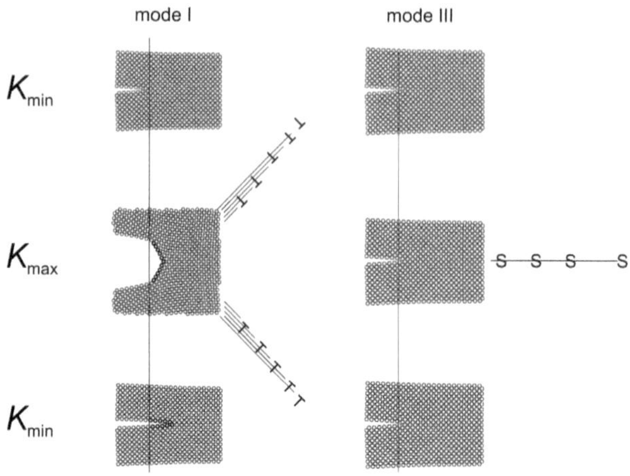

Figure 3.39 Atomistic scheme of fatigue crack advances in the mode I and the mode III during a single loading cycle

number of generated dislocations at the crack tip times the Burgers vector, i.e., to the cyclic crack tip "opening" displacement.

In the case of an idealized mode III loading screw dislocations are generated at (or near) the crack tip as shown in Figure 3.39. Although these dislocations form a plastic zone in a similar way to edge dislocations in the mode II case, they do not create new fracture surfaces along the crack front since their Burgers-vector is parallel to the crack front. Thus, in a planar sample, the movement of the screw dislocations generates ledges on the surface within the plastic zone. Each dislocation generates a surface ledge that is equal to the Burgers vector times the distance of its movement. If one assumes that the surface ledge cannot annihilate during the reverse motion of dislocations during unloading, a new "fracture surface" is generated only within the plastic zone at the surface of the sample. However, along the crack front no crack extension should take place. During further loading the surface crack (the ledge in the plastic zone) will grow in a mode II along the mode III crack front.

In summary, the cyclic movement of dislocations under idealized modes I and II generates a new fracture surface along the whole crack front in each cycle. On the other hand, the screw dislocations in a pure mode III generate only ledges on the surface of the sample which may propagate as "local" mode II cracks. The situation along the crack front is similar to the movement of dislocations near a free surface: the movement of the dislocation will not generate a ledge (a surface step) if its Burgers vector is parallel to the surface. One can imagine that, in reality, mutual interactions of dislocations (including those statistically stored) might generate some microcracks ahead of the crack front in the case of all modes I, II and III. This might lead to local advances of the main crack front by a coalescence of microcracks with

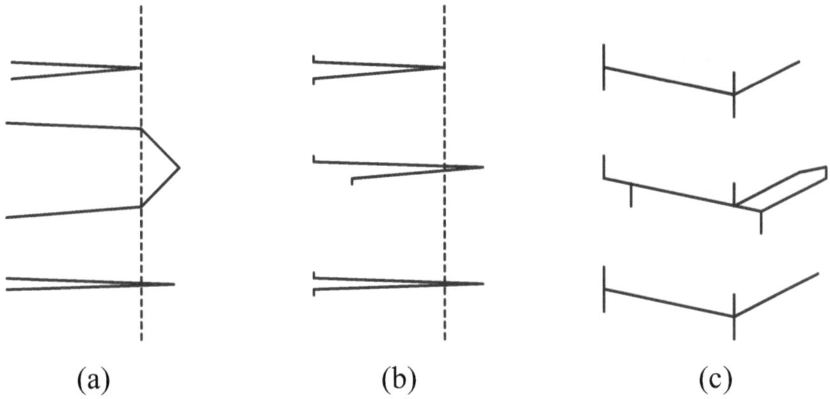

Figure 3.40 The geometric scheme of fatigue crack advance under: (a) mode I, (b) mode II, and (c) mode III during a single loading cycle

the main crack front. However, the crack front advance per cycle produced in such a way must be negligible when compared with the straightforward crack growth caused by the irreversible movement of geometrically necessary dislocations in modes I and II.

The basic difficulty with a pure mode III mechanism in homogeneous materials can also be simply understood when following the macroscopic crack growth schemes drawn in Figure 3.40. During one loading cycle, new surfaces are created ahead of both mode I and mode II fatigue crack fronts by non-zero components of shear displacements parallel to the crack growth direction. Environmental degradation of a newly created surface and irreversibility of dislocation movement are two commonly accepted reasons for an incomplete recovery of atomic bonds at the crack tip during the reversed loading. On the other hand, no shear displacements creating such new surfaces are produced by pure mode III loading.

3.3.1.2 In-plane Shear Models of Out-of-plane Shear Crack Growth

It should be emphasized that a macroscopic mode III crack front propagation does not necessarily need to be produced by pure mode III displacements. Indeed, pure mode II cracking micromechanisms can be exclusively responsible for such crack front advance. In homogeneous materials this demands only a natural assumption of a microscopically tortuous crack front. The crack fronts in engineering materials are never absolutely straight except for single crystals provided that the crack plane is identical with the shear plane and, simultaneously, the direction of the "straight" crack front is exactly parallel to the Burgers vector. In all other cases the crack fronts have a serrated shape at least at the atomic level. Therefore, a global out-of-plane shear deformation will cause fatigue crack propagation of the serrated flanks as depicted in Figure 3.41. In case (a) the small displacement ranges Δu_{III} mean that the sum of edge components of dislocations emitted from the more horizontal parts of the crack front does not exceed a few total Burgers vectors. This corresponds to a loading near the local mode II+III crack growth threshold. In case (b) higher displacement ranges Δu_{III} cause crack propagation along the whole crack front.

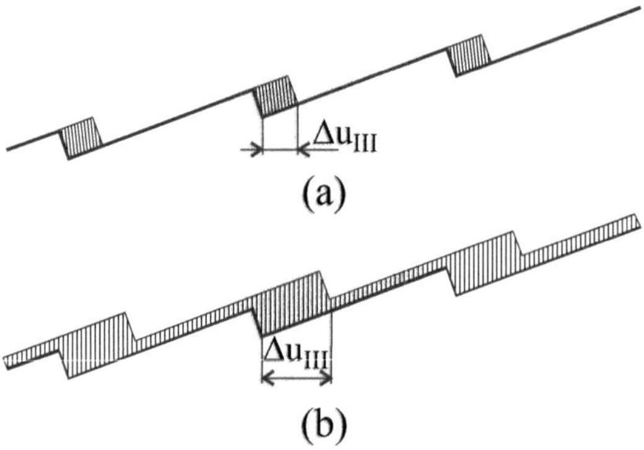

Figure 3.41 Scheme of deformation-controlled mode III growth of a serrated crack front: (a) small out-of-plane strain, and (b) large out-of-plane strain. Reprinted with permission from John Wiley & Sons, Inc. (see page 265)

Let us now consider a simple model of a macroscopically straight but microscopically tortuous crack front under a pure macroscopic mode III loading as shown in Figure 3.42. The triangular microledges are loaded in a mixed-mode II+III, but the out-of-plane shear stress vector can be resolved into two pure mode II (in-plane) components, perpendicular to the legs of the triangle. This enables an alternating step by step growth of the crack front segments under a pure mode II mechanism and leads to a gradual smoothing of the front. This effect may decelerate the macroscopic mode III crack growth or even cause its arrest. Such a behaviour, already reported by Tschegg [350–352], was originally attributed only to the friction and clinching of spatially tortuous crack flanks. The front propagates in the x direction and, indeed, it remains parallel to the direction of the macroscopic "mode III" crack front. This can elucidate a mode III-like fatigue crack growth from a circumferential mode I pre-crack under torsion [353, 354]. When applying the alternating mode II advance of the crack front segments to the circumferential pre-crack described above [348], the resulting shape of the growing crack always becomes qualitatively very similar to that experimentally observed by Murakami [355]. After a certain number of cycles, the shape of the crack front gives an impression of a mode III-like crack propagation (see [348] for more details).

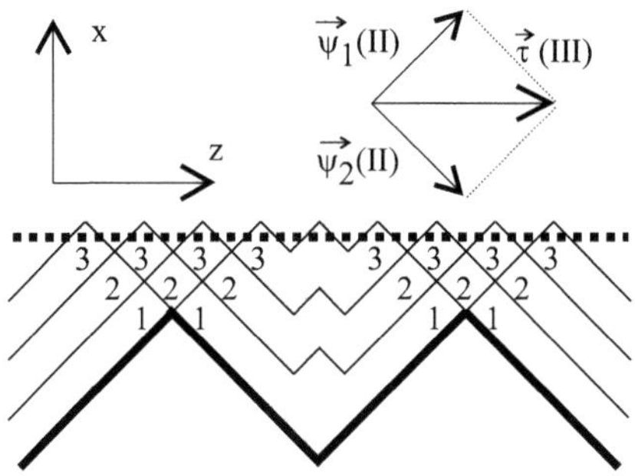

Figure 3.42 Scheme of the alternating pure mode II mechanism of a gradual advance of a microscopically tortuous (macroscopically straight) crack front in a macroscopic mode III. The *thin lines* indicate positions of the crack front after a specific number of fatigue cycles. Reprinted with permission from John Wiley & Sons, Inc. (see page 265)

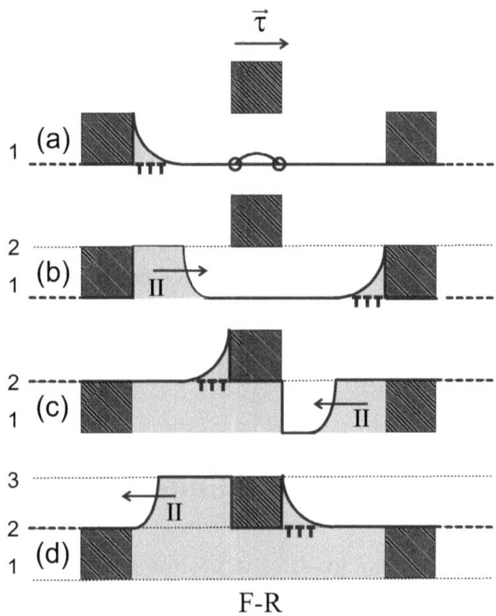

Figure 3.43 Scheme of a mode II mechanism ensuring a macroscopic mode III crack advance (*grey areas*) between secondary phase particles (*hatched squares*). Fracture of the particle-matrix interface is assumed when the crack front approaches the particles

Even a microscopically straight crack front can propagate in a macroscopically pure mode III when considering the assistance of cracks related to secondary phase particles [348, 349]. Indeed, the cracks formed by fracture of particles or particle-matrix interfaces can spread in a pure mode II along the "mode III" crack front as depicted in Figure 3.43. The particles in the vicinity of the crack front are depicted by hatched squares. Frank–Read (F-R) sources adjacent to the main crack front generate dislocation pile-ups at the particle-matrix interfaces and produce coplanar interface cracks with crack fronts nearly perpendicular to the main crack front. Consequently, these cracks can extend under mode II loading along the crack front. The mean crack propagation rate in such a model depends on the total length of the microscopic mode II crack front, given by both the size and the line concentration of particles, associated with a unit length of the "mode III" crack front.

Let us note, however, that another dominant damage mechanism might be responsible for fatigue crack propagation in some metallic materials. One can imagine that, inside the whole cyclic plastic zone ahead of the crack front, many secondary particles can be separated from the matrix during one loading cycle. Such microcracks can then be mutually jointed by shearing under the local mode II to coalesce with the main crack front by breaking remaining ligaments (in a local mixed-mode I+II+III). In this way, rather large crack advances in a nearly pure mode III through the whole process zone can be accomplished during each of several loading cycles and, consequently, the crack growth rate under mode III must not necessarily be lower than that under mode II. Although such a mechanism might be primarily expected to operate in a low cycle fatigue, one cannot exclude that it could also be efficient in a high-cycle fatigue.

3.3.2 Propagation of Cracks under Cyclic Torsion

The great majority of shear loading experiments was performed by cyclic torsion of cylindrical specimens. Many of these studies revealed that there is a big difference in the crack propagation mode in smooth and notched specimens (e.g., [192, 350, 356, 357]). In smooth specimens, the maximal shear and normal stresses are equal in magnitude. Therefore, there is competition between the planes of the maximal shear (parallel and perpendicular to the specimen axis) and those of the maximal tensile stress (main planes, inclined at 45° to the specimens axis) with respect to the fracture process. During the first few cycles, this competition is, as a rule, decided in a favour of the latter planes. Thus, the cracks start to propagate under an opening mode I.

In the circumferentially notched specimens, however, the concentration of shear stresses in the plane perpendicular to the specimen axis stabilizes the crack growth in this plane. Therefore, the cracks propagate under shear modes II+III at least for a certain portion of the fatigue life to an extent which increases with increasing level of torsion loading. In quasistatic and very low-cycle fatigue regions, the cracks can propagate under a dominant shear mode until the final fracture. In these cases, an intensive cyclic deformation within a large plastic zone (embracing many grains and microstructural elements) quickly produces a network of microcracks at grain boundaries or secondary-phase particles. This damage zone is extended ahead of the crack front rather coplanar with the notch plane (e.g., [358]). A coalescence of these microcracks and their shear interconnections with the main crack front keep the crack path in the plane perpendicular to the specimen axis. It is to be expected that the coalescence is controlled predominantly by local mode II shearing processes already described in the frame of the particle-assisted model of mode III crack growth (Figure 3.43). In the high-cycle region, on the other hand, the plastic zone size is comparable to the characteristic microstructural distance. Therefore, the microstructurally-induced tortuosity of the crack front soon causes the initial shear-mode propagation to transfer into the zig-zag mode I growth by formation of factory roofs (e.g., [359, 360]).

In the next subsections, the results of fractographical studies of both smooth and notched specimens made of low-alloyed high-strength steel are presented [348]. Moreover, a theoretical model of factory roof formation is briefly described according to a more extended analysis reported in [361].

3.3.2.1 Smooth Specimens

Initiation and propagation of fatigue surface cracks in smooth cylindrical specimens made of a low-alloy, Cr-Al-Mo steel BS 970/1-83 ($\sigma_y = 850\,\text{MPa}$), were investigated by means of optical and scanning electron microscopes. Pure torsion fatigue tests (number of cycles to fracture $N_f \in (10^4, 10^6)$ cycles) were interrupted after defined numbers of cycles, the specimens were statically fractured in liquid nitrogen and their fracture surfaces were analyzed in SEM.

A formation of a network of microcracks macroscopically perpendicular and parallel to the specimen axis was detected during the first loading stage. This initiation stage was followed by growth of the perpendicular microcracks preferentially in the mode II along the surface and their coalescence.

A typical shape of microcracks that macroscopically lie in the plane perpendicular to the specimen axis is shown in Figure 3.44. However, the real crack plane related to stage I is inclined to the macroscopic plane at an angle of 45°. The crack front at the end of the stage I exhibits, similarly to the scheme in Figure 3.42, a microscopically rough zig-zag geometry. The depth of stage I cracks was in the range of 10–30 μm. The plane of the stage II crack is inclined at 50° to the opposite direction and twisted at 20° so that it gets an additional mode I support. As a rule, the crack fronts at the end of stage II were smoother than those of stage I. All such cracks were rather

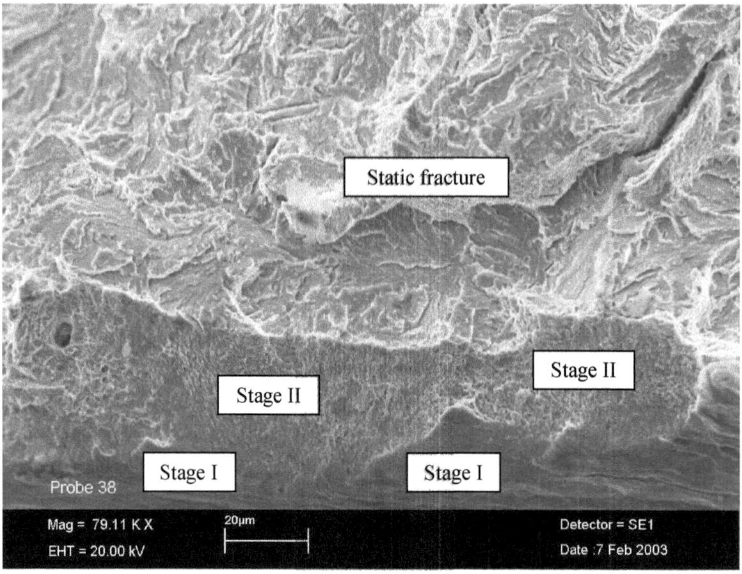

Figure 3.44 A typical shape of a surface microcrack on the fracture surface macroscopically perpendicular to the specimen axis. Both the stage I and stage II planes are inclined at about 45° to the macroscopical fracture plane. Reprinted with permission from John Wiley & Sons, Inc. (see page 265)

slowly propagating in a mode I+III to a depth of nearly 100 μm while simultaneously growing and coalescing along the specimen surface under modes II or I+II. As a result, a shallow circumferential macrocrack has developed round the whole specimen as depicted in Figure 3.45. A much higher growth rate of a mode II or I+II crack front segments in comparison with those of the mode I+III can, however, be partially attributed to the interaction and coalescence of surface microcracks. In some cases, however, long mode I branches were also observed leading to a deep and extremely tortuous surface macrocrack propagating into the interior of the specimen along 45° planes of the maximum tensile stress.

In general, the fractographical analysis revealed that:

1. mode III crack growth was always supported by a mode I component due to the propagation in planes of maximum tensile stress (short stage I cracks) or even with additional twisting of the crack plane (stage II cracks);
2. the mode II (or I+II) crack growth rate was much higher than that of the mode I+III which often resulted in a continuous circumferential crack.

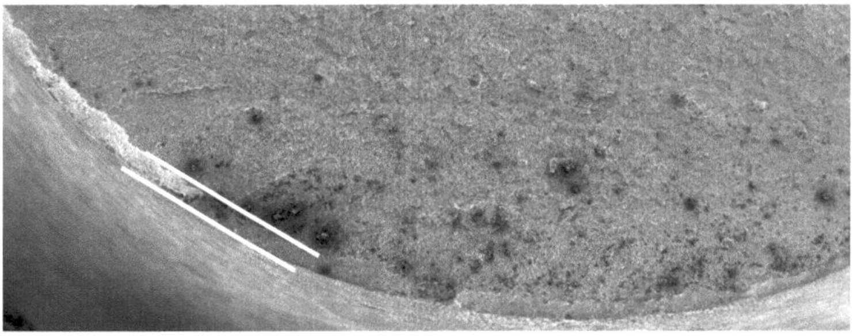

Figure 3.45 SEM picture of a circumferential crack developed after coalescence of surface microcracks in modes II or I +II (depth about 100 μm). White lines mark a segment of the continuous fatigue crack. Reprinted with permission from John Wiley & Sons, Inc. (see page 265)

3.3.2.2 Notched Specimens

A typical way in which the cracks emanating from circumferential notches of cylindrical specimens further propagate into the specimen interior under torsional loading is by formation of factory roofs. The factory roof (F-R) is one of the most extraordinary fractographical patterns ever observed in fatigue and fracture of metallic materials. The roughness (or visibility) of F-R particularly depends on the applied cyclic shear stress amplitude though a significant influence of both the material microstructure and the material yield strength was also observed (e.g., [349, 353, 356, 357, 362]). In spite of the fact that first reports on the F-R appeared in the early 1950s, their formation mechanism was only just beginning to be understood in 2006 when the experimental work of Matake *et al.* [363] appeared. This study revealed that there are three stages of F-R formation: (1) initiation and growth of surface semi-elliptical microcracks under shear loading modes II+III, (2) their interaction, coalescence and growth in the local mixed-mode I+II+III by forming mode I branches (tilted and twisted segments) and (3) growth of the periodic main crack under the prevalent mode I loading. The theoretical work [364] was focused on the problem of friction and shielding phenomena associated with a simple saw-tooth model of F-R patterns. However, neither a detailed geometry of F-R nor any quantitative rules of their formation were reported. Therefore, many principal questions concerning the phenomenon of F-R still remained unsolved:

1. What is the characteristic 3D picture of the F-R?
2. Which physically based relationships control the initiation and growth of the F-R?
3. What is the kinetics of the F-R formation?
4. Why the F-Rs are not observed in the region of a very low cycle fatigue?
5. Are the F-Rs observed only in the case of torsional loading?

Answers to these questions are the main subject of the paper [361]. Some of the most important results are mentioned hereafter.

3D Topography of Factory Roof

The three-dimensional micromorphology of F-R was investigated by stereophotogrammetry and optical chromatography. The fracture surfaces of V-notched cylindrical specimens made of a high-strength low-alloy steel were generated by a reversed torsion loading. Three applied values of the torsion moment $M_{t1} = 13\,\mathrm{Nm}$, $M_{t2} = 17.9\,\mathrm{Nm}$ and $M_{t3} = 22.7\,\mathrm{Nm}$ led to fatigue lives of $N_{f1} = 8.31 \times 10^5, N_{f2} = 2.44 \times 10^5$ and $N_{f3} = 1.58 \times 10^4$ cycles, respectively. Unlike in the latter case (low-cycle fatigue), distinct F-R patterns were observed in both high-cycle regimes. The macrophotograph of the fracture surface with highlighted region of investigated F-R patterns is shown in Figure 3.46.

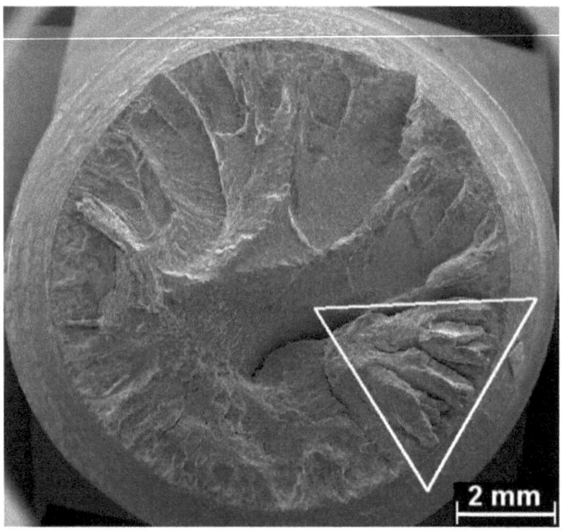

Figure 3.46 Fracture surface with highlighted region of investigated factory-roof patterns ($M_t = 17.9\,\mathrm{Nm}$, $N_f = 2.44 \times 10^5$ cycles)

Figure 3.46 documents not only a complexity but also a certain regularity of F-R patterns. Indeed, the lamellar-like F-R structure consists of nearly parallel elongated "mountain-like" massifs (segments) joined by rather narrow valleys with secondary cracks. Near the surface, the massif is usually split into two smaller segments which constitute the initial stage of the F-R formation. Note that there are several F-R patterns which significantly differ in their size: The smaller the F-R patterns the finer is their lamellar structure.

The profiles of F-R topology in nearly tangential directions are plotted in Figure 3.47 as obtained by stereophotogrammetry. The profiles well document the general geometrical features: the slopes (hillsides) of the embryonic

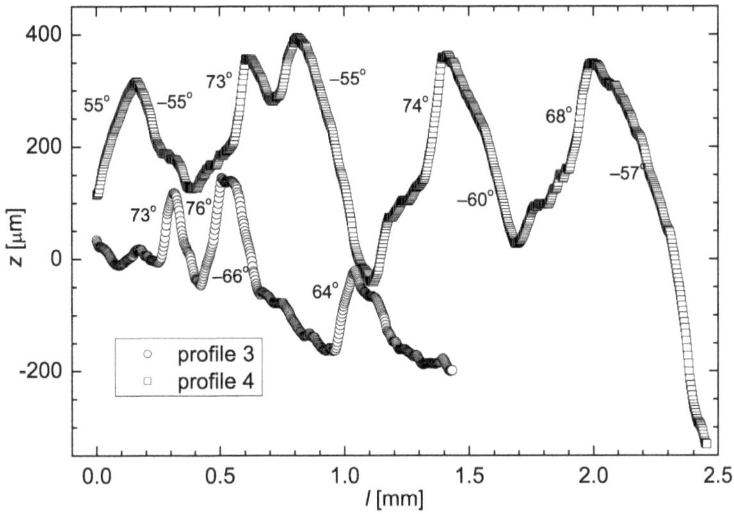

Figure 3.47 Topological profiles of F-R patterns in tangential direction

triangle-like segments gradually decrease starting from an initial angle α of 65–77° near the surface (profile 3), continuing to somewhat lower angles of 60–70° near their conjunction to one main segment (profile 4) to finally reach 40–60° at the top of the main ridge and near the centre of the fracture surface.

A repeated contact of fracture surfaces and the related bending loading initiates secondary cracks in the valleys and also contributes to a further mode I propagation of the main F-R crack inside the specimen. Therefore, numerous wear traces (fibrous patterns or tire tracks) could be found on the SEM pictures of the fracture surface.

Initiation of Factory-roof Patterns

Mixed-mode II+III exists at all points of the semi-elliptical crack front except for two points on the surface (pure mode II) and one point at its centre (pure mode III). However, nearly straight crack fronts of such semi-elliptical cracks change to a highly tortuous profile by inclinations towards a mode I loading rather soon. This change is accomplished by tilting (branching) and twisting of the crack front segments. There are several possible reasons for such behaviour that are particularly related to the crack advance in the radial direction. Indeed, the notch stress concentration decreases and, moreover, there is also a very limited ability of a pure mode III segment to propagate in that direction [348]. Interactions of the crack front with microstructural barriers are accompanied by an increase in the roughness of crack flanks. This leads to an increase in the friction stress (shear closure) and to the reduction of the mode II+III crack driving force.

In order to get a maximum support of the opening mode I to avoid the above-mentioned problems, mode II segments rotate around the axis parallel to the crack front. Such a rotation is relatively easy even for large tilted segments since their planes intersect the main crack plane along the line (curve). On the other hand, twisting around the axis perpendicular to the crack front provides mode III crack segments with mode I support. Since the planes of the twisted elements and the main crack intersect only at a single point, the size of the twisted elements is very limited and the twisting can occur only on microscopic ledges at the main crack front. This means that the formation of mode I branches at the mode II crack front is easier than that at the mode III crack front.

Thus, the F-R formation starts by the creation of mode I branches at particular sites along the elliptical mixed-mode II+III front. A detailed mathematical analysis of mode I branching based on LEFM is given in [361]. The main aim of this theoretical analysis was to predict the most probable sites of mode I branching at the semi-elliptical crack front. These sites are considered to be associated with branches of a pure local mode I stress intensity factor K_I at the semi-elliptical crack front. The angles X and Θ of such branches respectively define the twist and the tilt of the branch with respect to the semi-elliptical crack front as shown in Figure 3.48. Simultaneously, these branches were found to be loaded by a maximal effective stress intensity factor $K_{eff} = (K_I^2 + K_{II}^2 + (1-\nu)^{-1}K_{III}^2)^{1/2}$. The polar angle φ defines the position of the branch at the semi-elliptical crack front as also depicted in Figure 3.48.

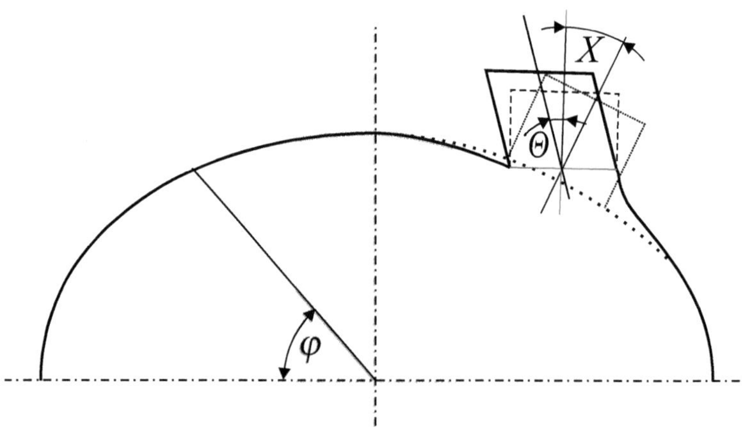

Figure 3.48 Scheme of a branched element at the semi-elliptical crack front

The analysis revealed that the tilt angle $\Theta = 71.6°$ keeps the same value within the whole range of $\varphi \in \langle 0°, 180° \rangle$ and the twist angle $X = 19.5°$ remains constant within the range $\varphi \in \langle 0°, 54° \rangle$ and $\varphi \in \langle 126°, 180° \rangle$. Starting from the critical polar angle $\varphi_c = 54°$, the twist angle X decreases to zero which corresponds to $\varphi = 90°$. This result holds well for all semi-elliptical cracks exhibiting aspect ratios in the range $b/a \in (0.6, 0.8)$ which contains a great majority of the experimentally observed semi-elliptical cracks. Only the critical polar angle changes in the range $\varphi_c \in \langle 45°, 60° \rangle$.

It should be emphasized that the superposition of tilting and twisting raises the values of $K_{I,\max}$ (or $K_{eff,\max}$) for the mode I branch significantly above the original K_{eff} values for the semi-elliptical crack front just before the initiation of the mode I branch. This very important result can be understood in terms of a synergy effect of both the mode II and the mode III loading on the creation of the mode I branch. For all the semi-elliptical cracks, the branches of maximal values of $K_{I,\max}$ (or $K_{eff,\max}$) lie in the ranges $\varphi_{\max} \in$ (20°, 30°) and (150°, 160°). This ranges determines two segments on the semi-elliptical crack front that correspond to the maximal probability of the creation of the F-R nuclei (mode I branches).

The initiation and further propagation of F-R nuclei is conditioned by exceeding the mode I threshold for the HSLA steel on the mode I branches. For cracks with $b/a \in (0.6, 0.8)$, the computed maximal values of K_I are in the range $K_{\max} \in (6.4, 7.1)\,\text{MPa}\,\text{m}^{1/2}$. These values are higher than the threshold amplitude $\Delta K_{th}/2 = 4.6\,\text{MPa}\,\text{m}^{1/2}$ but still sufficiently close. This result, along with the good prediction of initiation sites, confirms the plausibility of the theoretical approach. Moreover, it can be used to define generalized conditions of mode I branching (kinking) from the shear-mode propagation:

1. the first branch forms at that site of the crack front where the value of ΔK_I on its facet would be maximal;
2. the branching appears at the moment when this maximal ΔK_I-value exceeds that of the threshold ΔK_{Ith} related to the applied cyclic ratio.

The geometrical and microstructural parameters related to both the crack front and the crack wake (the level of the roughness-induced closure) of the mode I branch might be somewhat different from those of a crack in the standard specimen used for the determination of ΔK_{Ith}. This difference could be a reason for some deviations from the rule (2). Nevertheless, the above-mentioned conditions give a general frame to a quantitative understanding of the transition from the shear-mode propagation to that under the opening mode.

Model of Factory-roof Formation

On the basis of the above-mentioned results, the kinetics of F-R formation can be qualitatively assessed. Individual stages of F-R formation in terms of gradual positions of F-R crack front are shown in Figure 3.49, where a 3D image of the investigated F-R is depicted according to the stereophotogrammetric reconstruction. The process starts with the creation of mode I branches at well defined sites of semi-circular cracks. The positions of mode I branches corresponding to the ranges $\varphi_{\max} \in (20°,\ 30°)$ and $(150°,\ 160°)$ are marked on the crack fronts of all deduced semi-elliptical cracks. One can see that there is a good correspondence with practically all the real initiation sites of F-R nuclei. In the cases of considerable size difference of neighbouring semi-elliptical cracks, the mode I branch appears first on the larger semi-ellipse. When a large semi-ellipse is adjacent to a small one, a short stage of a backward growth towards both the critical site at the smaller semi-ellipse and the remaining semi-elliptical crack front may appear due to their mutual interaction. This accelerates both the initiation of the second mode I branch and the coalescence process.

After the coalescence, the initial F-R crack front consists of the local spatial ledges and branches that form the embryonic massifs connected by the remaining fronts of semi-elliptical cracks. The contact bending loading on these massifs starts to produce the secondary cracks adjacent to the nuclei and, later on, to the main valleys (or hilltops on the mating fracture surface). These mode I cracks spread in planes inclined at 45° to the macroscopic plane of the maximum shear stress, and eventually approach the advancing F-R crack front.

In order to reduce the line tension of such a tortuous front, the embryonic massifs expand in both the radial direction (inside the specimen bulk) and the tangential direction (along the semi-elliptical fronts). In this way, both the width and the height of the nuclei increase while forming the local U-shaped valleys of decreasing width. Finally, the embryonic segments link-up to form the main massifs and to terminate the local valleys. From that moment on, the main front of a nearly saw-tooth profile propagates further in the direction of a maximum increase of the crack driving force, i.e., more or less in the radial direction. For simple geometrical reasons, the F-R crack front has to contract during propagation towards the specimen centre. Consequently, the main massifs are brought mutually closer and their heights and widths decrease. The extinction of F-R patterns near the specimen centre precedes the final fracture of the specimen. A geometrical model of F-R patterns [361] also revealed that both the height h and the width w of the F-R patterns decrease with decreasing distance (or increasing density) of individual segments.

When considering all the presented theoretical and experimental results, the first three basic questions seem to be answered in a satisfactory manner. The last two questions can also be answered in a rather simple way. The density of initiated semi-elliptical cracks increases with decreasing number of cycles to failure (increasing applied stress range). For the above-mentioned geometrical reasons, the higher the density, the lower the size of the F-R patterns. In the very low-cycle and quasistatic regions, moreover, the microcrack coalescence in the damage zone start to dominate the fracture process. Both these facts mean that the F-R practically vanishes when approaching

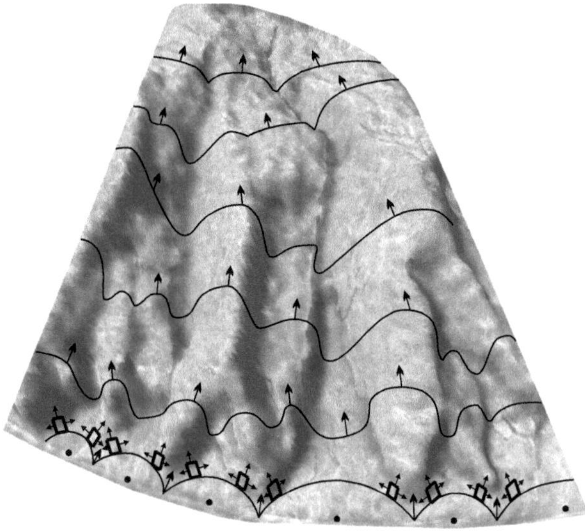

Figure 3.49 The 3D model of the F-R formation

these regions. It should be noted that some authors emphasize another reason for the vanishing of F-R: the shear displacements in the low-cycle regime become very large and the related strong wear can destroy the F-R morphology. However, the relevancy of this reason is substantially weakened by the theoretical analysis performed by Vaziri and Nayeb-Hashemi [359] that predicts sliding of F-R walls rather than their abrasive wear in the low-cycle region (see Section 3.3.3 for more details).

Let us finally remark that the biaxial stress state induced by torsion is not the only kind of loading that produces the F-R patterns. Indeed, these patterns are also developing under pure shear loading, i.e., under a uniaxial stress state [365] (see also the Section 3.3.3).

3.3.3 Propagation of Cracks under Cyclic Shear

An investigation of shear-mode cracks in the case of torsional loading is very difficult and ineffective especially in the high-cycle fatigue region. Indeed, there is only a very short initial period of shear mode crack growth during which almost the whole crack front grows either in the mixed-mode I+II+III (smooth specimens) or II+III (notched specimens). Therefore, it is practically impossible to study the pure modes II and III separately in order to distinguish their growth mechanisms. Under this kind of loading, moreover, an appropriate identification of threshold values ΔK_{IIth} and ΔK_{IIIth} is also impracticable. However, good experiments enabling the pure mode II or mode

III crack tip loading are very difficult to arrange. One of the main problems is to avoid a parasitic mode I vibrational loading. To our knowledge, the first experimental device of this kind was based on four-point bending of angular beams [366, 367]. However, pure mode II or mode III loadings could be realized only in the very centre of the beams so that the extent of the related crack growth was very limited. Diverse experimental arrangements were employed in [365, 368]. The basic idea was to realize simultaneous pure mode II and mode III crack propagations in one specimen. Two different devices were utilized for low-cycle and high-cycle fatigue experiments. Hereafter, a brief description of these experiments and obtained results are presented.

3.3.3.1 Low-cycle Fatigue

Experimental Arrangement

A loading scheme of a special cylindrical specimen utilized in the experiments is introduced in Figure 3.50. Two very sharp circumferential notches were machined by a lathe tool and additionally subjected to a compressive loading in order to produce a crack-like notch. Then, the specimen was annealed in order to remove residual stresses at the crack tip induced by compressive loading.

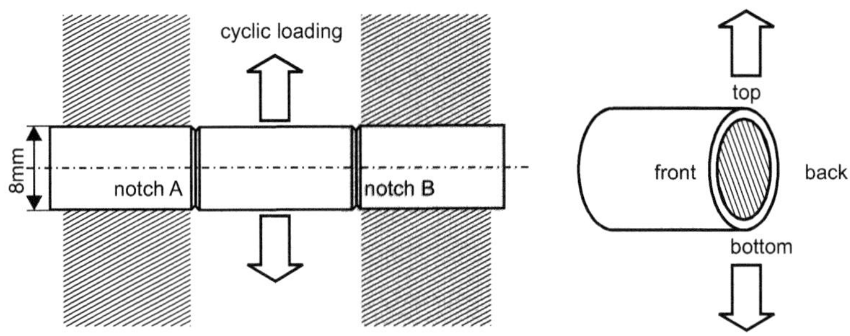

Figure 3.50 The loading scheme of the two-notch specimen. Reprinted with permission from Elsevier B.V. (see page 265)

Specimens were then placed into a fixed rigid holder in the machine, where both side parts of the specimen could be gripped tightly. The whole middle part of the specimen (in between both circumferential notches) was gripped by a moving part of the machine and loaded strictly uniaxial, in tension and compression. The clamping device was stabilized in order to prevent any vibrations and movements out of the tensile axis. A more detailed description of the testing device can be found in [368]. In this way, pure shear loading was transferred to the crack-like notches: At both the top and the bottom sites of two round notches a pure mode II loading was applied, whereas the cracks starting from both middle-site segments were subjected to a pure mode III loading. Thus, four sites of both mode II and mode III growing cracks could be analyzed in a single specimen. Other sites along the notch root experienced a mixed-mode II+III loading.

The constant applied cyclic displacement produced a large-scale cyclic yielding in the ligament that corresponded to a displacement of the crack flaws of the order of a few microns. Such a loading produced crack advance of several hundreds microns during about a hundred of cycles applied in the experiments.

An austenitic steel X5NiCrTi26-15, used e.g., in the aviation industry or as turbine blade material (yield strength of 600 MPa), was selected as an experimental material. This material was nearly free of inclusions or precipitates so that the particle-induced mechanism of mode III propagation (Figure 3.51) was not expected to be dominant.

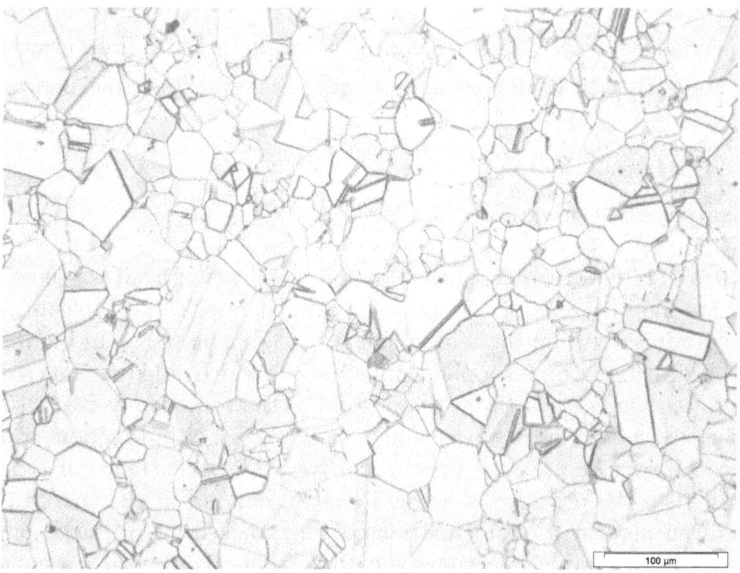

Figure 3.51 The microstructure of the austenitic steel along the crack growth plain. Reprinted with permission from Elsevier B.V. (see page 265)

Crack Growth Data

Two specimens were fatigued by 200 cycles and one specimen by 100 cycles. After the fatigue tests the specimens were broken by cyclic tensile loading. Fractographic observation permitted the determination of the crack extension caused by mode II and mode III cyclic loading, which was clearly distinguishable from the crack propagation caused by the subsequent cyclic tensile loading. The fatigue crack advance under macroscopically pure modes II and III was measured at appropriate sites of fracture surfaces along short crack-front segments of 100 μm. In spite of a relatively high scatter of individual measured fatigue crack lengths, a clear difference in averaged remote mode II and mode III final crack lengths was identified. In the case of 200 applied cycles, the averaged length of 420 μm corresponded to mode II but only 200 μm was determined for mode III. The respective lengths of 210 μm (mode II) and 120 μm (mode III) were measured in the case of 100 applied cycles. Since the fatigue crack path was only a tenth of the total crack length, a nearly constant crack growth rate could be assumed in individual tests. Thus, the remote mode II crack growth rate was approximately twice higher than that of the mode III. It should be emphasized that the numerical analysis for the circumferential crack subjected to pure remote shear revealed that the value of CTOD_{III} was more than 1.4 times higher than that of CTOD_{II} in the small-scale yielding case for the same applied strain range [369]. By assuming the dependence $\frac{\mathrm{d}a}{\mathrm{d}N} \propto \mathrm{CTOD}^2$ this means that, in fact, the crack growth rate under mode III must have been about five times lower than that under mode II.

Fractographical Observations

A detailed 2D fractographical investigation by means of SEM was performed in order to identify real local fracture modes. Examples of a typical fracture surface produced during the remote mode II loading is shown in Figure 3.52. Here the directions of both the applied shear stress and the macroscopic crack growth are vertical (from the bottom to the top). Many facets covered by striations, mostly nearly perpendicular to the growth direction, could be observed all over the fatigue fracture surface. The occurrence of striations was related to the presence of a small opening mode I at the crack tip that was induced here by local inclinations of the crack front from the plane of maximum shear strain. Sometimes the crack front advanced in pure mode II as can be easily deduced from tire tracks typical for shear mode presence that could be identified in many parts of the fracture surface (see Figure 3.52). The direction of these periodic patterns indicates the expected propagation of the shear crack front from bottom to top.

Many secondary cracks, mostly perpendicular to the growth direction, were also found all over the fracture surface [368]. These cracks are considered to be preferentially created at the corners of asperities left in the crack wake which, under the applied mode II shear, are loaded by cyclic bending.

All observed morphological features confirm that pure mode II and combined mode I+II are the dominating microscopical fracture mechanisms caus-

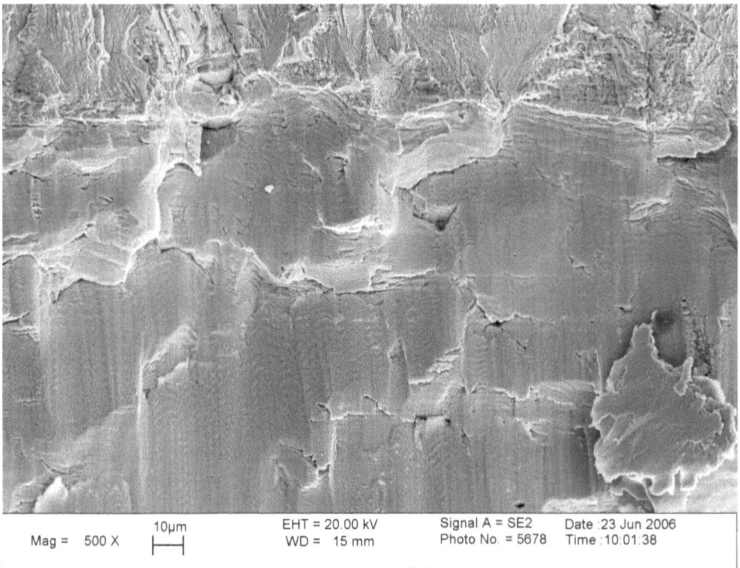

Figure 3.52 Tire tracks and striations (*at the top*) on the remote mode II fracture surface

ing rather straightforward advance of the crack front under remote mode II loading.

A typical example of fracture surfaces produced under remote mode III is displayed in Figure 3.53. Here the direction of the applied shear stress is horizontal. The fracture morphology was found to be completely different from that created by the remote mode II. Practically all striation fields confirmed the dominance of mode II or II+III propagation in accordance with the models in Figures 3.42 and 3.43, often supported by mode I due to kinking and twisting of the crack plane. Indeed, a prevalent direction of striations is nearly vertical which means that the crack front propagated horizontally, i.e., under local mode II. A careful examination of the mode III fractographs uncovered some regions looking like mode I+III crack propagation, where the crack formed striations parallel to the external applied mode III loading. Due to the crystallographic nature of plasticity, the macroscopic shear displacement parallel to the crack front is expected to be always locally associated with small mode I and mode II displacements in addition to the mode III ones. Indeed, there must practically always be a certain deviation of the Burgers vectors from the crack front direction.

The secondary cracks generated under remote mode III are more frequent and noticeable than those under remote mode II loading (see Figure 3.53). These cracks, mostly parallel to the growth direction, were created preferentially as corner cracks at the crack-wake asperities bent by external mode III shear. Coalescence of such cracks causes crumbling of asperities and falling

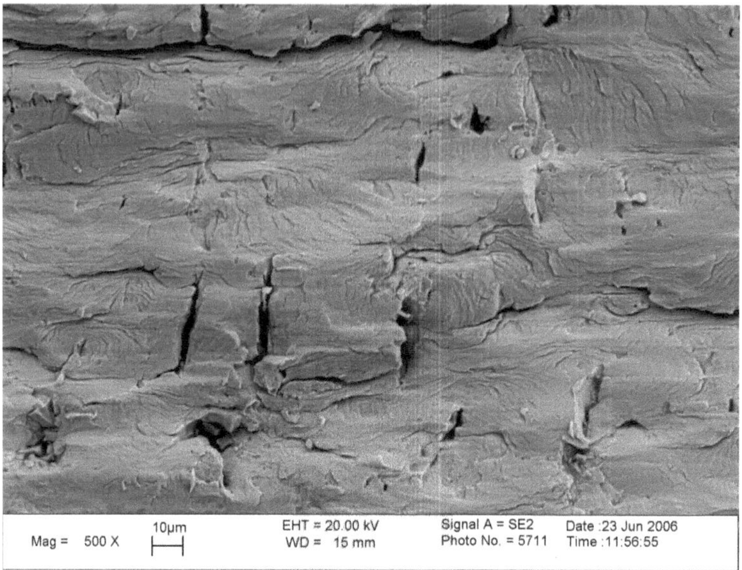

Figure 3.53 Striations on the remote mode III fracture surface revealing the dominance of local mode II or I+II crack propagation mechanisms. Reprinted with permission from Elsevier B.V. (see page 265)

of the fracture surface. As a result, large voids are left in between mating surfaces as was previously detected by Slámečka and Pokluda [370] by using stereophotogrammetrical methods after torsional fatigue fractures. Secondary cracks also sporadically initiated at particles of secondary phases near the main crack front. These cracks then propagated as mode II or I+II cracks in between the particles along the main crack front – see Figure 3.53 (left). Indeed, the striations emanating from the cracked particles indicate a horizontal direction of the crack front propagation in accordance with the model in Figure 3.43.

In summary, the fracture morphologies generated by remote mode II and mode III loadings are significantly different. However, the observed morphological features revealed that the dominating growth mechanisms in both cases are similar: pure mode II and the combined mode I+II. This is in agreement with the models in Figures 3.42 and 3.43. Some indication of the local mode I+III crack propagation under the remote mode III loading was also found. Thus, the local fracture modes caused a complicated zig-zag advance of the crack front unlike the rather straightforward growth under the remote mode II. As a consequence, the crack front propagation rate under remote mode III was significantly lower than that in the remote mode II case.

3.3.3.2 High-cycle Fatigue

Experimental Arrangements

In the high-cycle regime, the shear displacements are much lower than those in low-cycle fatigue and, consequently, small parasitic mode I vibrations and bendings could play a much more important role. Therefore, two other original testing setups (cells) with a higher stiffness were designed and utilized to assure a pure remote shear loading. The loading scheme of the first cell is depicted in Figure 3.54. The construction of the specimen holder and its orientation with respect to the loading axis provided pure mode II loading at the "top" and "bottom" sites of the specimen and pure mode III loading at the "front" and "back" sites. At all other points along the crack front mixed-mode II+III was applied. A circumferential V-notch was machined by a lathe tool at the specimen mid-length and a pre-crack was introduced by a blade mechanism. Finally the specimens were compressed by 20 kN to sharpen the pre-crack and subsequently annealed. The resulting pre-crack is depicted in Figure 3.55. Specimens made of the ferritic steel (0.01 %C) with outer diameter of 8 mm and inner diameter of 4 mm were loaded by different ranges of the nominal ligament shear stress τ_n (the cyclic ratio R = 0.1) After the shear mode tests, the specimens were rapidly fractured in liquid nitrogen. Specimens made of austenitic steel X5NiCrTi26-15 were also tested by means of this experimental set-up. These specimens were fractured by cyclic tensile loading.

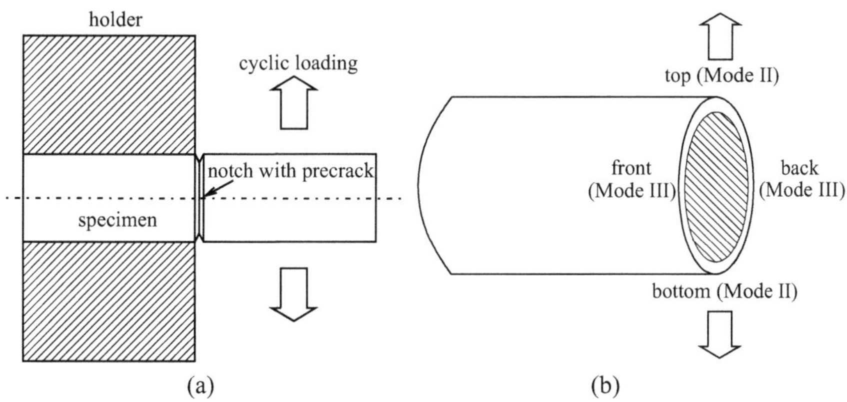

Figure 3.54 The experimental setup: (a) the loading scheme, and (b) the loading modes operating at different specimen sites

The second special cell for loading specimens made of austenitic steel was manufactured to enable higher loadings (see Figure 3.56). The pre-crack was introduced by means of a compressive load of 200 kN. After the shear mode tests, the specimens were also fractured by cyclic tensile loading.

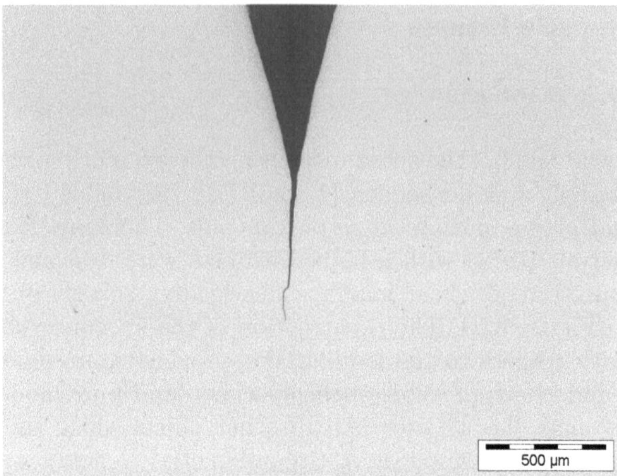

Figure 3.55 The shape of the pre-crack after compression

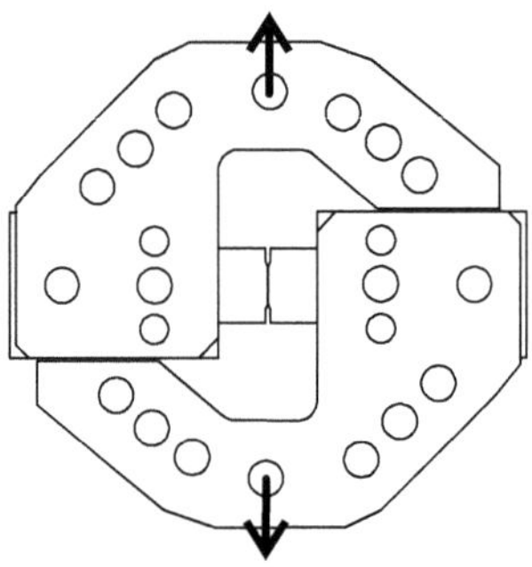

Figure 3.56 The scheme of the loading cell for higher loading forces

Numerical K-calibration of Specimens

In order to determine the mode II and mode III stress intensity factors at the crack tip, a numerical analysis was performed by means of the ANSYS code. Although the loaded specimen was modelled as rotationally symmetric, a full linear–elastic 3D solution had to be used owing to a different symmetry of the loading. In the first step, the stress-strain field along the crack front loaded by the remote shear stress of 180 MPa was determined by utilizing a rough finite-element network. This field was used to create a ring-like submodel with a very fine finite-element network that embraced only the pre-crack region. The submodel surface was loaded by the rimstrains computed for the same surface on the rough model.

In this way, very precise values of the stress intensity factors K_I, K_{II} and K_{III} could be determined for many points along the circular crack front. Mutual shear displacements of crack flanks were calculated at four points

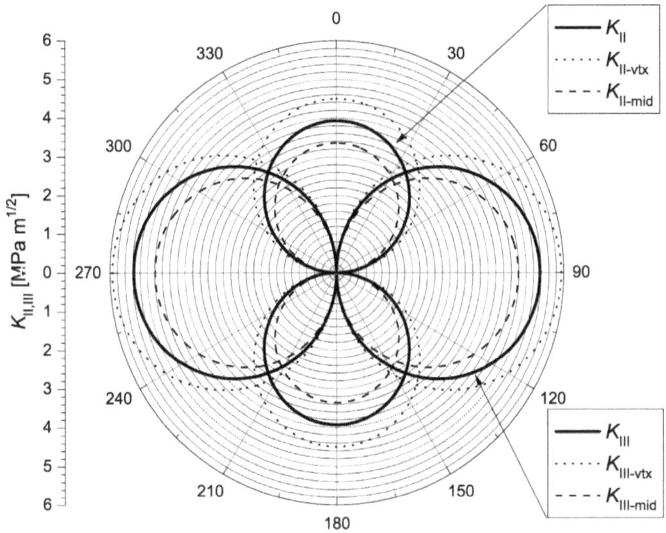

Figure 3.57 Values of K_{II} and K_{III} along the crack front in polar coordinates. The *dashed lines* correspond to the middle points of the finite elements, the *dotted lines* are related to their vertexes and the *full lines* represent the average values

near the crack front. The values of K_I, K_{II} and K_{III} were determined by an extrapolation to the crack front for all applied nominal shear stress ranges. The ratio of maximal values in pure shear modes II and III was found to be $K_{III\,\max}/K_{II\,\max} = 1.37$ while the values of $K_{I\,\max}$ were negligible (for more results see [365]). This can be seen from the polar diagram of maximal values of $K_{III\,\max}$ and $K_{II\,\max}$ in the loading cycle (for $\tau_n = 180\,\mathrm{MPa}$) that is plotted in Figure 3.57. It should be emphasized that such determined numerical values of $K_{III\,\max}$ were found to be in excellent agreement with calculations performed using the asymptotic method [371]. Let us mention that recent results of elasto-plastic analysis have shown that the ratio $\mathrm{CTOD}_{III\,\max}/\mathrm{CTOD}_{II\,\max}$ in the large-scale yielding case can be even higher than 1.4 [369].

Crack Growth Data

The spatial shear crack path was determined by an SEM identification of the fracture surface morphology in selected rectangular regions corresponding to pure mode II and mode III loading of both austenitic and ferritic steels. The fracture morphology of pure mode II and III shear cracks is shown in Figure 3.58 for the austenitic steel. The areas corresponding to the pre-crack, the shear crack propagation and the final tensile fracture are marked as well. Practically all the mode II shear cracks were globally inclined from the shear plane. Averaged deflection angles were found to be 60±16° (austenite) and 53±15° (ferrite). This means that just after the start of shear mode cycling, the mode II cracks branched to the opening mode I. In terms of the previously defined branching condition, the value of $\Delta K_{Ith} \approx 5.8$ for the austenitic steel at $R = 0.1$ must have already been exceeded at the mode I branches. Indeed, the crack flanks near the pre-crack tip are microscopically tortuous, as can be seen in Figure 3.55. Consequently, a bending moment at interlocked asperities has, most probably, produced an additional local mode I loading at the pre-crack tip. The fracture morphology of mode III cracks typically consisted of factory-roof patterns (Figure 3.59).

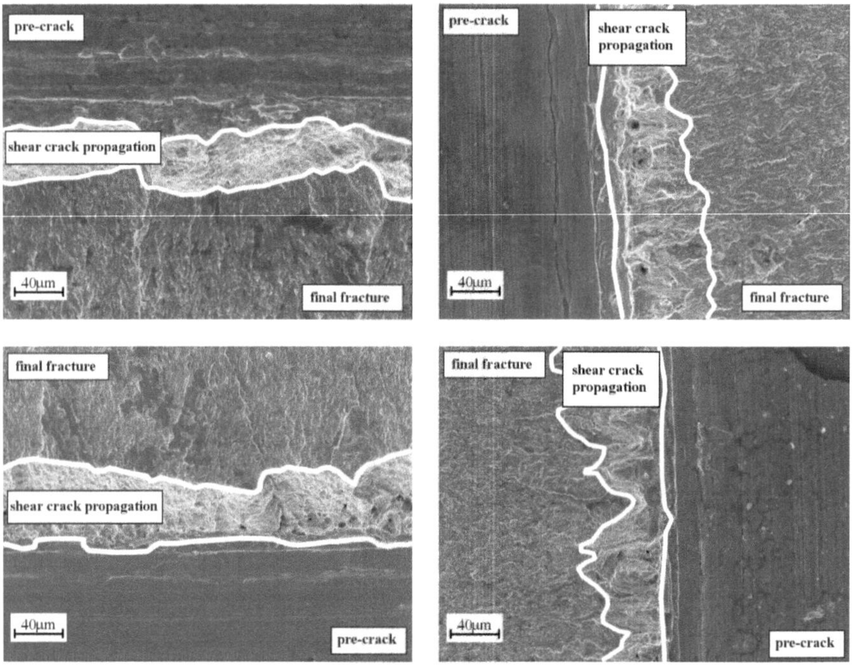

Figure 3.58 SEM pictures of fracture surfaces showing propagation of shear cracks. The *left* (*right*) pictures correspond to remote mode II (mode III) cracks

Profiles of fracture micromorphology in the direction of applied stress for both mode II and mode III crack propagation stages are depicted in Figure 3.60. As expected, the crack path in the remote mode III case clearly reveals the formation of factory roofs. The related interlocking of crack flanks caused by the F-R asperities must lead to a dramatic reduction of the crack tip driving force especially in the near-threshold crack growth region [359]. This is schematically shown in Figure 3.61 where the interlocked F-R asperities are associated with a low effective crack driving force $\Delta K_{eff,III}$ for low applied values of τ. When the applied shear stress or ΔK_{III} becomes sufficiently

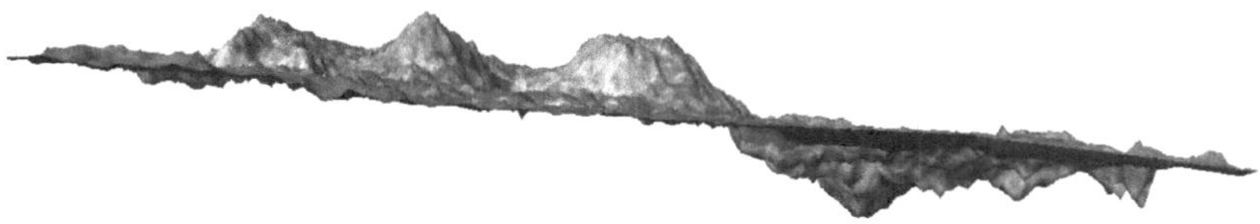

Figure 3.59 3D picture of the factory-roof morphology created by a pure mode III shear (viewed parallel to the fracture surface)

high (or the FR patterns sufficiently small) the sliding of fracture surfaces in the crack wake rapidly becomes dominant. This is accompanied by a sudden increase in the effective driving force in the near-fracture region of the crack growth rate diagram (or in the low cycle region). On the other hand, the microrelief of mode II crack growth is very smooth so that the effective driving force is high. Consequently, the crack growth rate under mode II is expected to be significantly higher than that under mode III.

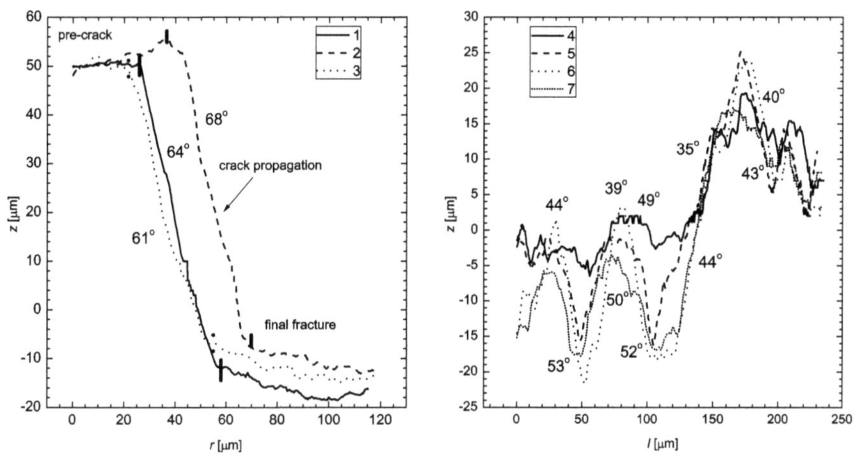

Figure 3.60 Profiles of fracture morphology in the direction of applied stress corresponding to mode II (*left*) and mode III (*right*) crack propagation. The smooth profiles 1, 2 and 3 show propagation paths of three elements of the mode II crack-front (extended in the tangential direction) along the radial direction r towards the specimen centre. The profiles 4, 5, 6 and 7 correspond to consecutive propagation stages of the mode III crack front extended in the tangential direction l

Since the length of the shear mode cracks was an order lower than that of the pre-crack, a nearly constant crack growth rate during the shear propagation could be assumed. Therefore, the crack growth rate was calculated simply by dividing the total length of shear cracks by the corresponding num-

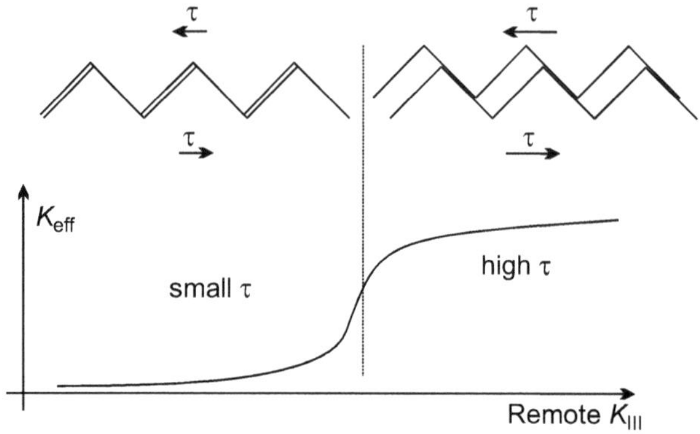

Figure 3.61 The scheme of the interlocked factory-roof asperities in the case of a small applied shear stress (*the left part*, wedging) and a high applied shear stress (*the right part*, sliding)

bers of cycles. The near-threshold crack growth curves for both the mode II and the mode III propagation in the austenitic steel are plotted in Figure 3.62. The related regression curves follow the Klesnil–Lukas relationship

$$\frac{\mathrm{d}a}{\mathrm{d}N} = A\left(\Delta K^n - \Delta K_{th}^n\right),$$

where A and n are material constants [289]. While the exponents $n_{II} = 5.6$ and $n_{III} = 5.1$ are nearly identical, the constants $A_{II} = 1.131$ and $A_{III} = 2.721$ [MPa m$^{1/2}$, m] are significantly different as well as the calculated thresholds $\Delta K_{IIth} = 3.5\,\mathrm{MPa\,m}^{1/2}$ and $\Delta K_{IIIth} = 4.7\,\mathrm{MPa\,m}^{1/2}$.

According to a general condition of mode I branching defined in Section 3.3.2, this effect can appear only when the mode I threshold value is exceeded at the mode I branch. With the help of the local bending moment induced by the crack-wake asperities near the crack tip, the value $\Delta K_{Ith} \approx 6\,\mathrm{MPa\,m}^{1/2}$ for the austenitic steel ($R = 0$) [372] could already be exceeded by applying ΔK_{II} values very close to the remote mode II threshold $\Delta K_{IIth} \approx 3.7\,\mathrm{MPa\,m}^{1/2}$. One can also clearly see that, for the same value of the applied stress intensity range, the crack growth rates for the mode II loading are about six times higher than those for the mode III loading. This is in agreement with results achieved in the low-cycle fatigue region and confirms the diversity of the mechanisms of mode II and mode III crack propagation in metals within the whole range of the fatigue life.

Also in the case of the ferritic steel, the threshold values $\Delta K_{IIth} = 1.2\,\mathrm{MPa\,m}^{1/2}$ and $\Delta K_{IIIth} \approx 2.0\,\mathrm{MPa\,m}^{1/2}$ were found to be different. This is documented in Figure 3.63, although further experiments must be performed in order to obtain more precise results.

To prevent mode I branching in order to observe a sufficiently long mode II and III crack growth near the threshold, the pre-crack flanks should be very smooth, particularly in the close vicinity of the pre-crack tip. This might be

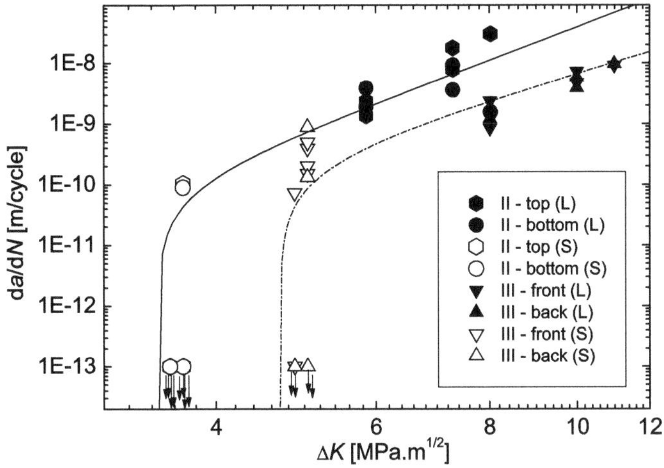

Figure 3.62 The crack growth curves for the austenitic steel in the near-threshold region

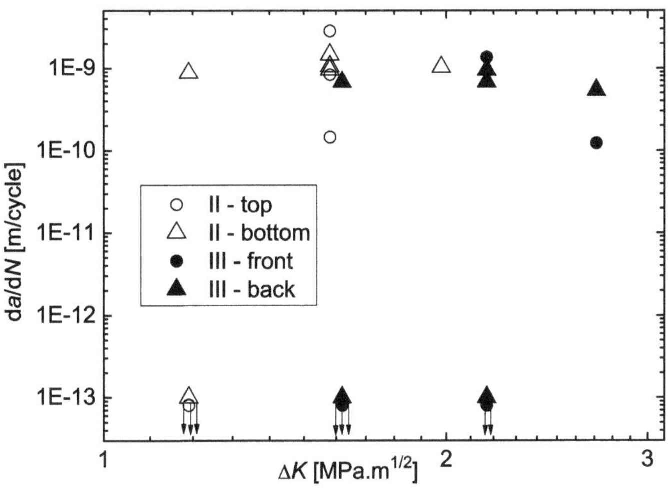

Figure 3.63 The crack growth data for the ferritic steel close to the fatigue threshold

achieved by generation of an additional short pre-crack by cyclic compressive loading. In this case, the intrinsic values $\Delta K_{IIth,in}$ and $\Delta K_{IIIth,in}$ could be determined by a sequential cyclic shear mode loading. Results of first experiments of this kind are plotted in Figure 3.64. For mode II loading, open stars and squares are associated with planar shear crack propagation with the deflection angle less than 15°, while solid stars and squares correspond to branching angles higher than 40°. The latter symbols can be found only for $\Delta K_{II} = 7.3\,\mathrm{MPa\,m}^{1/2}$ which is a higher value than that of ΔK_{Ith}. Due to the smoothing and sharpening of the crack tip by cyclic compression, there is also a significant shift of both mode II and mode III thresholds towards lower values ($\Delta K_{IIth,in} \approx 2.9\,\mathrm{MPa\,m^{1/2}}$ and $\Delta K_{IIIth,in} \approx 4\,\mathrm{MPa\,m^{1/2}}$).

These results do not confirm the identity $\Delta K_{IIth,in} = \Delta K_{IIIth,in} \approx 9.6\,\mathrm{MPa\,m^{1/2}}$ as obtained by Murakami *et al.* [354] for carbon steel under cyclic torsion. It should be noted, however, that the methodology used in the latter case was rather complicated and, in our opinion, not correct. Indeed, the extremely high threshold values were the result of a wrong presumption that the branching of the mode II+III cracks occurred at the first joint sites of the factory-roof massifs. In fact, the branching starts much sooner: at sites of embryonic semi-ellipses that are well defined by the theoretical analysis in Section 3.3.2. Consequently, the real threshold values should be substantially lower.

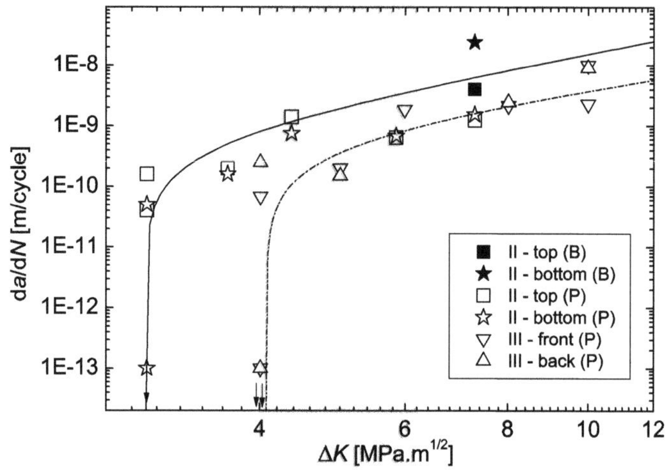

Figure 3.64 The crack growth curves for the specimens with an additional pre-crack made by cyclic compression in the austenitic steel

The results achieved in both the low and high-cycle region show that the crack growth rate under mode III is about five times lower than that under the mode II for the same applied strain or stress ranges. This fact was already respected in empirical codes for mixed-mode crack growth in the small scale yielding case [373, 374] by utilizing the effective stress intensity factor approximately as

$$K_{\mathit{eff}}^2 = K_I^2 + \xi K_{II}^2 + \eta K_{III}^2, \qquad (\eta < \xi < 1).$$

Thus, it seems that a similar assumption can be also extended to the large scale yielding regime by using the effective crack opening displacement

$$\mathrm{COD}_{\mathit{eff}} = \mathrm{COD}_I + \xi^* \mathrm{COD}_{II} + \eta^* \mathrm{COD}_{III}, \qquad (\eta^* < \xi^* < 1).$$

Let us finally emphasize that, up to now, the experiments on mode II and mode III crack propagation were accomplished only for a relatively small number of engineering materials. In some metallic materials, the dominant fatigue crack growth mechanism must not necessarily be based on the environmentally prevented recovery of the newly created surfaces ahead of the crack front. Indeed, the accumulation of damage due to the creation of microcracks inside the cyclic plastic zone could also control the crack propagation rate in some metallic materials. This could lead to nearly identical crack growth rates under both mode II and III loading cases. Therefore, the problem of shear-mode fatigue crack propagation still remains a challenge for further research.

3.3.4 Crack Growth and Fatigue Life under Combined Bending-torsion Loading

Combined cyclic bending-torsion (CCBT) is a kind of multiaxial fatigue loading that is probably mostly employed in scientific experiments. At the same time, many engineering components such as, for example, shafts, piston rods or gear wheels also operate under CCBT. Basically, there are two types of CCBT observed in practice and utilized in experiments: in-phase and out-of-phase loadings. In the next subsections, some results achieved by us and our co-workers using the in-phase variant of CCBT are presented.

3.3.4.1 Fatigue Life

Stress State under In-phase Bending-torsion Loading

In experimental investigations sinusoidal loading is usually applied. In this case, the time-dependence of non-zero components of the stress tensor can generally be expressed as

$$\sigma_{ij}(t) = \sigma_{ijm} + \sigma_{ija}\sin(\omega_{ij}t - \varphi_{ij}), \qquad i,j = x,y,z, \tag{3.33}$$

where σ_{ijm} is the mean stress, σ_{ija} is the stress amplitude, ω_{ij} is the angular velocity and φ_{ij} is the initiation phase.

Because of the biaxial character of bending-torsion, the loading trajectories can be plotted in 2D diagrams [312]. In the cases of pure bending, pure torsion and, generally, in-phase loading, the directions of principal stresses remain time-independent. This means that the ratio of torsion/bending stresses is constant along the whole loading trajectory (τ/σ = const). In the case of out-of-phase loading, on the other hand, only the amplitude ratio remains constant ($\tau_a/\sigma_a = \text{const}$) and the principal stress directions are time dependent.

The stress tensor at an arbitrary point inside the cylindrical specimen under in-phase bending-torsion (Figure 3.65) can be described as

$$\hat{\sigma}(t) = \begin{bmatrix} 0 & 0 & \tau_{xz}(t) \\ 0 & 0 & \tau_{yz}(t) \\ \tau_{zx}(t) & \tau_{zy}(t) & \sigma_{zz}(t) \end{bmatrix} = \begin{bmatrix} 0 & 0 & -\frac{M_t(t)}{I_p}y \\ 0 & 0 & \frac{M_t(t)}{I_p}x \\ -\frac{M_t(t)}{I_p}y & \frac{M_t(t)}{I_p}x & \frac{M_b(t)}{I_y}x \end{bmatrix},$$

$$I_p = \frac{\pi d^4}{32}, \qquad I_y = \frac{\pi d^4}{64},$$

where the components denoted τ (σ) represent shear (normal) stresses, M_t (M_b) is the torsion (bending) moment, d is the specimen diameter, I_p is the polar inertial moment and I_y is the inertial moment round the y-axis ([279]).

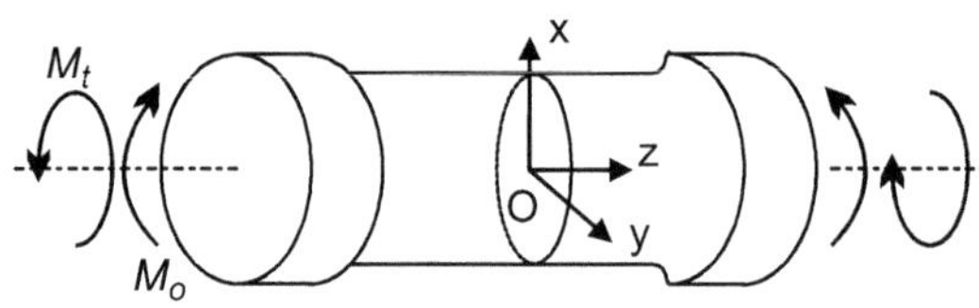

Figure 3.65 A scheme of the cylindrical specimen loaded by bending-torsion moments

In the case of in-phase symmetric loading one can consider $\omega_\sigma = \omega_\tau = \omega$, $\sigma_m = \tau_m = 0$ and $\varphi_\sigma = \varphi_\tau = 0$ so that Equation 3.33 becomes very simple:

$$\sigma_{ij}(t) = \sigma_{ija}\sin(\omega t), \qquad i,j = x,y,z.$$

This means that only stress amplitudes are sufficient for further reasoning and the time dependence can be neglected.

Since the short crack growth near the surface of rather smooth specimens occupies a significant part of their fatigue life, the stress components at the surface ($x = \pm d_2$, $y = 0$) are the most important parameters, especially in the case of high-cycle fatigue. At these points the principal stresses can be determined by solving the equation

$$\begin{vmatrix} -\sigma_{ia} & 0 & 0 \\ 0 & -\sigma_{ia} & \tau_a \\ 0 & \tau_a & \sigma_a - \sigma_{ia} \end{vmatrix} = 0,$$

where $\sigma_{zza} = \sigma_a$ and $\tau_{yza} = \tau_{zya} = \tau_a$. Hence, in accordance with the convention $\sigma_1 > \sigma_2 > \sigma_3$, one obtains

$$\sigma_{1a} = \frac{1}{2}\left(\sigma_a + \sqrt{\sigma_a^2 + 4\tau_a^2}\right), \quad \sigma_{2a} = 0, \quad \sigma_{3a} = \frac{1}{2}\left(\sigma_a - \sqrt{\sigma_a^2 + 4\tau_a^2}\right),$$

which means that the combined bending-torsion loading induces the biaxial state of stress. One can easily show that the directional cosines c_{ij} related to i and j axes of the principal coordinate system can be expressed as

$$c_{ji}^2 = \frac{(\sigma_{kk} - \sigma_j)(\sigma_{mm} - \sigma_j) - \sigma_{km}^2}{3\sigma_j^2 - 2\sigma_j I_1 + I_2}, \; i \neq k \neq m,$$

where J_1 and J_2 are the first and second invariant of the stress tensor, respectively:

$$J_1 = \sigma_{xx} + \sigma_{yy} + \sigma_{zz}, \qquad J_2 = \sigma_{xx}\sigma_{yy} + \sigma_{xx}\sigma_{zz} + \sigma_{yy}\sigma_{zz} - (\tau_{xy}^2 + \tau_{xz}^2 + \tau_{yz}^2).$$

The maximal shear stresses, crucial for the short crack initiation, act in planes inclined at $\pm 45°$ towards the principal axis, and their magnitude is

$$\tau_{12} = \frac{1}{2}\left|\sigma_1 - \sigma_2\right|, \quad \tau_{13} = \frac{1}{2}\left|\sigma_1 - \sigma_3\right|, \quad \tau_{23} = \frac{1}{2}\left|\sigma_2 - \sigma_3\right|.$$

For selected bending-torsion loading cases, the orientation of principal stresses and maximal shear stresses at surface points is depicted in Figure 3.66.

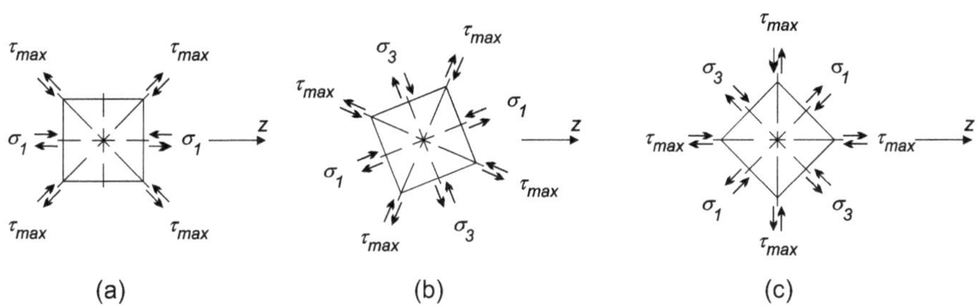

Figure 3.66 Principal stresses and maximal shear stresses at surface points of the cylindrical specimen loaded under: (a) pure bending, (b) in-phase combined bending-torsion ($\tau_a = \sigma_a$), and (c) pure torsion. The z-axis is parallel to the specimen axis

Let us finally emphasize that, in general, the stresses in planes not too much inclined at those in Figure 3.66 are very similar in magnitude. In the case of uniaxial tension, for example, the stresses in planes inclined at 10° are reduced by only 3% [375]. This is particularly important with respect to the crystallographic nature of short crack propagation in fatigue.

Stress-based Criteria of Multiaxial Fatigue Life

The experimental determination of fatigue life curves for complicated multiaxial loading cases represents, as a rule, a very difficult and expensive task. Therefore, many criteria of multiaxial fatigue life based on data available from uniaxial push-pull, bending or torsion tests were developed during the second half of the last century. In general, these criteria can be categorized as based on stress, strain or energy [375]. The stress-based criteria are appropriate for the large class of materials and components that operate near or below the fatigue limit (high-cycle fatigue).

The stress-based criteria can be written in the general form

$$af(\tau_a) + bg(\sigma_a, \sigma_n, \sigma_m) = 1, \tag{3.34}$$

where a and b are parameters that can be determined from two independent simple fatigue tests and $f(\tau_a)$ and $g(\sigma_a, \sigma_n, \sigma_m)$ are functions of applied shear and normal stresses, respectively. In the case of combined bending-torsion loading, for example, the fatigue limits in fully reversed pure torsion t_{-1} and pure bending f_{-1}, corresponding to a prescribed number of cycles to failure, are preferably used to obtain parameters a and b. In order to assess the efficiency of a particular criterion, the error index I is widely utilized as

$$I = 100(\mathrm{LHS} - 1) \quad [\%],$$

where LHS means the left-hand side of Equation 3.34. The ideal prediction leads to $I = 0$. If $I > 0$, the criterion gives conservative (safe) results since it predicts the specimen (or component) failure under lower loads than observed in the experiment. In other words, the applied stress amplitudes τ_a and σ_a lead to higher than prescribed number of cycles to failure.

The oldest criterion was suggested by Gough and Pollard [376, 377]. This criterion is based on the Huber–Hencky–Mises equivalent stress and its simplest form can be written as

$$a\tau_a^2 + b\sigma_a^2 = 1,$$

where $a = 1/t_{-1}^2$ and $b = 1/f_{-1}^2$.

Another classical empirical criterion, as proposed by Sines [378], uses the amplitude of the second stress invariant $J_{2,a}$ and the mean hydrostatic stress σ_m:

$$a\sqrt{J_{2,a}} + b\sigma_m = 1.$$

In the case of bending-torsion fatigue, $a = 1/t_{-1}$ and $b = \sqrt{3}f_{-1}/(\sigma_u t_{-1})$, where σ_u is the ultimate tensile strength. The criterion reported by Crossland [379] is of the same form while $b = 3/f_{-1} - \sqrt{3}/t_{-1}$ and $\sigma_m \to \sigma_{ma}$, where σ_{ma} is the amplitude of the mean hydrostatic stress.

More sophisticated criteria suggested by Findley, McDiarmid, Matake, Dang Van and Spagnoli are based on identification of the stress acting on specific planes within the component bulk. These planes are termed critical planes and are defined as one or more planes subjected to maximum damage. This concept respects the fact that the fatigue crack initiates in slip systems of maximal shear stresses and, in the short crack stage, it propagates along these planes with an enhancing support of normal stresses that open the crack tip. Particularly in smooth specimens or components, these stages occupy a majority of the fatigue life [247]. Thus, the fatigue life is controlled by the combination of shear and normal stresses acting on a critical plane. Findley [380] suggested that the maximal normal stress σ_n (the sum of the mean stress σ_m and the stress amplitude σ_a) on the critical plane might have a linear influence on the allowable alternating shear stress τ_a:

$$(a\tau_a + b\sigma_n)_{\max} = 1, \tag{3.35}$$

where $a = 1/\sigma_a$ and $b = k/\sigma_a$. The damage leading to failure is expected to be related to the critical plane that is associated with the largest term in the left-hand side of Equation 3.35. The constant k must be determined experimentally by performing tests under pure bending and pure torsion. In the case of in-phase proportional bending-torsion loading, a simple relation $\tan 2/\Theta = k^{-1}$ incorporates the angle Θ between the direction of the first principal stress σ_1 and the critical plane [375]. However, the Findley criterion incorrectly predicts a dependence of the torsion fatigue limit when the mean torsion stress is superimposed [381].

In the McDiarmid criterion [382], the critical plane is defined as the plane on which the amplitude of the shear stress reaches its maximum (not the plane on which the damage quantity is maximized). The torsion fatigue strength $t_{A,B}$ is introduced either for case A (cracking parallel to the surface) or case B (cracking inwards from the surface). The bending-torsion loading refers to case A and the McDiarmid criterion becomes

$$\frac{\tau_{\mathrm{a,max}}}{t_{-1}} + \frac{\sigma_n}{2\sigma_u} = 1.$$

The Matake criterion [363] can be formally expressed in the form of the Findley relation at Equation 3.35. However, the critical plane is assumed to be the same as that in the McDiarmid criterion (the maximum shear stress amplitude) which leads to $a = 1/t_{-1}$ and $b = 2/f_{-1} - 1/t_{-1}$ (see also Equation 3.36).

The criterion of Carpinteri and Spagnoli [383] is an example of a quadratic form

$$a\tau^2_{\mathrm{a,max}} + b\sigma^2_n = 1,$$

where $a = 1/t^2_{-1}$ and $b = 1/f^2_{-1}$.

Criteria of Dang Van [384], Papadopoulos (integral approach) [385] and Gonçalves *et al.* [386] are examples of advanced approaches. These criteria also take, besides the critical plane, the distribution of local microstructural stresses and possible slip systems into account. Dang Van has proposed an endurance limit criterion based on the concept of microstresses within a critical volume of the material. This criterion can be written as a time-dependent cumulative combination of the local shear stress $\tau_L(t)$ and the local hydrostatic stress $\sigma_{h,L}(t)$:

$$d = \max\left[\frac{\tau_L(t)}{b - a\sigma_{h,L}(t)}\right].$$

The critical damage is reached when the condition $d \geq 1$ becomes fulfilled.

In the Papadopoulos criterion both the shear stress and the normal stress are integrated over all slip planes:

$$\sqrt{\frac{5\kappa^2}{8\pi^2}\int\limits_{\varphi=0}^{2\pi}\int\limits_{\psi=0}^{\pi}\int\limits_{\chi=0}^{2\pi}\left(T_a\left(\varphi,\psi,\chi\right)\right)^2 \mathrm{d}\chi \sin\psi \mathrm{d}\psi \mathrm{d}\varphi} + \left(3-\sqrt{3}\kappa\right)\cdot\sigma_{h,\max} \leq f_{-1},$$

where T_a is the amplitude of resolved stress, φ, ψ and χ are the Euler angles and $\sigma_{h,\max}$ is the maximum value of the mean hydrostatic stress.

The Gonçalves criterion is expressed as

$$\frac{\kappa-1}{\sqrt{2}\left(1-\frac{1}{\sqrt{3}}\right)}\sqrt{\sum_{i=1}^{5} d_i} + \frac{\sqrt{3}-\kappa}{\sqrt{3}-1}\sigma_{1,\max} \leq f_{-1},$$

where the parameters d_i can be determined from minimum and maximum values of the deviatoric stress tensor as

$$d_i = \frac{1}{2}\left(\max s_i(t) - \min s_i(t)\right).$$

Fatigue Life of Steel Specimens under Bending-torsion

Comparison of prediction efficiency of individual multiaxial criteria in the case of combined bending-torsion fatigue of smooth specimens was reported in the work of Major *et al.* [387]. One set of specimens was subjected to a nitriding procedure, a technological treatment that results in a harder surface layer containing compressive residual stresses. Hereafter, these specimens will be called "nitrided" ones unlike the "virgin" ones without the surface layer.

The cylindrical specimens were made of high-strength low-alloy Cr-Al-Mo steel (equivalent to EN 37CrAlMo6) of the chemical composition given in Table 3.7. When compared with commonly used nitrided steels, the investigated material contains a slightly higher percentage of carbon, which offers higher strength but, at the same time, it introduces larger differences in microstructure caused by local variations in the nitriding process. After the heat treatment consisting of annealing (920°C, 25 min, air), quenching (930°C, 25 min, oil), and tempering (650°C, 25 min, air), the microstructure consisted of the sorbitic phase. The heat treatment resulted in yield strength $\sigma_y = 840\,\text{MPa}$ and ultimate tensile strength $\sigma_u = 950\,\text{MPa}$.

Table 3.7 Chemical composition of EN 37CrAlMo6 steel (wt%)

C	Mn	Cr	Mo	V	Cu	Al	W	Si	P	S
0.357	0.468	1.49	0.194	0.01	0.072	1.4	0.032	0.292	0.006	0.006

Nitriding is a prominent industrial technology that is used to enhance key properties of engineering components, especially surface hardness, fatigue strength and wear and corrosion resistance [388–390]. The plasma nitriding process is characterized by adsorption of nitrogen in the form of N^+, NH^+, NH^{2+} and NH^{3+} ions. The most important characteristics determining the depth of the nitrided layer are the nitriding time and the nitriding temperature. The nitrided layer consists of (1) the thicker subsurface diffusion layer and (2) the thin brittle surface compound layer composed of different iron nitride phases (often called the white layer). The higher strength of the nitrided layer together with associated compressive residual stresses causes subsurface crack nucleation and, therefore, improvement of fatigue strength. The micropulse plasma nitriding procedure was applied in two steps – cleaning (30 min) and nitriding (8 h) resulting in the depth of the diffusion layer $h_{dl} = 200\,\mu\text{m}$ and the thickness of the white layer $h_{wl} = 3\,\mu\text{m}$; see Table 3.8 for the details of this procedure. The basic tensile mechanical properties of the plasma nitrided specimens were measured as follows: yield strength $\sigma_y = 870\,\text{MPa}$ and ultimate strength $\sigma_u = 1020\,\text{MPa}$. This means that a slight improvement of both characteristics was achieved.

Table 3.8 Parameters of the nitriding process

Step	Temperature	Time	Atmosphere			Pressure	U	Pulse
	[°C]	[h]	N_2	H_2	CH_4	[mbar]	[V]	[μs]
Cleaning	510	0:30	20	2	-	0.7	800	120
Nitriding	515	32:00	21	7	0.4	2.6	530	150

The fatigue life of nitrided specimens was found to be significantly longer than that of virgin specimens, as can be clearly seen in Figure 3.67. The curves are plotted by using the McDiarmid criterion for the number of cycles

Table 3.9 Average error indices I_{avr} and their absolute values $I_{ABS,avr}$ of selected multiaxial criteria for virgin and nitrided specimens

Criterion / Specimens Indices [%]	Virgin $I_{ABS,avr}$	Virgin I_{avr}	Nitrided $I_{ABS,avr}$	Nitrided I_{avr}
Dang Van	7.35	−3.03	8.45	−0.35
Crossland	7.28	−5.42	8.65	−2.47
Sines	11.85	−11.23	12.32	12.82
McDiarmid	7.08	−3.02	8.65	1.72
Findley	7.08	−3.38	8.62	−0.87
Matake	7.12	−3.04	8.32	−0.23
Spagnoli	10.5	4.33	8.56	3.35
Papadopoulos (integral approach)	7.36	−3.04	8.46	−0.35
Papadopoulos (critical plane)	11.08	−6.30	16.56	−6.70
Gonçalves *et al.*	8.62	4.14	13.92	5.20

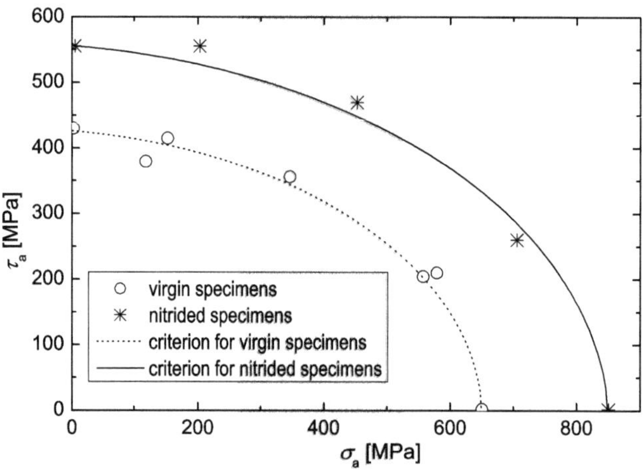

Figure 3.67 The constant fatigue life diagram ($N_f = 5 \times 10^5$ cycles) for bending-torsion loading of virgin and nitrided specimens according to the McDiarmid criterion

$N_f = 5 \times 10^5$ and the experimental data correspond to a fatigue life of $N_f = 5 \times 10^5 \pm 2 \times 10^5$. This conclusion also holds for push-pull fatigue tests reported elsewhere [390].

The error indices, calculated for selected multiaxial criteria by averaging the experimental data, are shown in Table 3.9. They reveal that the McDiarmid criterion was the most successful in fatigue life prediction for virgin specimens whereas the Matake criterion was the best for nitrided ones, although both criteria provide slightly non-conservative results. Nevertheless, the related errors are less than 10% which is safely below the safety factors commonly utilized in fatigue strength analyses applied in engineering practice.

3.3.4.2 Topology of Bending-torsion Fracture Surfaces

Cyclic torsion loading leads to a very complicated crack path exhibiting local crack arrests, a branch/twist crack morphology or the factory roof [347,361,375,391,392]. An interaction between both mating fracture surfaces (some combination of sliding, climbing, sticking, slipping and deforming) often makes even a qualitative understanding of fatigue crack propagation difficult [347,375]. On the other hand, a high amount of opening loading mode I or sometimes high shear amplitudes generate a fracture surface that appears macroscopically flat. Despite all mentioned studies, the crucial problem in biaxial fatigue topography is still a significant lack of experimental data. This shortcoming could be partially reduced by a study devoted to the fracture surface topology formed by a combined bending-torsion loading in both the low-cycle fatigue (LCF) and the high-cycle fatigue (HCF) regimes [393].

Experimental Procedure and Results

Experimental settings were based on Matake's critical plane criterion [381, 383]. According to fatigue life N_f, the virgin cylindrical specimens made of the EN 37CrAlMo6 steel were divided into LCF and HCF, respectively. The differences in N_f within both LCF ($N_f \approx 10^4$) and HCF ($N_f \approx 10^6$) sets were well within an order of magnitude, which is a scatter typical for fatigue life data.

Topological 3D data of 0.5×0.5 mm regions, selected on the fracture surfaces at a distance of 0.8 mm from the specimen surface (the crack initiation site), were obtained by means of stereophotogrammetry. Using the Delaunay triangulation, two sets of profiles were traced for all analyzed regions: the first in the crack propagation direction (referred to in the following as the y-direction) and the second in the perpendicular direction (referred to as the x-direction) thus representing different positions of the progressing fatigue crack front.

The following parameters were found to be sufficiently sensitive to different topological aspects of the extracted fracture surface profiles, corresponding to a different loading mixture: the standard deviation R_q of vertical z-coordinates, the number of peaks of the profile per unit length m_0 and the Hurst exponent H. It was expected that these parameters might be used for a wide range of fracture surfaces of engineering components made of materials with different microstructures. It should be emphasized that the arithmetic roughness R_a exhibited similar trends as R_q, and the fractal dimension D_D did not reveal any significant susceptibility to differentiate the loading cases.

Average values of analyzed parameters, together with their standard deviations (error bars), are plotted in Figure 3.68 for profiles oriented in the

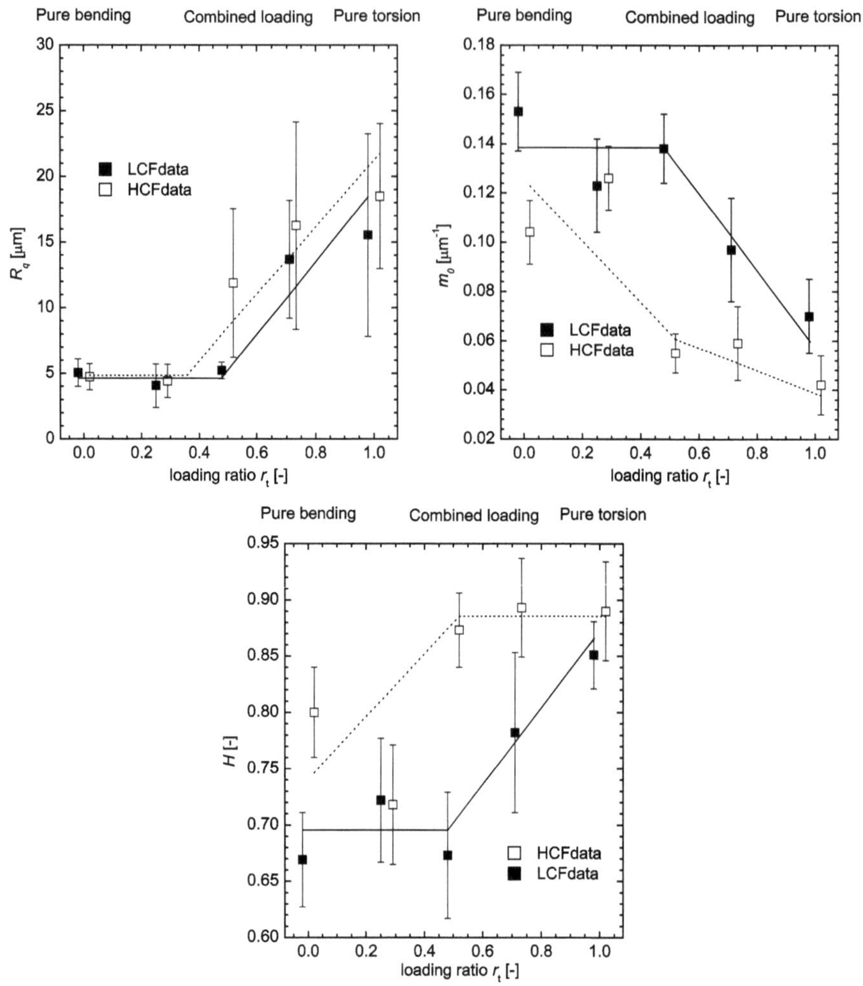

Figure 3.68 Averaged values of topological parameters as functions of the loading ratio r_t. Reprinted with permission from Elsevier B.V. (see page 265)

y-direction as functions of the loading ratio r_t. This ratio is defined as

$$r_t = \tau_a/(\tau_a + \sigma_a).$$

Discussion of Results

Since the parameters R_q and R_a describe both macroscopic and microscopic levels of the surface roughness, it is not surprising that lower values $R_q <$ 10μm correspond to low torsion components of the loading ($r_t < 0.5$) in both LCF and HCF. Fracture surfaces corresponding to a higher torsion component ($r_t > 0.5$) exhibited substantially higher R_q values.

Calculated values of the parameter m_0 decrease with an increase in the torsion component which is likely associated with frictional contact of both fracture surfaces during the fatigue process. Indeed, the contact friction must be more intensive for loading regimes of high r_t. Generally higher values of m_0 for LCF can be understood in terms of a higher roughness level. On the other hand, the values of H increase with increasing loading ratio r_t. Moreover, they are generally higher in the case of HCF. This behaviour, the opposite of that of m_0, can also be attributed to the frictional contact of fracture surfaces.

The most important result lies in the existence of a critical value of the loading ratio $r_t \approx 0.5$ that corresponds to a significant change of all examined topological characteristics. This means that the topography of analyzed fracture surfaces substantially changes when the torsion component becomes higher than the bending one. This result might be very important for the assessment of the loading type in failure analysis, provided that it remains true for other metallic materials.

Let us finally note that the experimental results obtained for the x-direction were similar to those for the y-direction. The only additional trend, observed only in the case of LCF, was a slight decrease (increase) in m_0 (H) with measured distance. In terms of the Hurst exponent, for example, this means that the fractality of the crack front decreases during its propagation. Such behaviour might be elucidated by a gradual merging of many initial surface microcracks that nucleate in the case of LCF.

3.3.5 Formation of Fish-eye Cracks under Combined Bending-torsion Loading

The high strength of the nitrided layer and the compressive residual stresses introduced within the diffusion zone are the main causes of an improvement of fatigue strength. Since such a layer hinders dislocation motion, the predominant failure mechanism in the high-cycle fatigue (HCF) region is subsurface fatigue crack growth (e.g., [394,395]). As a rule, the cracks initiate at internal inclusions within the core region and propagate in a near vacuum by forming so-called fish-eye cracks. The interior of these cracks looks bright to the naked eye or in the optical microscope, whereas the outside region seems to be grey. The difference in colour is, most probably, caused by different fracture micromechanisms. The bright morphology of the fish-eye crack is produced by a cyclic contact of the mating fracture surfaces under conditions of suppressed atmospheric effects. When the crack front approaches the low-toughness nitrided layer, local through-the-layer brittle cracking creates a connection to the surface and, subsequently, a penetration of the atmosphere to the inside of the fish-eye [394]. This changes the growth mechanism to an environmentally assisted one which leads to a different surface roughness and to the optical contrast.

Fish-eye cracks are also a typical feature of fractures in the ultra-high-cycle fatigue (UHCF) region [396–398]. The transition of the crack initiation site from the surface into the interior causes the stepped, double or multi-stage fatigue curve appearance. The initiation sites in the UHCF region are often non-metallic inclusions, although the crack nucleation near finely dispersed segregations or phase interfaces was sometimes observed. In the former case, the dark-looking fracture morphology in the vicinity of the inclusion, termed as the "optically dark area" (ODA), is usually present. The formation of the ODA is, most probably, attributed to slow intermittent fatigue crack growth assisted by the internal hydrogen trapped by the inclusion [399].

When the maximum tensile loading exceeds the yield strength of the material, the low-toughness nitrided layer breaks during the first few loading cycles due to a high strain mismatch at the layer/matrix interface [390]. Such initiated surface cracks can easily penetrate to the specimen bulk owing to the related hard/soft transition direction [198]. Thus, this damage mechanism is typical for the quasi-static region.

The geometric characteristics of the fish-eye cracks were, to some extent, studied by several researchers (e.g., [388, 389, 396, 400]). In general, the fish-eye cracks were found to be approximately of either a circular or an elliptical shape depending on the type of loading (push-pull, rotating bending, plane bending) and the presence of high compressive residual stresses introduced by some of the surface hardening procedures. A restricted propagation of the crack towards the nitrided layer is considered to be the reason why there is an increase of the fish-eye crack radial size with increasing distance of the inclusion from the specimen surface, which in turn is reported to be a function of the level of applied stress inside the nitrided layer and the compressive residual stress profile. In the UHCF region, the ODA is of particular interest due to the still insufficiently understood interplay of cyclic strain and internal hydrogen. The size of the ODA was found to increase with decreasing stress level or increasing fatigue life.

The residual stress effect on the crack initiation site under high-cycle push-pull loading was studied in [390]. The size of the inclusion required for fatigue crack initiation was computed as a function of inclusion depth. The local shift of the loading asymmetry caused by the presence of a compressive residual stress field was accounted for by using an equation describing the dependence of the fatigue threshold ΔK_{th} on the cyclic ratio R [401]. For the investigated nitrided steel, this dependence was determined by using pre-cracked flat specimens with nitrided layers of the thickness of approximately a quarter of the specimen width and, therefore, the measured fatigue thresholds represented averaged layer/core values. When identifying the inclusion with a critical crack at the fatigue limit, the critical inclusion size d_c could be estimated as $d_{c1} = 3.79\,\mathrm{mm}$ near the specimen surface, $d_{c2} = 0.87\,\mathrm{mm}$

at depth $h = 0.3$ mm inside the diffusion zone and $d_{c3} = 0.068$ mm at depth $h = 0.4$ mm [390] at the outer layer/matrix boundary. In spite of the fact that the averaged values of ΔK_{th} might have been inaccurate, the calculated critical size of the inclusion near the surface exceeded by two orders of magnitude the inclusion size of tens of microns found in the investigated steel as well as in most commercial steels. This is a clear reason why there was no crack initiation observed at (or near) the surface.

All the above-mentioned studies were, however, conducted on fish-eye cracks generated by the uniaxial loading regime or torsion and no information is currently available for the case of multiaxial loading. Therefore, the motivation for further research was to study the fish-eye cracks developing under symmetrical bending, symmetrical torsion and their synchronous in-phase combinations. The results of such an investigation [402] are presented in the following subsection.

3.3.5.1 Experimental

The specimens were made of high-strength low-alloy Cr-Al-Mo steel specified in the previous subsection. In the HCF regime, both the compressive residual stresses and the hardness of the nitrided layer induce the internal fatigue fish-eye crack initiation at the inclusion-matrix interface and influence a subsequent crack propagation. The dependence of the residual stress level inside the layer on the distance from the free surface is depicted in Figure 3.69. This dependence was obtained by means of precise X-ray measurement performed on samples made of an equivalent high-strength steel that was nitrided using the technology described above [403]. One can see that the compressive residual stresses reach a maximum of 750 MPa close to the surface and, up to about 0.2 mm depth, they start to decay steeply. Beyond the depth of 0.2 mm corresponding to the microstructurally distinguishable diffusion layer, a rather slow decrease of residual stresses continues to reach a zero value at 0.7 mm. The maximum value of the residual compressive stress of about 800 MPa was also confirmed by less precise measurement on the investigated nitrided steel [404].

Plasma nitrided specimens were tested at room temperature using a Polish resonance testing machine MZGS-200. The sinusoidal symmetrical bending and torsion as well as the synchronous in-phase bending-torsion were applied at a frequency $f = 29$ Hz. Experimental settings were based on the Matake critical plane criterion

$$\tau_a + \left(\frac{2\tau_c}{f_{-1}} - 1\right)\sigma_n = t_{-1}. \tag{3.36}$$

This criterion was used to estimate both bending and torsion loading components in order to constrain the fatigue life into the HCF region. The details of experimental settings and fatigue life data are published elsewhere [393].

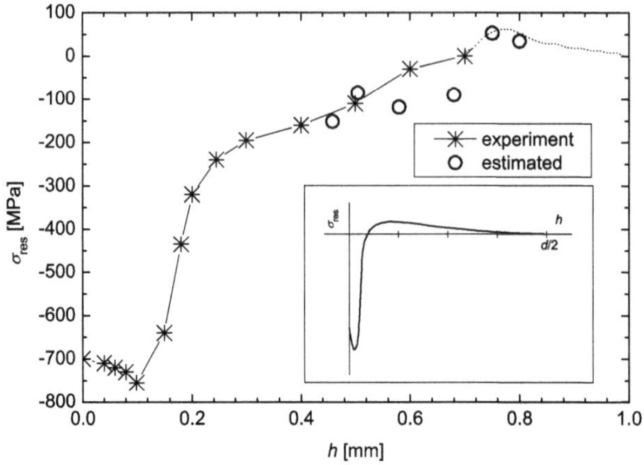

Figure 3.69 The experimental profile and the theoretical estimation of residual stresses in the nitrided layer. The scheme in the *inset* shows a general shape of the residual stress curve along the specimen radius. Reprinted with permission from Elsevier B.V. (see page 265)

3.3.5.2 Fish-eye Crack Geometry

Fish-eye cracks were studied using a chromatic optical profilometer FRT MicroProf 100, Fries Research & Technology GmbH. In the first step, the distance h of the centre of the inclusion from the free specimen surface in the radial direction was measured. After that, the maximal projection plane was found by tilting the specimen mounted on the moving x-y table and fish-eye crack sizes in radial (R_{tl}, R_{tr}) and tangential (R_{ru}, R_{rd}) directions were recorded; see Figure 3.70 for the nomenclature. While the inclination of the crack plane was almost negligible in the radial direction, the inclination angles in the tangential direction varied in the range of $(0°, 45°)$, where 0° nearly corresponds to pure bending ($r_t = 0$) and 45° approximately stands for pure torsion ($r_t = 1$). This actually means that in all cases of non-zero torsion component the crack front propagates under the local mode I loading. In several cases, the diameter of the inclusion was also estimated in the scanning electron microscope (SEM) by analyzing either the inclusion itself or its imprint found on the fracture surface. The average diameter was found to be $37 \pm 14\,\mu\text{m}$.

The observed fish-eye cracks were, in general, of an elliptical shape; see Figure 3.71. In the case of fish-eyes initiated close to the nitrided layer, the upper radial dimension R_{ru} (crack propagation towards the free specimen surface) was found to be smaller than the lower radial dimension R_{rd} (crack growth into the specimen core). This behaviour, which is understandable in terms of the retardation effect of the surface nitrided layer, is consistent with observations reported in the work of DelaCruz *et al.* [388] where, however, only fish-eye cracks generated under reversed bending were studied.

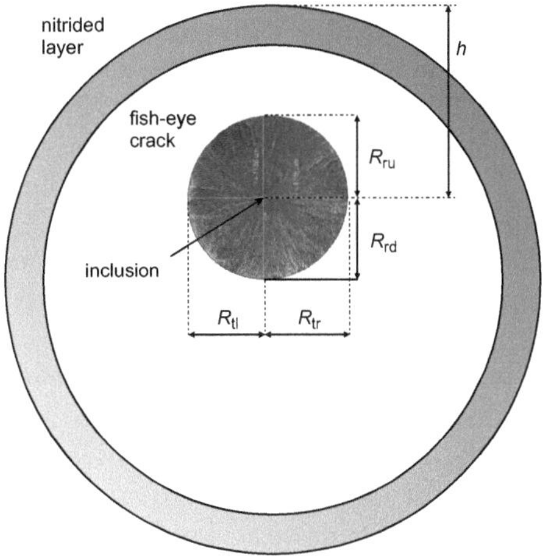

Figure 3.70 A scheme showing the nomenclature for the fish-eye crack characterization

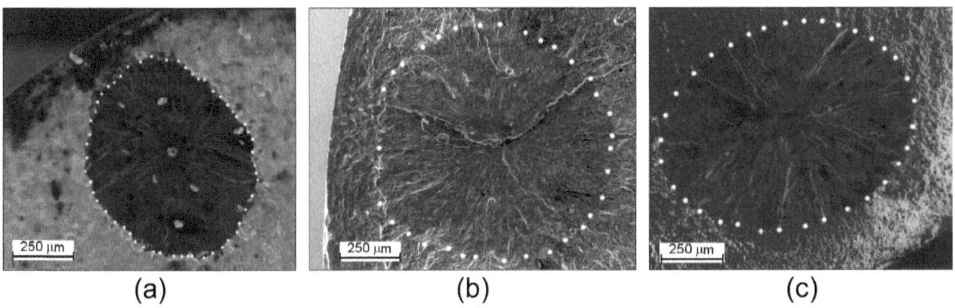

Figure 3.71 An example of the SEM microphotographs of fish-eye cracks for different loading regimes: (a) $r_t = 0$, (b) $r_t = 0.5$, and (c) $r_t = 1$. Reprinted with permission from Elsevier B.V. (see page 265)

The fish-eye crack shape was characterized by the radial asymmetry, S_r, the average size of the fish-eye crack (free of the influence of the nitrided layer), R_{avr}, and the elliptical coefficient Q. These quantities were defined as

$$S_{\mathrm{r}} = \frac{R_{\mathrm{ru}}}{R_{\mathrm{rd}}},$$

$$R_{\mathrm{avr}} = \frac{1}{4}\left(2R_{\mathrm{rd}} + R_{\mathrm{tl}} + R_{\mathrm{tr}}\right),$$

$$Q = \frac{R_{\mathrm{tl}} + R_{\mathrm{tr}}}{2R_{\mathrm{rd}}}.$$

All of the fish-eye cracks studied were nucleated at internal non-metallic inclusions. The distance h of the inclusion centre from the specimen free surface was in the range of $420 - 1040\,\mu\mathrm{m}$. The crack growth towards the free surface was always, at least partially, influenced by the compressive residual stress field ($h - R_{ru} < 0.6\,\mathrm{mm}$). Moreover, some fish-eye cracks (5 out of 26) grew into the compound layer ($h - R_{ru} < 0.2\,\mathrm{mm}$).

The effect of the free surface and the bending stress gradient on the fish-eye crack growth in the radial direction without the influence of the nitrided layer was analyzed using the analytical expressions for K_{IA} and K_{IB} at two opposite radial crack fronts A and B of the elliptical crack in a shaft under bending. The results revealed that, almost independently of the depth h in the range of (0.6, 1.2) mm, the radial asymmetry, S_r, was found to be in the range of (1.20, 1.22) [402]. Note that the influence of the nitrided zone on fish-eye cracks initiating outside the range of compressive stresses ($h > 0.7\,\mathrm{mm}$) must be nearly negligible. Because the stress gradient is also similar in the case of torsion loading, most of the measured values of S_r for the fish-eye cracks initiated at the depth $h > 0.7\,\mathrm{mm}$ should lie close to $S_r = 1.2$ for all loading regimes.

3.3.5.3 Results

The dependence of the radial asymmetry, S_r, on the distance of the inclusion from the specimen surface, h, is shown in Figure 3.72. As mentioned above, the fatigue crack growth in the radial direction is influenced by bending-torsion loading component gradients, the free surface and the residual compressive stresses, which are prominent especially for the fish-eye cracks initiated near the nitrided layer. Obviously, when the inclusion depth is not very large ($h < 0.7\,\mathrm{mm}$), the residual compressive stresses decelerate the crack front growing towards the free surface as evidenced by the radial asymmetry values $S_r < 1$. On the other hand, as was predicted by the growth simulation, the radial asymmetry reaches $S_r = 1.2$ in the range $h \in (0.7, 0.8)\,\mathrm{mm}$. With an increasing inclusion depth $h > 0.8\,\mathrm{mm}$, however, the general tendency seems to be the growth in a rather symmetrical fashion $S_r \approx 1$. This might be a consequence of small tensile residual stresses that must balance the compressive ones beyond the depth $h = 0.7\,\mathrm{mm}$ (see also hereafter).

An increase in the average fish-eye size, R_{avr}, with increasing inclusion depth and fatigue life, N_f, is depicted in Figure 3.73. The first tendency is in agreement with the result reported in [388] and follows from a simple geometrical consideration. Indeed, the fish-eye dimension R_{ru} is, more or less, determined by the distance from the initiation site to the nitrided layer. The second dependence can be understood in terms of the increasing equivalent stress amplitude with the decreasing number of cycles to failure N_f.

The elliptical coefficient, Q, exhibits no correlation with the number of cycles to failure (Figure 3.74(a)) in accordance with the result reported in [400]. On the other hand, the loading regime has a certain effect, as is demonstrated in Figure 3.74(b). Despite a large scatter in the data, the elliptical coefficient increases with increasing loading ratio (the regression line). In Figure 3.75, a

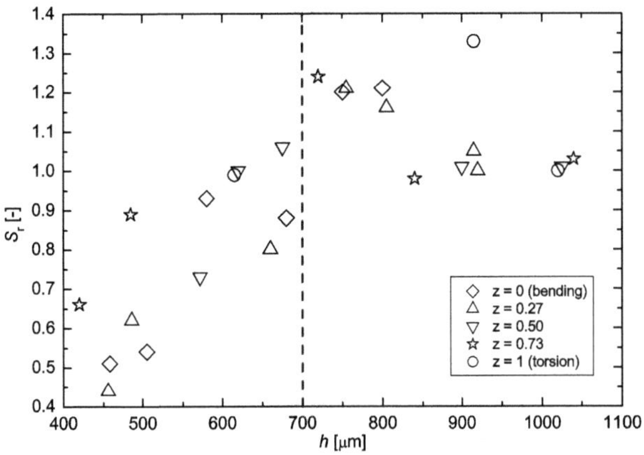

Figure 3.72 The dependence of the radial asymmetry on the inclusion depth. Reprinted with permission from Elsevier B.V. (see page 265)

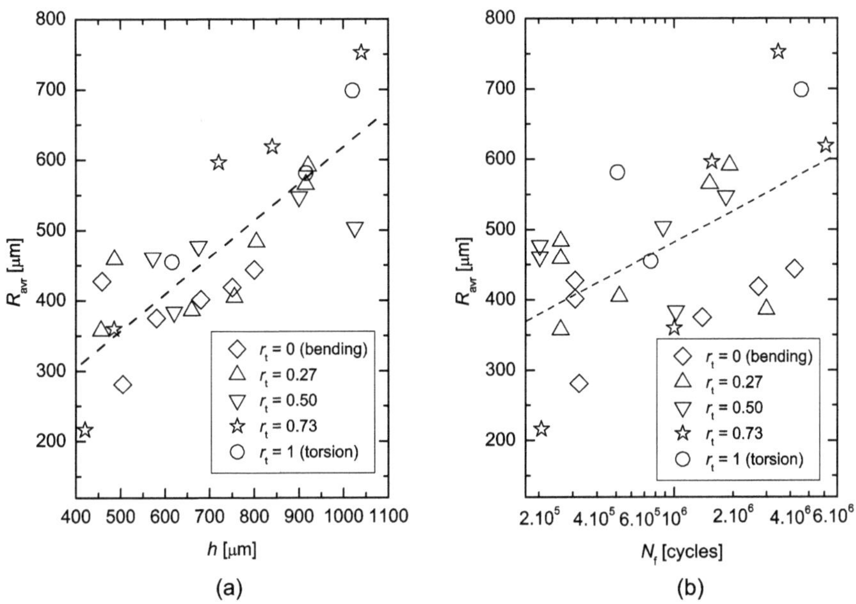

Figure 3.73 The dependence of average fish-eye crack size on: (a) the inclusion depth, and (b) the fatigue life

corrected coefficient $Q' = (R_{tl} + R_{tr})/(2R_{rd})$, where the lower radial dimension R_{rd} (free of the influence of the nitrided layer) replaced the upper radial dimension R_{ru}. This correction eliminates the effect of the nitrided layer on the fish-eye crack growth, diminishes the data scatter and accentuates the trend observed in Figure 3.74b. The dependence Q *vs* r_t means that with a higher portion of the torsion loading component the crack growth rate in the tangential direction becomes more rapid than that in the radial direction. This effect is, most probably, caused by a different influence of the stress gradients on the crack growth in tangential and radial directions.

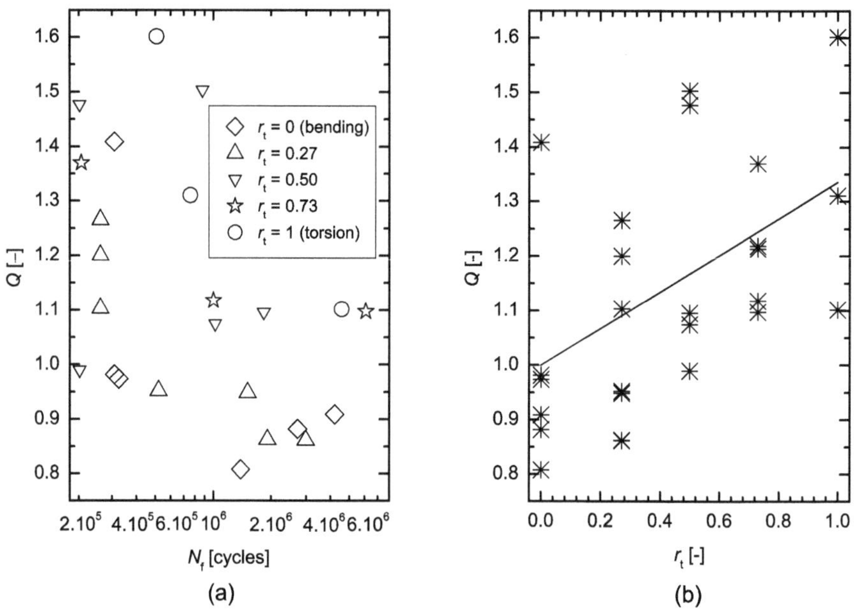

Figure 3.74 The dependence of the elliptical coefficient on: (a) the fatigue life, and (b) the loading ratio

3.3.5.4 Estimation of Residual Stresses

The experimental data obtained under pure bending could be used for an approximate assessment of the residual stresses in the innermost part of the nitrided layer and the adjacent core region. In the pure bending case, the direction of internal stresses is equal to that of the principal stress. The orientation of the residual stress vector is opposite to that of the main stress during the tensile loading half-cycle and identical to it during the compressive half-cycle. This means that the residual stress σ_{res} shifts the S-N (Wöhler) curve for a symmetric loading ($R = -1$) of the virgin material (without

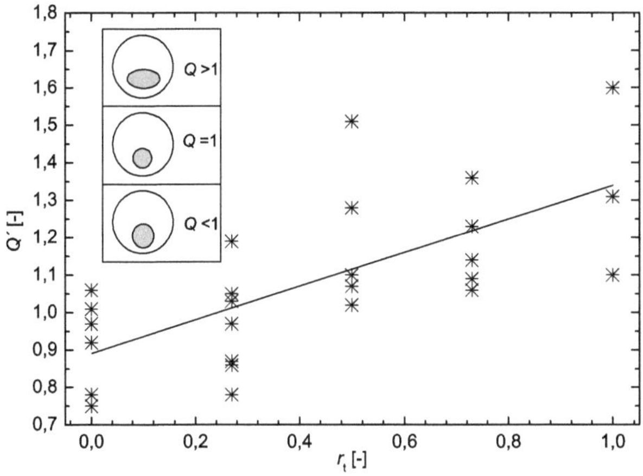

Figure 3.75 The dependence of the corrected elliptical coefficient Q' on the loading ratio

the nitrided layer) to a curve for a lower cyclic ratio $R < -1$. This fact is employed in the proposed method described as follows:

1. measurement of the initiation depth h_k of k-th fish-eye;
2. calculation of the amplitude of the bending stress σ_{ak} that corresponds to the depth h_k;
3. identification of the number of cycles to failure N_{fk} corresponding to the k-th bending experiment;
4. determination of the bending amplitude σ_{avk} (the surface stress amplitude) corresponding to N_{fk} on the Wöhler curve for virgin specimens;
5. utilization of both the ratio σ_{ak}/σ_{avk} and the relationship generally describing the shift of S-N curves to a determination of the mean stress $\sigma_m = \sigma_{res}$ corresponding to the depth h_k;
6. application of this procedure to fish-eyes of various depths h in order to obtain the dependence $\sigma_{res}(h)$.

This procedure was applied to obtain an approximate dependence $\sigma_{res}(h)$ for comparison with the measured values in Figure 3.69. In Figure 3.76, the S-N curve for virgin specimens [390] is plotted by a solid line along with that for nitrided specimens (the dashed line). The solid stars refer to the calculated stresses σ_{ak} at the depths h_k. For depths $h_k < 0.7\,\text{mm}$, these points lie in between the S-N curves for virgin and nitrided specimens while, for $h_k > 0.7\,\text{mm}$, they lie slightly below the virgin S-N curve. The well known Soderberg ($\sigma_a/\sigma_{av} = (1 - \sigma_m/\sigma_y)$) and Goodman ($\sigma_a/\sigma_{av} = (1 - \sigma_m/\sigma_u)$) relations for the S-N curve shift were adopted for calculation of residual stresses. All the experimental and computed data are displayed in Table 3.10. Since the Soderberg relation generally gives conservative results unlike the Goodman approximation that provides nonconservative estimates [192], the mean values of both approaches are plotted in Figure 3.69 (open symbols) as the theoretical assessment of residual stresses. One can see a rather good agreement between the calculated and measured data within the innermost part of the nitrided layer. In the adjacent core region, the computed data predicted small tensile residual stresses. This is to be expected owing to a necessary balance of residual stresses through the specimen radius $d/2$. Indeed, the integral of the residual stresses along the radius must be zero, as is schematically shown in the inset of Figure 3.69.

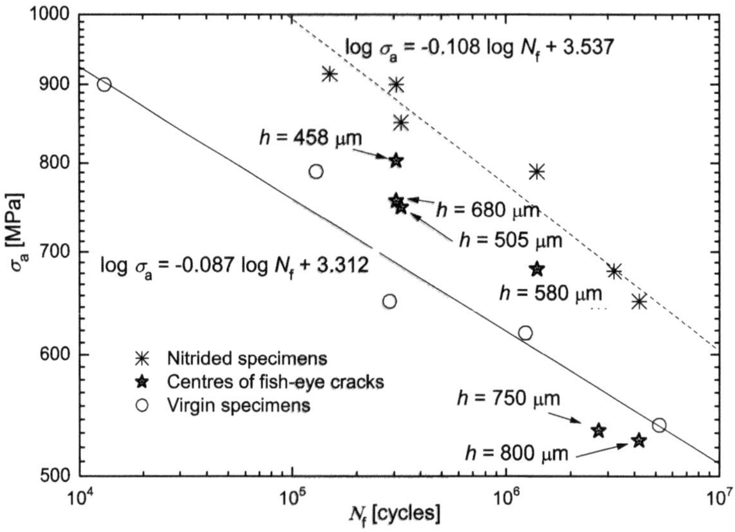

Figure 3.76 Wöhler curves for virgin and nitrided specimens and the data points corresponding to the analyzed fish-eye cracks. Reprinted with permission from Elsevier B.V. (see page 265)

Table 3.10 Results of the residual stress analysis

N_f [cycles]	h_k [μm]	σ_{ak} [MPa]	σ_{avk} [MPa]	σ_m [MPa] Soderberg	σ_m [MPa] Goodman
4.20×10^6	800	527.6	548.0	31.2	35.3
2.71×10^6	750	535.3	569.2	50.0	56.5
1.39×10^6	580	682.2	602.9	−110.5	−125.0
3.24×10^5	505	749.0	684.0	−79.9	−90.3
3.08×10^5	680	756.0	687.0	−84.3	−95.4
3.08×10^5	458	803.0	687.0	−141.8	−160.4

Although the estimation of the residual stresses seems to be plausible, one should take some limitations of the proposed method into account. First, the fatigue resistance of the layer increases with decreasing depth h due to the hardening effect of the nitriding process. This means that the virgin S-N curve certainly does not apply for $h_k < 0.2\,\text{mm}$, i.e., for the depth region corresponding to the hard diffusion zone. For the values $h_k > 0.3\,\text{mm}$, however, the microstructurally induced hardening effect becomes rather negligible and the hardening component induced by residual stresses starts to dominate. Consequently, the applicability of the method is restricted to fish-eyes initiating within that region. Since the initiation depth decreases with increasing σ_a, the depth condition $h_k > 0.3\,\text{mm}$ also implies a limitation of the applied loading. Another load limit is given by the yield stress σ_y due to the related breaking of the nitrided zone. On the other hand, a different geometry of the crack growth stage in the virgin and nitrided specimens does not represent any serious problem. Indeed, the number of cycles to failure in both types of specimens are determined by the crack initiation stages where the inclusion-assisted nucleation mechanisms are identical. However, this is not necessarily true for low-strength materials, where the crack initiation mechanism in virgin specimens could be different (persistent slip bands, grain boundaries, etc.). This is a further limitation of the proposed method. Let us finally mention that the data scatter of S-N curves for both the virgin and nitrided specimens does not cause a significant complication, as one can observe in Figure 3.76.

The main results of the fish-eye analysis can be summarized in the following points:

1. The nitrided layer reduces the crack growth rate. The retardation effect of the residual compressive stresses is prominent especially for the fish-eye cracks initiated on inclusions at depths $h < 0.7\,\text{mm}$, as evidenced by the radial asymmetry $S_r < 1$.
2. The average size of the fish-eye crack increases with increasing distance from the free surface and the number of cycles to failure.
3. The elliptical coefficient Q increases with an increasing portion of the torsion loading component. The transient value of the loading ratio can be estimated to be $r_t \approx 0.4$.
4. An assessment of residual stresses can be made by a combination of fractographical and strength analyses related to the fish-eye centre. The applicability of this theoretical method is restricted to the innermost part of the nitrided layer and to the adjacent core region.

3.4 Failure Analysis

Linear–elastic fracture mechanics can be successfully applied in numerical procedures predicting the residual fatigue life of structural components containing cracks longer than about 0.5 mm. According to defect-tolerant design approaches to fatigue, the useful fatigue life is the number of cycles to propagate the largest undetected crack to an unstable fracture. The discovery of fatigue striations in the early 1960s has promoted a development of quantitative retrogressive methods that can be very useful for the reconstruction of conditions under which the failure process occurred.

In failure analysis, the re-estimation of the stress amplitude, the cyclic ratio and the number of cycles to failure is of main practical interest [246]. First attempts to assess those parameters from the fracture surface were published in the early 1960s [405, 406]. These simple approaches allowed the assessment of applied loading amplitude (assumed to be a constant) but the value of cyclic ratio had to be anticipated. Similar concepts were described in the early 1970s [407, 408]. Several years later two different methods enabling an assessment of both $\Delta\sigma$ and R were reported by Uchimoto *et al.* [409] and Pokluda and Staněk [410, 411]. The method proposed by the latter authors is applicable even to cases of stationary random loading. Nowadays, failure analysis can be supported by laboratory devices that are able to simulate the fatigue process on real components under realistic loading spectra. Methods based on overload markings or image analysis can better identify the function $s(\mathrm{d}a/\mathrm{d}N)$ that defines the difference between striation spacings and real crack growth rate during the whole period of stable crack propagation [412, 413]. However, these time consuming and expensive methods do not necessarily provide much more accurate results. The reason lies in a lack of knowledge about a precise fracture mechanical description of crack growth rate with respect to the complexity of both applied loading spectra and fatigue crack paths in structural components [192, 347].

As a useful example, the method [411] which was developed by one of the authors during his employment in the military research institute is presented here. This method was widely used in case studies in order to assess fatigue loading parameters as well as the number of cycles spent by a long fatigue crack during its propagation from a small initial size to the final fracture (e.g., [414, 415]. The original method was recently improved by utilizing a more sophisticated relation describing the whole range of the long crack growth rate [416]. The applicability of the method is demonstrated in a case study concerning the fatigue failure of a compressor blade.

3.4.1 Theoretical Background

As was shown in Section 3.1, both the subcritical (stage II) crack propagation and the unstable fracture are usually reflected on the fracture surface in a characteristic way. In the reconstruction procedure, the following assumptions are utilized: (1) an approximate 1:1 relation between the mean projected striation spacing $\bar{s}$ and the crack growth rate in the Paris–Erdogan region of stage II crack growth; (2) the morphological boundary between stable/unstable crack growth, associated with the the crack length a_c, can be simply related to the cyclic fracture toughness K_c. Moreover, the basic crack-growth rate equation proposed by Forman *et al.* [417] is used in a modified form allowing a better accommodation to the near-threshold region:

$$\frac{\mathrm{d}a}{\mathrm{d}N} = A\frac{\Delta K^n - \Delta K_{th}^n (1-R)^{nm}}{K_c(1-R) - \Delta K}, \tag{3.37}$$

where ΔK is the range of applied stress intensity factor, ΔK_{th} is the threshold factor for $R = -1$, R is the cyclic ratio and A, m, n are experimental constants. Equation 3.37 reduces to the Klesnil–Lukas relation [418] for $\Delta K_{th} \approx \Delta K \ll K_c$. As a rule, $n \in (2,5)$ and $m \in (0.3, 0.5)$ for metallic materials. The left-hand side of Equation 3.37 may be replaced by the mean striation spacing $\bar{s}$ ($\mathrm{d}a/\mathrm{d}N \approx \bar{s}$) that should be measured closely in the range of $a/a_c \in (0.3, 0.6)$ on the fracture surface (usually corresponding to the Paris–Erdogan region). The value $\bar{s}$ should be determined according to the relation at Equation 3.4 at k points corresponding to the same crack length a on the fracture surface. It should be emphasized, however, that the limit $k = 1$ is also permissible which substantially extends the applicability of the method. For many reasons, indeed, one can find only a single facet covered by striations. On the other hand, the accuracy of such an assessment might be lower.

The moment of unstable fracture corresponds to the cyclic fracture toughness K_c and to the critical crack length $a = a_c$. Therefore, one can write

$$K_c = \frac{\Delta K(a_c)}{1-R}. \tag{3.38}$$

Since $\Delta K = \Delta\sigma\sqrt{\pi a} f(a, a/W)$ (W is the width of the component and $f(a, a/W)$ is the shape function), a numerical solution of Equations 3.37 and 3.38 gives the estimates of loading parameters $\Delta\sigma$ and R that caused failure of the structural component. It should be noted that comprehensive handbooks of stress intensity factors are available today (e.g., [419]).

Substituting the obtained values $\Delta\sigma$ and R into the relation

$$N_c = \int_0^{N_c} \mathrm{d}N = \frac{1}{A}\int_{a_0}^{a_c} \frac{K_c(1-R) - \Delta K}{\Delta K^n - \Delta K_{th}^n (1-R)^m}\, \mathrm{d}a, \tag{3.39}$$

one can calculate the number of cycles N_c, associated with the fatigue crack propagation in the range of $\langle a_0, a_c \rangle$, where a_0 is the initial crack length (usually a minimum detectable length of inspection methods).

For the method described above, nearly constant values of $\Delta\sigma$ and R during the fatigue process are assumed. In engineering practice, however, this kind of loading occurs rather rarely (only about 10%). Roughly 40% of applied loading cases are stationary or quasi-stationary spectra (highly variable ΔK and nearly constant R). In these cases, the striation spacings $\bar{s}_1$ and $\bar{s}_2$ for at least two different crack lengths a_1 and a_2 are to be measured on the fracture surface within the Paris–Erdogan region [411]. The reason lies in an uncertain stress range at the moment of the final fracture corresponding to the crack length a_c. The values $\Delta\sigma_{rms}$ and R can then be obtained by a simultaneous numerical solution of two Equations 3.37 for a_1 ($\bar{s}_1$) and a_2 ($\bar{s}_2$), where $\Delta\sigma_{rms}$ means the root mean square value of the applied spectrum loading. About 50% of all exploitation loading spectra exhibit a highly variable R, i.e., a non-stationary behaviour. In these cases the method presented becomes useless, similar to most other approaches.

Thus, the following steps should be performed to accomplish the failure analysis:

1. experimental determination of Equation 3.37 in the laboratory by using samples made of the material of the fractured component;
2. selection of appropriate shape function(s) $f(a, a/W)$ for the fractured component and crack lengths a and a_c;
3. measurement of $\bar{s}$ for the selected length(s) a on the fracture surface;
4. measurement of the critical length a_c on the fracture surface;
5. numerical solution of Equations 3.37 and 3.38 to estimate $\Delta\sigma$ and R;
6. anticipation of the shape function $f(a, a/W)$ in the whole range of (a_0, a_c);
7. numerical solution of Equation 3.39 to assess N_c.

3.4.2 Case Study

Apart from many other applications, this method was used to reconstitute conditions of a surprisingly quick fatigue failure of a compressor blade in an aircraft engine after a general repair. The engine was tested in a stand placed on the ground inside a special semi-natural cave. Blocks of a constant loading amplitude were applied by changing engine frequencies to simulate the service regime. The blade was made of the high-strength alloy Ti-4Al-3Mo-1.5Zr ($\sigma_y = 1060\,\text{MPa}$). A careful investigation in the scanning electron microscope revealed that the fatigue crack was initiated at the leading edge of the blade. Some silicon micro-particles were found to be stuck in the blade surface along with many micro-craters near the crack initiation site. Thus, the crack initiation was a result of impacting silicon particles coming, most probably, from a dust whirled up from the ground.

The shape function for the cracked compressor blade can be written as

$$\begin{aligned} f(a) = {} & 0.56419(1 - 1.1 \times 10^{-4}a^3) \times \\ & \times (1.67687 - 0.43573a + 0.0819338a^2 - 0.0065158a^3 + 0.00018858a^4). \end{aligned}$$

This form was obtained by compliance measurements performed on an identically designed blade containing crack growing from the leading edge [420]. The influence of a variable blade width W is already respected by coefficients in the shape function (a in [mm]).

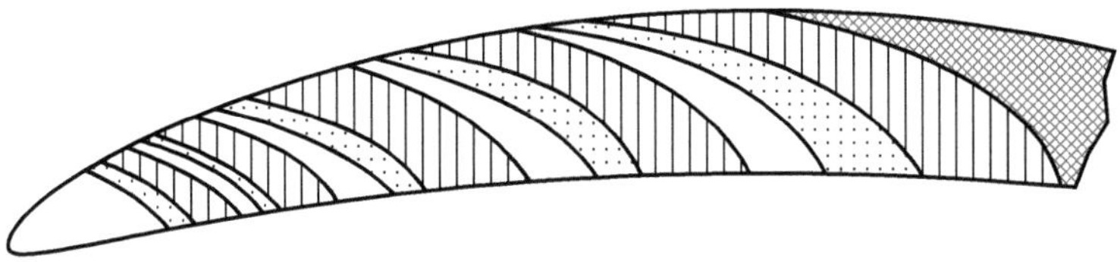

Figure 3.77 The scheme of growth curves marking different loading regimes on the fracture surface of the compressor blade

Table 3.11 Measured and calculated values

a [mm]	$\bar{s}$ [nm]	R	$\Delta\sigma$ [MPa]	N_c
8	45	0.345	518	2.01×10^6
10	120	0.265	582	1.28×10^6

The macromorphology of the fracture surface exhibited a sequence of characteristic regions and growth curves corresponding to different loading blocks; see Figure 3.77. Bioden-carbon replicas of fracture surfaces had to be prepared for observations using the transmission electron microscope in order to find some facets covered by fatigue striations. The critical crack length $a_c = 17\,\text{mm}$ corresponded to the final fracture. Three-point bending specimens made of the blade material were used for measuring parameters in Equation 3.37: $\Delta K_{th} = 5\,\text{MPa}\,\text{m}^{1/2}$, $K_c = 80\,\text{MPa}\,\text{m}^{1/2}$, $A = 4.582 \times 10^{-36}$ [Pa,m], $n = 4.9$ and $m = 0.3$. Measured mean striation spacing $\bar{s}$ related to two different crack lengths is shown in Table 3.11 along with calculated values of R, $\Delta\sigma$ and N_c.

Computed values of $\Delta\sigma$ lie below the fatigue limit $\Delta\sigma_c \approx 600\,\text{MPa}$ (for $R \approx 0.3$) of the blade material. The initial crack length $a_0 = 0.5\,\text{mm}$ was chosen to be nearly equal to the size of silicon particles. The computed residual fatigue life $N_c \approx 10^6$ cycles also shows that the stable crack propagation ran at a stress below the fatigue limit of the material. The crack initiation stage was very low due to the impact of silicon particles. When neglecting the initiation stage and assuming the blade vibration frequency of several hundreds of Hz, the duration of the whole fracture process was no longer than several days.

It was realized later on that the test was performed during a week when, due to a service cut-off, the ground floor in the cave remained uncleaned.

Chapter 4
Final Reflections

The book has been finished and the authors would like to thank in particular those tolerant and patient readers who have read it to the very end. One can be quite sure that such readers would not mind further brief reflection. In spite of the fact that these final remarks will bring no additional scientific information, the authors believe that selected pieces of knowledge that are assembled in the main text might be worth mentioning again. Because this book preferentially refers to the scientific work of the authors, only the results that arose from their own research will be highlighted. On the other hand, many important things still remain open for further investigations. Some of these tasks related to our research will also be recalled.

4.1 Useful Results

1. The highest achievable tensile strength of a solid of a particular chemical composition, the ideal tensile strength of the perfect crystal, strongly depends on the stress triaxiality. For the majority of metallic crystals, the ideal strength related to the volumetric instability increases almost linearly with increasing transverse biaxial stresses. On the other hand, ceramic crystals with a diamond structure exhibit a sharp maximum either close to the zero biaxial stress (Si, Ge) or in the compressive biaxial region (C).
2. Under uniaxial tension, however, perfect metallic crystals usually fail when reaching the first shear instability. Therefore, their uniaxial ideal strength σ_{iut} does not exceed 10 GPa which is far below their volumetric ideal uniaxial strength and only several times higher than that of the strongest related engineering metallic materials. This is not necessarily true for the case of isotropic (hydrostatic) tension in which the ideal strength σ_{iht} usually reaches values as high as several tens of GPa.

3. The ranking of tensile strengths σ_u of the strongest grades of engineering materials with chemically different matrices nearly follows that of σ_{iht} of related perfect crystals. The same is true for the ductile/brittle response of engineering materials and that of their perfect crystals. Therefore, these mechanical properties are, to a considerable extent, predetermined by those of a perfect lattice. On the other hand, such a correspondence is not apparent for the fracture strain and its dependence on the stress triaxiality. This property is determined instead by crystal defects and secondary phases.
4. Crystals of metals and diamond exhibit a nearly linear decrease in ideal shear strength with increasing superimposed isotropic (hydrostatic) stress σ_h. On the other hand, the dependence $\tau_{is}(\sigma_h)$ of covalent crystals Si, Ge and SiC reveals an opposite trend. These trends also refer to the dependence of ideal shear strength on the superimposed normal stress. This is in agreement with the normal stress influence found for the dislocation nucleation stress (non-Schmid behaviour). However, the ideal shear strength of Cu and Ni crystals becomes lowered by both tensile and compressive normal stresses.
5. By considering the coupling of shear and normal stresses, the values of σ_{iut} can be calculated from those of τ_{is}. This method, proven on a variety of fcc metals, avoids cumbersome examinations of the stability conditions necessary for a standard computation of σ_{iut}.
6. Nanoindentation appears to be the only efficient method for an experimental determination of ideal shear strength values. This method is based on a quantitative interpretation of the pop-in effect on the load–penetration diagram.
7. The geometrically-induced shielding of the crack tip due to crack kinking and branching can substantially increase fracture toughness values of brittle and quasi-brittle engineering materials. This is the reason why, even in the case of severely segregated grain boundaries of very low fracture energy, the fracture toughness of such defect containing steels can reach values higher than 50 MPa m$^{1/2}$. This effect also elucidates the anomalous fracture behaviour of ultra-high-strength low-alloy steels when considering an influence of the size ratio d/r_p (the characteristic microstructural parameter/the plastic zone size). The size ratio is a very important parameter that couples micro (d) and macro (r_p) scales of fracture and fatigue processes. Decreasing size ratio means a decreasing influence of microstructure and thus a decreasing level of geometrical shielding.
8. Enhanced dislocation activity around growing voids gives a steep increase of the strain rate inside the neck during the tensile test of ductile metallic materials. The related model reasonably predicts the value of fracture strain when using the Brown–Emburry or percolation based void-coalescence models.
9. The fracture toughness of steels exhibiting a ductile fracture morphology can be reasonably predicted by means of a simple model based on the frac-

ture strain diagram. This diagram can be easily determined from tensile and compressive tests.

10. The ratcheting process in metallic materials starts when reaching a critical plastic strain range during cyclic softening. This threshold value decreases with increasing cyclic ratio. The initial ratcheting rate can be well described by a model based on discrete dislocation theory. This model explains the role of cyclic softening (hardening) in the acceleration (deceleration) of the ratcheting process.
11. Besides geometrically-induced shielding, an increasing size ratio also raises the effect of the roughness-induced crack closure in fatigue. A generalized model of crack tip shielding and closure under mode I loading, based on the concepts of discrete dislocations and size ratio, allows one to separate the intrinsic component from the measured fatigue threshold value. This model correctly predicts a decrease in roughness-induced crack closure with increasing range of applied stress intensity factor as well as the low fatigue threshold in nanostructured materials.
12. Micromechanisms of mode II and mode III crack propagation in metallic materials are generally different. Unlike for the mode II case, the commonly accepted idea of fatigue crack front advance due to an oxygen-assisted creation of new fracture surfaces is rather irrelevant for the mode III case. Therefore, mode II based micromechanisms of the remote mode III crack growth were proposed and experimentally verified for austenitic steel. In notched specimens, the mode I growth by formation of factory roofs (F-Rs) is a typical micromechanism under the remote mode III crack tip loading in the high-cycle fatigue region. These rather complicated micromechanisms are, most probably, the main reason for the low mode III crack growth rates observed in some steels.
13. In the high-cycle fatigue region, the size ratio is rather high and shear mode cracks incline (or branch) to the mode I loading case very easily. This is mainly caused by an increase in both the roughness-induced friction and the microstructurally affected tortuosity of the crack path. Moreover, there is a synergy of modes II and III by a formation of the mode I branch in terms of the crack driving force. The generalized conditions of mode I branching from the shear-mode crack propagation can be defined as follows: (1) the first branch forms at that site of the shear crack front where the value of ΔK_I on a facet of a potential mode I branch becomes maximal; (2) the branching appears at the moment when this maximal value exceeds that of the threshold ΔK_{Ith} for the applied cyclic ratio.
14. A textbook example of the branching behaviour is a formation of the factory roof from mode II+III semielliptical cracks on the surface. The relevancy of a quantitative model of F-R formation based on the synergy effect was proven by stereogrammetrical measurements of the three-dimensional topography of factory roofs. There are two main reasons why the formation of factory roofs is not observed in the low-cycle fatigue region. First, the size of factory-roof patterns decreases with increasing density of crack nu-

clei on the surface which becomes very high in the low-cycle fatigue region. Second, in this region a network of microcracks at microstructural heterogeneities is quickly produced inside a large plastic zone that is coplanar with the notch plane and embraces many characteristic microstructural elements (low size ratio). Coalescence of these microcracks and the main crack front keeps the crack path in their common (shear) plane.

15. Plasma nitriding can substantially raise not only the fatigue limit of mild steels but also that of high-strength steels under various kinds of loading such as push-pull, bending, torsion and combined bending-torsion. Both the improved hardness and the compressive residual stresses in the nitrided layer cause fish-eye cracks to initiate at inclusions inside the specimen and propagate in a near vacuum towards the surface. Both the level and the sign of residual stresses that are present in locations of fish-eye centres can be assessed by utilizing a method based on a determination of bending (or push-pull) stress amplitude corresponding to these centres on the fracture surface.
16. A rather simple method can be employed to reconstitute the applied stress, the cyclic ratio and the number of cycles spent for stable crack propagation during the fatigue process. This method couples quantitative fractography on both microscopic and macroscopic levels and might be very useful for the analysis of failure cases of engineering components and structures.

4.2 Open Tasks

17. Calculations of ideal strength should be corrected by thorough analyses of phonon spectra under various loadings and temperatures. This should finally explain the gaps between the strength of the strongest solids and their upper theoretical limits.
18. Verification of the simple method for recalculating the ideal tensile strength from the ideal shear-strength data of bcc metals and ceramics.
19. The synergy effect found in W-V and Mo-V nanocomposites should be verified for more composite components. A physically justified reason for this phenomenon might possibly be found by means of a precise analysis of stress coupling effects.
20. Nanoindentation experiments should be performed on as many single crystals as possible in order to establish an extended database of experimental values of ideal shear strengths.
21. The dependence of the fracture energy on the grain boundary concentration of phosphorus and other detrimental elements should be established by fracture experiments on well defined segregated bicrystal boundaries. This would finally solve the problem of the severity of individual impurities and set up physically justified impurity limits.

22. Correct measurements of crack growth rates and crack growth thresholds in mode II and mode III should be performed for an extended range of metallic materials. This would solve the practically important question concerning the resistance of materials to crack propagation under both kinds of shear mode loading.
23. The generalized conditions of mode I branching from shear-mode crack propagation should be verified for more metallic materials under various external shear loading modes.
24. Extended experiments on fish-eye crack initiation and growth should be realized for materials with reinforcing surface layers in both standard and gigacycle fatigue regimes. This would decisively complete our knowledge about micromechanisms of interior nucleation and propagation of fatigue cracks.

The authors hope that performing a majority of these tasks might be attempted in a rather near future.

Appendix A
Ab initio Computational Methods

The objective of *ab initio* approaches based on applied quantum theory is to calculate stationary states for electrons in the electrostatic field of nuclei, i.e., the electronic structure (ES). The energy of this ground state can then serve as a potential energy for displacements of nuclei. From the point of view of ideal-strength (IS) calculations, the total energy of the system is the most important output of *ab initio* methods.

The main advantage of *ab initio* methods is their independence of experimental data. Unlike in the case of empirical and semi-empirical methods, there is no need for calibration parameters. Thus, they can also be used for calculation of some structural and mechanical characteristics of hypothetical systems (prediction of properties of materials that have not yet been developed) or study of materials behaviour under large deformations (far from the equilibrium state) that can give a better understanding of micromechanisms of materials failure.

First attempts to develop applicable theories were made in the late 1920s [421, 422], a few years after the foundations of modern quantum theory were laid (derivation of the Schrödinger equation). A very successful step forward was the Hartree–Fock method [422, 423]. This method yields very accurate bond lengths in molecules. On the other hand, the binding energies are generally not in good agreement with experimentally obtained energies. Moreover, for solids, the Hartree–Fock method has problems with a description of band structures. The density functional theory (DFT) [37, 38] was invented to include correlation effects without using the very costly wave function methods. All the methods used within this book are based on the DFT.

In DFT the energy is not obtained as eigenvalues of a wave function, but rather as a functional of the electron density. The complex problem of many interacting electrons is transformed into a much simpler study of single electron interactions with an effective potential U_{eff} created by other electrons and all nuclei. This is expressed by the Kohn–Sham equation (one-electron Schrödinger equation)

$$(-\Delta + U_{eff}(\mathbf{r}) - \varepsilon_i)\psi_i(\mathbf{r}) = 0,$$

where ε_i represents the one-electron energies and ψ_i are the one-electron wave functions.

The wave functions are then occupied in accordance with the Pauli principle and a new field is obtained by solving the Poisson equation for point nuclei shielded by the electronic density

$$\rho(\mathbf{r}) = \sum_{i,occ} |\psi_i(\mathbf{r})|^2 .$$

In the case of periodic crystalline materials, the one-electron wave functions are expanded into appropriate basis sets and satisfy the Bloch theorem. The Kohn–Sham equation is solved iteratively until the solution becomes self-consistent. This means that the electron density, determined from the effective one-electron potential, must generate the same effective potential (which is again a functional of the electron density). The self-consistent cycle usually starts with a guess of the effective potential (superposition of atomic-like potentials) and then the input and output potentials are appropriately mixed before starting a new iteration. The quality and speed of the convergence of such calculations is related not only to the choice of a suitable basis, but also to the sophistication of the iterative process. The necessary corrections for exchange and correlation are also included in the effective potential U_{eff}. This seems to be the crucial point of *ab initio* calculations because the exchange-correlation (XC) functional is not known exactly and must be approximated. The first (and the simplest) attempt to build an approximation of the XC energy functional in the DFT is the local density approximation (LDA) [424]. The LDA is local in the sense that the electron exchange and correlation energy at any point in space is a functional of the electron density at that point only. As a consequence of this, LDA fails in situations where the density undergoes rapid changes (molecules). An improvement to this can be made by considering the gradient of the electron density. The density gradient corrections are implemented in the so-called Generalized Gradient Approximation (GGA). While there is only one LDA there are several different parameterizations of the GGA [425–427].

Nevertheless, it is the choice of the basis wave functions that makes the main difference between various methods used for the ES calculations. The better we choose them (according to the character of the problem), the smaller number of them is needed for the description of one-electron wave functions. Commonly used bases are augmented (APW) and orthogonalized (OPW) plane waves, linear muffin-tin orbitals (LMTO), linear combination of atomic orbitals (LCAO), of Gaussian orbitals (LCGO) and linear augmented Slater-type orbitals (LASTO), augmented spherical waves (ASW), etc. The Korringa–Kohn–Rostoker (KKR) method proceeds by the use of the Green function of the Kohn–Sham equation and is also called Green's function (GF) method. The pseudopotential approach applied mostly to solids containing no d- or f-electrons is also widely used. A detailed description of these methods may be found in many books and articles, e.g., in [86, 88, 89]. The atomic configurations corresponding to deformed structures usually have lower symmetries and, at the strength limit, they are very far from the lowest-energy equilibrium state. Therefore, to obtain reliable structural energy differences, the full-potential methods (i.e., without any shape approximation of the crystal potential and the electronic charge density) have to be utilized in such studies. At present, several codes are available, e.g., WIEN, VASP, FHI, FLEUR, FPLO, FPLMTO, ABINIT, SIESTA, etc.

All the *ab initio* calculations of IS presented in this book were performed by using the following three computational codes: LMTO-ASA, WIEN and VASP.

A.1 TB-LMTO-ASA Code

Linearized *ab initio* methods have been successfully utilized for solving many problems in materials science [428, 429]. One of the most effective approaches for early first principles calculations was the LMTO formalism which has been continuously developed since 1980 [428–430]. This method is very appropriate for self-consistent calculations.

In the LMTO-ASA code [431], the crystal potential U is approximated by a muffin-tin (MT) shape potential which is composed of a set of spherically symmetric potentials inside slightly overlapping spheres around individual nuclei and a constant potential in the interstitial region outside the spheres (Figure A.1). Atomic-like orbitals derived for the MT potential constitute a suitable basis set. The tight binding approximation, which is also implemented into the code [432], assumes that the full Hamiltonian may be approximated by that of an isolated atom centred at each lattice point. The atomic orbitals (eigenfunctions of the single-atom Hamiltonian) are assumed to be negligible at distances exceeding the lattice constant.

In all presented calculations, the LMTO method is used in the framework of an atomic sphere approximation (ASA) which is particularly suitable for closely packed structures like fcc, hcp or bcc [428]. The size of a spherically symmetric potential is assumed to be equal to that of the Wigner–Seitz cell (Figure A.1(b)). This suppresses the interstitial region and neglects the kinetic energy of related free electrons. Owing to the necessary space-filling condition, ASA represents a physically plausible model only for a description of an infinite system of atomic spheres.

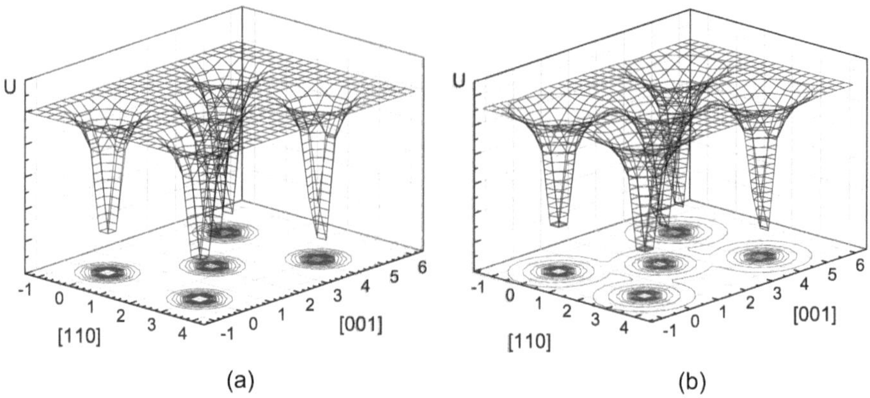

Figure A.1 Muffin-tin shape potential in (110) plane of a general bcc crystal of the lattice constant $a = 3\,\text{au}$ with spherical radii: (a) $r_{MT} = 1\,\text{au}$, and (b) $r_{MT} = r_{WS} = 1.48\,\text{au}$

A.2 Wien 95 – w2k Codes

The program package WIEN [433] does not use any shape approximation to the potential. The crystal environment is divided into a region of non-overlapping atomic spheres (centred at individual atomic sites) and an interstitial region as can be seen in Figure A.2. In order to describe ES reliably and effectively, two different basis sets are employed: a product of linear combination of radial functions and spherical harmonics is used inside the spheres whereas the wave functions in the interstitial region are expanded into a linear combination of plane waves. The solution in both regions must be continuous at the spherical boundaries. Each basis function is then defined as a plane-wave in the interstitial region connected smoothly to a linear combination of atomic-like functions in the spheres, thus providing an efficient representation throughout the space. A similar representation is used for potentials and charge densities. The method is called the linear augmented plane wave (LAPW) method [428, 434].

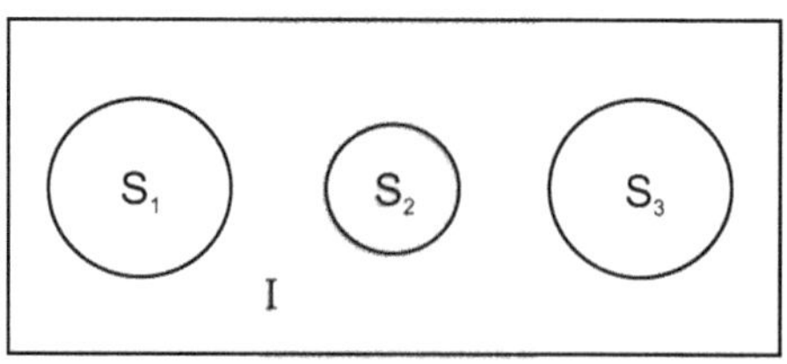

Figure A.2 Illustration of a crystal model – three atomic spheres (S_1 – S_3) with potential $U_S(\mathbf{r}) = \sum_{lm} U_{lm}(r) Y_{lm}(\hat{r})$ embedded in the interstitial region I with $U_I(\mathbf{r}) = \sum_K U_{\mathbf{K}}(r) \mathrm{e}^{i\mathbf{Kr}}$

In order to increase the flexibility of the basis (to improve upon the linearization of wave functions) and to make possible a consistent treatment of semicore and valence states in one energy window (to ensure orthogonality) additional basis functions can be added. They are called local orbitals [435] and consist of a linear combination of two radial functions at two different energies (e.g., at the 3s and 4s energy) and one energy derivative (at one of these energies). The local orbitals are normalized and have zero value and slope at the spherical boundaries.

A.3 VASP Code

Another way of avoiding a problem with plane wave basis set in the vicinity of atomic nuclei, where the number of plane waves would exceed any practical limits (perhaps except for H or Li), is to substitute the exact potential by a pseudopotential.

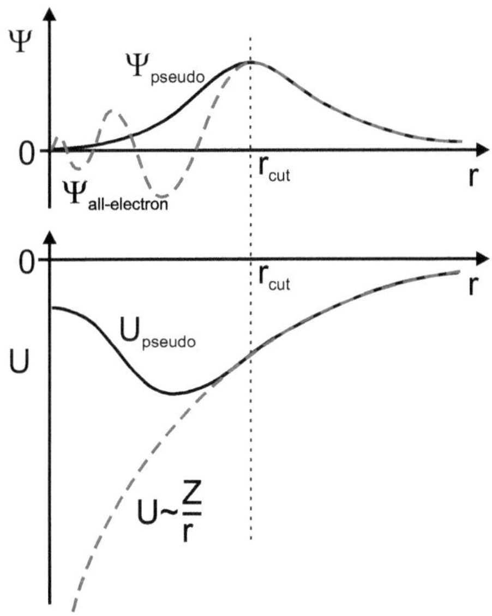

Figure A.3 Comparison of a wavefunction in the Coulomb potential of the nucleus (*dashed lines*) to that in the pseudopotential (*solid lines*)

Construction of the pseudo-wavefunctions is schematically described in Figure A.3. The Coulomb potential of the nucleus and the corresponding wavefunction are represented by dashed lines. The solid line displays the pseudopotential and the pseudo-wavefunction. The real and the pseudo-wavefunction (and also the potentials) match above a certain cutoff radius r_{cut}.

The pseudopotential approach has been implemented in the VASP (Vienna *ab initio* Simulation Package) code. This code was developed at the Institut für Materialphysik, Universität Wien. The VASP currently supports three types of pseudopotentials: norm-conserving pseudopotentials, ultrasoft pseudopotentials and projector augmented wave pseudopotentials [436–438]. In all three cases, the core electrons (at lower energy levels than valence electrons) are precalculated in an atomic environment and kept "frozen" during the remaining calculations.

Appendix B
Mixed-mode Criteria of Crack Stability

In order to describe the crack stability under mixed-mode loading, various concepts within the framework of LEFM were proposed (e.g., [162–164]). In this brief overview, the stress intensity factors are denoted only by capital letters K_I, K_{II} and K_{III}, as is common for remote quantities. However, the criteria are relevant also to local stress intensity factors usually denoted as k_1, k_2 and k_3.

B.1 Energy Criterion

The criterion postulates that the total energy $G_I+G_{II}+G_{III}$, released by the system to create a unit of a new surface, equals the crack growth resistance G_c:

$$G_c = G_I + G_{II} + G_{III}. \tag{B.1}$$

When the crack grows in its original plane (or propagates in a self-similar manner), the energy criterion can be expressed in terms of the effective K-factor as

$$K_c = \sqrt{K_I^2 + K_{II}^2 + \frac{4}{\kappa+1}K_{III}^2}, \tag{B.2}$$

where K_c is the critical stress intensity factor (fracture toughness), $\kappa = 3-4\nu$ for plane strain, $\kappa = (3-\nu)/(1+\nu)$ for plane stress and ν is the Poisson's ratio.

B.2 Criterion of Linear Damage Accumulation

Each of Equations B.1 and B.2 can be understood as a condition of subsequently reaching the critical level of material damage during loading. If the partial damages accumulate independently, the mixed-mode criterion

$$\frac{G_I}{G_{Ic}} + \frac{G_{II}}{G_{IIc}} + \frac{G_{III}}{G_{IIIc}} \leq 1$$

can be accepted. Clearly, for the special case $G_{Ic} = G_{IIc} = G_{IIIc} = G_c$, the criterion of linear damage accumulation transfers to the energy criterion. Under the dominant mode I loading, the problem of shear friction in modes II and III on the rough fracture surfaces is not crucial due to sufficiently large opening displacements of crack flanks. Consequently, the equality of critical crack driving forces in all modes might be assumed and the energy criterion can be accepted.

Since the assumption of self-similar crack propagation does not hold under a general remote mixed-mode loading, one must be careful when applying both the above-mentioned criteria. They appear to be useful only for a prediction of the moment of unstable fracture or, generally, in the case of a dominant external mode I loading of microscopically tortuous cracks. The description of long fatigue crack propagation under external mixed-mode loading conditions by means of these criteria is rather limited in the sense that the growth direction cannot be predicted.

B.3 Criterion of Minimal Deformation Energy

The deformation energy density S is minimal in those elements near the crack front, where the ratio of the hydrostatic tensile stress and the octahedral stress is maximal [439]. In such elements, therefore, the first local fracture is expected to occur and the crack propagates towards such damaged sites in the matrix. In the plane mixed-mode I+II, the condition of minimal S yields the stability condition as

$$K_{\mathit{eff}}^2 = \frac{16G}{2\,(\kappa - 1)} \left(a_{11}(\theta_m) K_I^2 + 2a_{12}(\theta_m) K_I K_{II} + a_{22}(\theta_m) K_{II}^2 \right),$$

where $a_{ij}(\theta_m)$ are angular functions and G is the shear modulus. The criterion generally enables us to predict the moment of crack nucleation as well as the crack growth direction in the 3D homogeneous continuum. Its disadvantage is a rather low transparency.

B.4 Criterion of Maximal Principal Stress

This criterion can be derived from the energy criterion by expressing the effective crack driving force G_{eff} as a function of growth angle θ and by searching its maximal value. It can be shown that this maximum corresponds to a crack perpendicular to the maximum principal stress. Considering the plane model, stresses in close vicinity of the crack tip can be expressed as follows:

$$\sigma_\theta = \frac{1}{\sqrt{2\pi r}} \cos\frac{\theta}{2}\left(K_I \cos^2\frac{\theta}{2} - \frac{3}{2}K_{II}\sin\theta\right),$$
$$\tau_{r\theta} = \frac{1}{2\sqrt{2\pi r}} \cos\frac{\theta}{2}\left(K_I \sin\theta + K_{II}\left(3\cos\theta - 1\right)\right).$$

If $\tau_{r\theta} = 0$, then the hoop stress σ_θ becomes the principal stress. This condition determines the growth angle θ_m as well as the related effective stress intensity value:

$$\tan\frac{\theta_m}{2} = \frac{K_I}{4K_{II}} \pm \frac{1}{4}\sqrt{\left(\frac{K_I}{K_{II}}\right)^2 + 8},$$

$$K_{eff} = K_I \cos^3\frac{\theta_m}{2} - 3K_{II}\cos^2\frac{\theta_m}{2}\sin\frac{\theta_m}{2}.$$

The criterion can be used for predictions of crack stability and growth in homogeneous materials. The criterion does not involve the antiplane mode III and, consequently, it can be applied only in the framework of 2D models. Nevertheless, a very good applicability of the maximum principal stress criterion to propagation of long fatigue cracks was already proven (e.g., [399,440]). This success can be understood from the micromechanical point of view. Indeed, the fatigue crack tries to maximize the opening as well as the friction between the flanks. Consequently, it inclines to the maximum opening mode I perpendicularly to the maximal main stress direction, thereby also minimizing the shear modes (friction).

Appendix C
Plastic Flow Rate Inside the Neck

C.1 Ideal Model

When taking Equations 2.25, 2.26 and ranges $\lambda_1 \in \langle 0.8, 2\rangle$, $\lambda_2 \in \langle 0, 1\rangle$ into account, the growth of the semi-axes of elliptical voids at the onset of necking can be described as

$$a = a_0 \mathrm{e}^{0.8\varepsilon_p}, \qquad b \approx b_0 \equiv a_0 \tag{C.1}$$

while

$$a = a_0 \mathrm{e}^{2\varepsilon_p}, \qquad b \approx a_0 \mathrm{e}^{\varepsilon_p} \tag{C.2}$$

holds in the final stages of necking just before the final fracture. In order to calculate the active volume V_2, the surface area of the elliptical void S_{el} has to be considered as

$$S_{el} \cong 2\pi \left(b^2 + ab\right). \tag{C.3}$$

By inserting Equations C.1 and C.2 into Equation C.3 and transferring the surface of the ellipsoid to that of the equivalent sphere of the same surface area one obtains

$$^{a}S_{el} = 4\pi a_0^2 \mathrm{e}^{0,4\varepsilon_p}, \qquad ^{b}S_{el} = 4\pi a_0^2 \mathrm{e}^{2,5\varepsilon_p}.$$

A corresponding error of this transfer is less than 10% for $0 < \varepsilon_p < 1$. Thus, the variation of the void surface during the necking process can be expressed as

$$S_{el} \approx 4\pi a_0^2 \mathrm{e}^{\kappa\varepsilon_p}, \tag{C.4}$$

where $0.4 < \kappa < 2.5$. Let us denote δ the mean distance from the void surface to the outer boundary of the volume V_2 (see Figure 2.30) and n the number of voids in V_1. With respect to Equations C.4 and 2.30, where $V_0 = \pi r_0^2 h$, one obtains

$$V_2 \cong 4\pi a_0^2 \delta n e^{\kappa \varepsilon_p}.$$

By considering Equation 2.30 ($V = V_1$) and denoting $\Theta = \frac{4\pi\gamma a_0^2 \delta n}{V_0}$, $\gamma = 1+\kappa$ we get

$$\mathrm{d}\left(\frac{V_2}{V_1}\right) \cong \mathrm{d}\left(\frac{4\pi a_0^2 \delta n}{V_0} \mathrm{e}^{(1+\kappa)\varepsilon_p}\right) = \Theta \mathrm{e}^{\gamma \varepsilon_p} \mathrm{d}\varepsilon_p. \tag{C.5}$$

When inserting Equation C.5 into Equation 2.33 one obtains

$$\frac{\mathrm{d}f_m}{\mathrm{d}\varepsilon_p} \cong -\Phi f_m + \Theta \mathrm{e}^{\gamma \varepsilon_p},$$

which is a linear differential equation. By solving this equation, a change of the fraction of mobile dislocations during the unstable deformation inside the neck can be expressed in the following form:

$$f_m \cong \exp\left\{\int_0^{\varepsilon_p} -\Phi \mathrm{d}t\right\} \left[f_m^u + \int_0^{\varepsilon_p} \mathrm{e}^{\gamma s} \exp\left\{\int_0^{s} \Phi \mathrm{d}t\right\} \mathrm{d}s\right].$$

Here f_m^u is the relative density of mobile dislocations at the onset of necking. After integration and some algebraic re-arrangement

$$f_m \cong \left(f_m^u \frac{\Theta}{\gamma + \Phi}\right) \mathrm{e}^{-\Phi \varepsilon_p} + \frac{\Theta}{\gamma + \Phi} \mathrm{e}^{\gamma \varepsilon_p}. \tag{C.6}$$

Equation C.6 shows the dependence of the relative density of mobile dislocations on the plastic deformation during the necking process. By introducing Equation C.6 into the general relationship at Equation 2.29 one obtains

$$\dot{\varepsilon}_p = \mu b v_c \left(\rho_u + M \varepsilon_p\right) \exp\left\{-\frac{D_u}{\langle \sigma_s \rangle_k}\right\} \left(\Omega_1 \mathrm{e}^{-\Phi \varepsilon_p} + \Omega_2 \mathrm{e}^{\gamma \varepsilon_p}\right), \tag{C.7}$$

where $\Omega_1 = f_m^u - \frac{\Theta}{\gamma + \bar{\Phi}}$, $\Omega_2 = \frac{\Theta}{\gamma + \bar{\Phi}}$ and $\bar{\Phi} = \frac{H}{\langle \sigma_s \rangle_k}$. Moreover, ρ_u and D_u are the density of dislocations and the friction stress, respectively, at the onset of necking, $\langle \sigma_s \rangle_k$ and $\bar{\Phi}$ are the mean effective stress and the mean value of Φ during the necking process, respectively ($\langle \sigma_s \rangle_k$ is nearly constant). Finally, an appropriate algebraic re-arrangement of Equation C.7 leads to Equation 2.36.

C.2 Real Model

The formation of cavities at secondary phase particles can be described by assuming either the energetic (matrix/particle interphase) or fracture-mechanics (breaking of the particle) criteria. In both cases, the analysis leads to the relation $\sigma_{cr} \propto a^{-1/2}$, where σ_{cr} is the critical applied stress for the cavity nucleation and a is the particle size (e.g., [157,441,442]. Therefore, the total initial surface S_P of nucleated cavities can be determined as

$$S_P \approx 4\pi K_1 \int\limits_{(K_2/\sigma_{cr})^2}^{\infty} a^{-2}\mathrm{d}a. \tag{C.8}$$

Beyond the ultimate stress σ_u, the true stress $\sigma(\varepsilon_p)$ is nearly a linear function of ε_p [236, 237]:

$$\sigma_{cr} \approx \frac{4}{3}\sigma_u + K_3\varepsilon_p, \tag{C.9}$$

where K_3 is a constant. A combination of Equations C.9 and C.8 gives

$$S_p \approx 4\pi K_1 \int\limits_{\left[\frac{K_2}{(4\sigma_u/3+K_3\varepsilon_p)}\right]^2}^{\infty} a^{-2}\mathrm{d}a. \tag{C.10}$$

Nucleated voids grow according to Equation 2.24 and their total surface area increases in correspondence with Equation C.4. At the moment of reaching the ultimate stress, the total surface area can be estimated by means of Equation C.10 as

$$S_p^u \approx 4\pi K_1 \int\limits_{(3K_2/4\sigma_u)^2}^{\infty} a^{-2}\mathrm{d}a. \tag{C.11}$$

Let us assume small increments of plastic strain $\Delta\varepsilon_p$. Then, with respect to Equation C.4, the total surface area at $\varepsilon_p^* \gg \Delta\varepsilon_p$ can be approximated as follows:

$$\begin{aligned} S_{el} &\approx S_p^u \mathrm{e}^{\kappa\varepsilon_p} + \left(\frac{\mathrm{d}s_p}{\mathrm{d}\varepsilon_p}\right)_0 \mathrm{e}^{\kappa\varepsilon_p}\Delta\varepsilon_p + \ldots \\ &+ \left(\frac{\mathrm{d}S_p}{\mathrm{d}\varepsilon_p}\right)_{\varepsilon_p/2} \mathrm{e}^{\kappa\left(\varepsilon_p^*/2\right)}\Delta\varepsilon_p + \ldots + \left(\frac{\mathrm{d}S_p}{\mathrm{d}\varepsilon_p}\right)_{\varepsilon_p^*} \Delta\varepsilon_p \\ &= S_p^u \mathrm{e}^{\kappa\varepsilon_p^*} + \sum_{k=0}^{\varepsilon_p/\Delta\varepsilon_p} \left(\frac{\mathrm{d}S_p}{\mathrm{d}\varepsilon_p}\right)_{k\Delta\varepsilon_p} \mathrm{e}^{\kappa\left(\varepsilon_p^*-k\Delta\varepsilon_p\right)}\Delta\varepsilon_p. \end{aligned}$$

For $\Delta\varepsilon_p \to 0$ the sum can be replaced by an integral:

$$S_{el} \approx S_p^u e^{\kappa \varepsilon_p^*} + \int_0^{\varepsilon_p^*} \left(\frac{dS_p}{d\varepsilon_p} \right) e^{\kappa \left(\varepsilon_p^* - \varepsilon_p \right)} d\varepsilon_p . \tag{C.12}$$

By combining Equations C.11, C.10 and Equation C.12, the total surface area related to the plastic strain ε_p reads

$$S_{el} \approx 4\pi K_1 \left[e^{\kappa \varepsilon_p} \int_{\left(\frac{3K_2}{4\sigma_u}\right)^2}^{\infty} a^{-2} da + \int_0^{\varepsilon_p} e^{\kappa(\varepsilon_p - e)} \frac{d}{de} \left(\int_{\left[\frac{K_2}{(4/3)\sigma_u + K_3 e}\right]^2}^{\infty} a^{-2} da \right) de \right]$$

and, after integration, one obtains

$$\begin{aligned} S_{el} \approx 4\pi \frac{K_1}{K_2^2} \times \\ \times \left[\left(\frac{4}{3}\sigma_u \right)^2 e^{\kappa \varepsilon_p} + \frac{8}{3\kappa} \sigma_u K_3 \left(e^{\kappa \varepsilon_p} - 1 \right) + \frac{2}{\kappa} \left(K_3 \right)^2 \left(\frac{1}{\kappa} e^{\kappa \varepsilon_p} - \frac{1}{\kappa} - \varepsilon_p \right) \right] . \end{aligned} \tag{C.13}$$

Since, for metallic materials, $K_3 < \frac{4}{3}\sigma_u$, $0.4 < \kappa < 2.5$ and

$$\frac{1}{\kappa} e^{\kappa \varepsilon_p} = \frac{1}{\kappa} + \varepsilon_p + \frac{1}{2} \kappa \varepsilon_p^2 + \ldots ,$$

the last term in the brackets can be neglected ($\varepsilon_p < 1$). After a simple re-arrangement of Equation C.13 (the constant $\frac{2}{3}\sigma_u + \frac{K_3}{\kappa}$ is not too much different from σ_u) one finally obtains

$$S_{el} = 2\pi K_1 \frac{1}{a_{0p}} \left(3e^{\kappa \varepsilon_p} - 1 \right) , \tag{C.14}$$

where $a_{0p} = \left(\frac{3K_2}{4\sigma_u} \right)^2$ is the size of voids that nucleate when reaching the ultimate stress.

Let us further consider that the initial void size is determined by a distribution function of particle sizes and, during the deformation, the number of nucleated voids increases. According to Bergh [443] and other authors (e.g., [444, 445]), the distribution function can be written as

$$g(a) = \frac{1}{a^4}, \qquad a \geq a^* ,$$

where a^* is a critical size of particles which start to nucleate voids. Consequently, the total number of particles which, at a given strain, have already nucleated voids reads

$$h(a) = K_1 \int_{a_n}^{\infty} \frac{1}{a^4} \mathrm{d}a, \tag{C.15}$$

where K_1 is a constant, a_n is the smallest size of particles that nucleate voids in a particular deformation stage ε_p. With respect to Equations C.14 and C.15, the dependence of the volume V_2 on the plastic strain is ε_p:

$$V_2 \approx 2\pi K_1 \frac{\delta}{a_{0p}} \left(3\mathrm{e}^{\kappa\varepsilon_p} - 1\right). \tag{C.16}$$

When inserting Equation C.16 into Equation 2.33, and following the same reasoning as in the case of the ideal model, one finally obtains the strain rate formula

$$\begin{aligned} \dot{\varepsilon}_p \approx\ & \mu b v_c \left(\rho_u + M\varepsilon_p\right) \exp\left\{-\frac{D_u}{\langle\sigma_s\rangle_k}\right\} \\ & \left[f_{om}^u - \Theta^* \left(\frac{1}{\gamma + \bar{\Phi}} - \frac{1}{3\gamma\left(1+\bar{\Phi}\right)}\right) \mathrm{e}^{-\Phi\varepsilon_p} + \right. \\ & \left. + \Theta^* \left(\frac{1}{\gamma + \bar{\Phi}} \mathrm{e}^{\kappa\varepsilon_p} - \frac{1}{3\gamma\left(1+\bar{\Phi}\right)}\right) \mathrm{e}^{\varepsilon_p}\right], \end{aligned} \tag{C.17}$$

where $\Theta^* = \frac{6\pi K_1 \delta\gamma}{a_{0p} V_0}$.

Since $\bar{\Phi} > 1$ (H in units of GPa and $\langle\sigma_s\rangle_k$ in hundreds of MPa) and $1.4 < \gamma < 3.5$, the terms $\frac{1}{3\gamma(1+\bar{\Phi})}$ in both brackets can be neglected and Equation C.17 reduces to Equation C.7 (both the ideal and the real model lead to a similar result).

List of Reprinted Figures

Fatigue & Fracture Engineering Materials & Structures 27:141–157, Fig. 10, © 2004 John Wiley & Sons, Inc.

- Figure 2.13 on page 95 is reprinted from the paper Pokluda J, Šandera P, Horníková J, Statistical Approach to Roughness-Induced Shielding Effects. Fatigue & Fracture Engineering Materials & Structures 27:141–157, Fig. 11, © 2004 John Wiley & Sons, Inc.
- Figure 2.20 on page 103 is reprinted from the paper Šandera P, Pokluda J, Horníková J, Vlach B, Lejček P, Jenko M, Fracture of Polycrystalline Fe-2.3%V-0.12%P Alloy. Engineering Fracture Mechanics 77:385–392, Fig. 2, © 2010 Elsevier B.V.
- Figure 2.22 on page 105 is reprinted from the paper Šandera P, Pokluda J, Horníková J, Vlach B, Lejček P, Jenko M, Fracture of Polycrystalline Fe-2.3%V-0.12%P Alloy. Engineering Fracture Mechanics 77:385–392, Fig. 5, © 2010 Elsevier B.V.
- Figure 2.24 on page 106 is reprinted from the paper Šandera P, Pokluda J, Horníková J, Vlach B, Lejček P, Jenko M, Fracture of Polycrystalline Fe-2.3%V-0.12%P Alloy. Engineering Fracture Mechanics 77:385–392, Fig. 7, © 2010 Elsevier B.V.
- Figure 2.25 on page 107 is reprinted from the paper Šandera P, Pokluda J, Horníková J, Vlach B, Lejček P, Jenko M, Fracture of Polycrystalline Fe-2.3%V-0.12%P Alloy. Engineering Fracture Mechanics 77:385–392, Fig. 3, © 2010 Elsevier B.V.
- Figure 3.35 on page 183 is reprinted from the paper Pokluda J, Šandera P, Horníková J, Analysis of Roughness-Induced Crack-Tip Shielding in Terms of Size Ratio Effect. Journal of ASTM International 2:1–15 (JAI11997), Fig. 5, © 2005 ASTM International.
- Figure 3.41 on page 192 is reprinted from the paper Pokluda J, Pippan R, Can a Pure Mode III Fatigue Loading Contribute to Crack Propagation in Metallic Materials? Fatigue & Fracture Engineering Materials & Structures 28:179–186, Fig. 3, © 2005 John Wiley & Sons, Inc.
- Figure 3.42 on page 193 is reprinted from the paper Pokluda J, Pippan R, Can a Pure Mode III Fatigue Loading Contribute to Crack Propagation in Metallic Materials? Fatigue & Fracture Engineering Materials & Structures 28:179–186, Fig. 4, © 2005 John Wiley & Sons, Inc.
- Figure 3.44 on page 196 is reprinted from the paper Pokluda J, Pippan R, Can a Pure Mode III Fatigue Loading Contribute to Crack Propagation in Metallic Materials? Fatigue & Fracture Engineering Materials & Structures 28:179–186, Fig. 7, © 2005 John Wiley & Sons, Inc.
- Figure 3.45 on page 197 is reprinted from the paper Pokluda J, Pippan R, Can a Pure Mode III Fatigue Loading Contribute to Crack Propagation in Metallic Materials? Fatigue & Fracture Engineering Materials & Structures 28:179–186, Fig. 8, © 2005 John Wiley & Sons, Inc.
- Figure 3.50 on page 204 is reprinted from the paper Pokluda J, Trattnig G, Martinschitz C, Pippan R, Straightforward Comparison of Fatigue Crack

References

1. Kelly A, Macmillan M H (1986) Strong Solids. Clarendon Press, Oxford
2. Raabe D (1998) Computational Materials Science: The simulation of Materials, Microstructures and Properties. Willey-VCH Verlag GmbH, Weinheim
3. Vitek V, Perrin R C, Bowen D K (1970) Phil Mag 21:1049–1073
4. Vitek V (1974) Cryst Lat Defects 5:1–34
5. Gröger R, Bailey A G, Vitek V (2008) Acta Mater 56:5401–5411
6. Paidar V, Pope D P, Vitek V (1984) Acta Metall 32:435–448
7. Gröger R, Vitek V (2008) Acta Mater 56:5426–5439
8. Weertman J (1996) Dislocation Based Fracture Mechanics. World Scientific, Singapore
9. Riemelmoser F O, Pippan R, Stüwe H P (1997) Int J Fract 85:157–168
10. Pippan R, Strobl G, Kreuzer H, Motz C (2004) Acta Mater 52:4493–4502
11. Pokluda J, Pippan R (2007) Mat Sci Eng A462:355–358
12. Pippan R, Kreuzer H G M, Kiener D, Rester M, Motz C (2007) Size Effect in Plastic Deformation and Fracture. In: Šandera P (ed) Materials Structure & Micromechanics of Fracture (MSMF-5). VUTIUM, Brno
13. Brenner S S, (1956) J Appl Phys 27:1484–1492
14. Nadgorny E M, (1962) Sov Phys Usp 5:462–469
15. Bahr D F, Kramer D E, Gerberich W W (1998) Acta Mater 46:3605–3612
16. Woodcock C L, Bahr D F (2000) Scripta Mater 43:783–788
17. Gouldstone A, Koh H J, Zeng K Y, Giannakopoulos A E, Suresh S (2000) Acta Mater 48:2277–2296
18. De la Fuente O R, Zimmerman J A, Gonzales M A, De la Figuera J, Hamilton J C, Pai W W, Rojo J M (2002) Phys Rev Lett 88:036101
19. Jokl M L, Vitek V, McMahon C J Jr (1980) Acta Metall 28:1479–1488
20. Kroupa F (1981) Theoretical strength. In: Pokluda J, Staněk P (eds) Prediction of Mechanical Properties of Metallic Materials by means of Structural Characteristics. ČSVTS, Brno (in Czech)
21. Thomson R (1986) Solid State Phys 39:1–129
22. Huang H, Gerberich W W (1992) Acta Metall Mater 40:2873–2881
23. Kelly A, Tyson W R, Cottrell A H (1967) Phil Mag 15:567–585
24. Pokluda J, Šandera P (1991) Phys Stat Solidi b167:543–50
25. Pokluda J, Šandera P (1994) Key Eng Mater 97–98:467–481
26. Goldschmidt D (1994) Single crystal blades. In: Coutsouradis D *et al.* (eds) Materials for Advanced Power Engineering: Part I. Kluwer Academic Publisher, Dodrecht
27. Frenkel J (1926) Z Phys 37:572–609
28. Polany M (1921) Z Phys 7:323–327

29. Orowan E (1949) Rep Progr Phys 12:185–232
30. Kelly A (1973) Strong Solids. Clarendon Press, Oxford
31. Šandera P, Pokluda J (1993) Scr Metall Mater 29:1445–1451
32. Paxton A T, Gumbsch P, Methfessel M (1991) Phil Mag Lett 63:267–274
33. Macmillan N H (1983) Ideal Strength. In: Latanision R M, Pickens J R (eds) Atomistics of Fracture. Plenum Press, New York
34. Pokluda J, Šandera P (1989) Phys Stat Solidi b151:85–91
35. Pokluda J, Šandera P (1990) Phys Stat Solidi b160:89–95
36. Šandera P, Pokluda J (1994) Metall Mater 32:180–193
37. Hohenberg P, Kohn W (1964) Phys Rev B136:864–871
38. Kohn W, Sham L J (1965) Phys Rev A140:1133–1139
39. Esposito E, Carlson A E, Ling B D, Ehrenreich H, Gelatt C D Jr (1980) Phil Mag A41:251–259
40. Price D L, Cooper B R, Wills J M (1992) Phys Rev B46:11368–11375
41. Šob M, Wang L G, Vitek V (1997) Mat Sci Eng A234–236:1075–1089
42. Šob M, Wang L G, Vitek V (1998) Metall Mater 36:145–158
43. Friák M, Šob M, Vitek V (2003) Phil Mag 83A:3529–3537
44. Kitagawa H, Ogata S (1999) Key Eng Mater 161–163:443
45. Ogata S, Kitagawa H (1999) Comp Mat Sci 15:435–440
46. Li W X, Wang T C(1999) Phys Rev B59:3993–4001
47. Telling R H, Pickard C J, Payne M C, Field J E (2000) Phys Rev Lett 84:5160–5163
48. Roundy D, Cohen M L (2001) Phys Rev B64:212103–212105
49. Luo W, Roundy D, Cohen M L, Morris J W Jr (2002) Phys Rev B66:094110
50. Ogata S, Hirosaki N, Kocer C, Kitagawa H (2001) Phys Rev B64:172102
51. Kocer C, Hirosaki N, Ogata S (2003) Phys Rev B67:035210
52. Xu W, Moriarty J A (1996) Phys Rev B54:6941–6951
53. Söderlind P, Moriarty J A (1998) Phys Rev B57:10340–10350
54. Jhi S H, Louie S G, Cohen M L, Morris J W Jr (2001) Phys Rev Lett 87:075503
55. Ogata S, Li J, Yip S, (2002) Science 298:807–811
56. Roundy D, Krenn C R, Cohen M L, Morris J W Jr (1999) Phys Rev Lett 82:2713–2716
57. Roundy D, Krenn C R, Cohen M L, Morris J W Jr (2001) Phil Mag A81:1725–1747
58. Černý M (2007) Mater Sci Eng A462:432–435
59. Černý M, Pokluda J (2008) Comp Mater Sci 44:127–130
60. Šandera P, Pokluda J, Wang L G, Šob M (1997) Mater Sci Eng A234/236:370–372
61. Černý M, Pokluda J, Šob M, Friák M, Šandera P (2003) Phys Rev B67:035116
62. Černý M, Pokluda J (2004) J Alloys Comp 378:159–162
63. Son Y, Yang R, Li D, Wu W T, Guo Z X (1999) Phys Rev B59:14220–14225
64. Born M (1940) Proc Camb Philos Soc 36:160–172
65. Born M, Fürth R (1940) Proc Camb Phil Soc 36:454–465
66. Hill R, Milstein F (1977) Phys Rev B15:3087–3096
67. Li W X, Wang T C (1998) J Phys: Cond Matter 10:9889–9904
68. Černý M, Šob M, Pokluda J, Šandera P (2004) J Phys: Cond Matter 16:1045–1052
69. Finnis M W, Sinclair J E (1984) Phil Mag A50:45–55
70. Aoki M, Nguyen-Manh D, Pettifor D, Vitek V (2007) Progr Mater Sci 52:154–195
71. Chantasiriwan S, Milstein F (1996) Phys Rev B53:1408–14088
72. Cawkwell M, Woodward C, Nguyen-Manh D, Pettifor D, Vitek V (2007) Acta Mater 5:161–169
73. Ackland G J, Mendelev M I, Stolovitý D J, Han S, Barashev A V (2004) J Phys Cond Matter 16:S2629–S2642

74. Clatterbuck D M, Chrzan D C, Morris J W Jr (2002) Phil Mag Lett 82:141–147
75. Dean D W, Wentzcovitch R M, Keskar N, Chelikowsky J R, Binggeli N (2000) Phys Rev B61:3303–3309
76. Ma Y, Tse J S, Klug D D (2003) Phys Rev B67:140301
77. Horsfield A P *et al.* (2008) Comp Mater Sci 44:16–20
78. Milstein F, Chantasiriwan S (1998) Phys Rev B58:6006–6018
79. Milstein F, Hill R (1979) Phys Rev Lett 43:1411–1413
80. Milstein F, Rasky D J (1996) Phys Rev B54:7016–7025
81. Murnaghan F D (1951) Finite deformation of an elastic solid. John Willey & Sons, New York
82. Nye J F (1985) Physical Properties of Crystals: Their Representation by Tensors and Matrices. Oxford University Press, USA
83. Pokluda J, Černy M, Šandera P, Šob M (2004) J Comp Aid Mater Design 11:1–28
84. Barron T H K, Klein M L (1965) Proc Phys Soc Lond 85:523–532
85. Morris J W, Krenn C R (2000) Phil Mag A80:2827–2840
86. Mehl J M, Klein B M, Papaconstantopoulos D (1994) First principles calculations of elastic properties of metals. In: Westbrook J H, Fleisher R L (eds) Intermetallic Compounds. John Willey & Sons, New York
87. Kim K Y (1999) Phys Rev B54:6245–6254
88. Deutsch T, Lancon F (2003) J Phys Cond Matter 15:1813–1826
89. Harrison W A (2004) Elementary Electronic Structure. World Scientific Publishing, Singapore
90. Craievich P J, Weinert M, Sanchez J M, Watson R E (1994) Phys Rev Lett 72:3076–3079
91. Paidar V, Wang L G, Šob M, Vitek V (1999) Modell Simul Mater Sci 7:369–381
92. Wang J, Li J, Yip S, Phillpot S R, Wolf D (1995) Phys Rev B52:12627–12635
93. Zhou Z, Joós B (1996) Phys Rev B54:3841–3850
94. Yip S, Li J, Tang M, Wang J (2001) Mater Sci Eng A317:236–240
95. Wallace D C (1972) Thermodynamics of Crystals. John Wiley & Sons, New York-London-Sydney-Toronto
96. Li J, Zhua T, Yip S, Van Vliet K J, Suresh S (2004) Mater Sci Eng A365:25–30
97. Černý M, Šandera P and Pokluda J (2001) *Ab initio* Modelling Deformation of Perfect Cubic Crystals under Triaxial Tension. In: Šandera P (ed) MSMF3. VUTIUM, Brno
98. Černý M, Pokluda J (2007) Phys Rev B76:024115
99. Ogata S, Li J, Hirosaki N, Shibutani Y, Yip S (2004) Phys Rev B70:104104
100. Song Y, Xu D S, Yang R, Li D, Wu W T, Guo Z X (1999) Mater Sci Eng A260:269
101. Clatterbuck D M, Krenn C R, Cohen M L, Morris J W Jr (2003) Phys Rev Lett 91:135501
102. Černý M, Pokluda J (2009) J Phys: Cond Matter 21:145406
103. Černý M, Pokluda J (2007) Mater Sci Eng A483–484:692–695
104. Černý M, Šandera P and Pokluda J (1999) Czech J Phys B49:1495–1501
105. Morris J W Jr, Krenn D, Roundy D, Cohen M L (2000) Phase transformations and evolution in materials In: Turchi P E, Gonis A (eds) Hume-Rothery Symp in honor of A G Khatchaturyan. TMS, Warrendale PA
106. Umeno Y, Černý M (2008) Phys Rev B77:100101
107. Černý M, Pokluda J, Šandera P (2004) Mater Sci Eng A 387/389:923–925
108. Šob M, Friák M, Legut D, Fiala J, Vitek V (2004) Mat Sci Eng A387/389:148–157
109. Šob M, Wang L G, Vitek V (1998) Phil Mag B78:653–662
110. Söderlind P, Moriarty J A, Wills J M (1996) Phys Rev B53:14063–14072
111. Wang F M, Ingalls R (1998) Phys Rev B57:5647–5654

112. Friák M, Šob M, Vitek V (2003) Phys Rev B68:184101
113. Milstein F, Farber B, (1980) Phil Mag A42:19–29
114. Mikhailovskii I M, Poltinin I Y, Fedorova L I (1981) Sov Phys Solid State 23:757
115. Šesták P, Černý M, Pokluda J (2007) Mater Sci Eng A481–482:247–249
116. Černý M, Pokluda J (2006) Influence of Superimposed Normal Loading on the Shear Strength in bcc Metals. In: Multiscale Materials Modelling. Freiburg Germany
117. Černý M, Pokluda J (2007) Phys Rev B76:024115
118. Černý M, Pokluda J (2008) A Simple Prediction of the Theoretical Tensile Strength of Cubic Crystals based on the Shear Strength Calculations. In: Pokluda J, Lukáš P, Šandera P, Dlouhý I (eds) Multilevel Approach to Fracture of Materials, Components and Structures (ECF17). Vutium, Brno
119. Černý M, Pokluda J (2010) Acta Mater 58:3117–3123
120. Krenn C R, Roundy D, Morris J W Jr, Cohen M L (2001) Mater Sci Eng A317:44–49
121. Holiday L (1966) Composite Materials. Elsevier, Amsterdam-London-NewYork
122. Černý M, Pokluda J (2006) *Ab initio* study of elasticity and strength of nanofibre reinforced composites. In: Gdoutos E (ed) 16th European Conference of Fracture. Springer
123. Černý M, Pokluda J (2006) Elasticity and strength of nanofibre reinforced composites from first principles. In: 12th European Conference on Composite Materials. Biarritz, France
124. Černý M, Pokluda J (2006) First principles study of of vanadium based composites reinforced by tungsten nanofibres. In: Šandera P (ed) Nano 06. Vutium, Brno
125. Vosko S H, Wilk L, Nusair M (1980) Can J Phys 58:1200–1207
126. Kittel C (1976) Introduction to Solid State Physics. John Wiley & Sons, New York
127. Simmons G, Wang H (1971) Single Crystal Elastic Constants and Calculated Aggregate Properties: A Handbook. MIT Press, Cambridge
128. Černý M, Pokluda J (2009) Elasticity and Strength of Nano-Fibre Reinforced Composites from First Principles. In: Elboujdaini M (ed) 12th International Conference on Fracture (ICF12). NRC, Ottawa, Canada
129. Goodwin L, Needs R, Heine V (1988) Phys Rev Lett 60:2050–2053
130. Huang Y M, Spence J C H, Sankey O F (1994) Phil Mag 70:53–59
131. Haydock R (1981) J Phys 14C:3807–3816
132. Šesták P (2009) Microstructural and mechanical characteristics of NiTi alloy from first principles. PhD Thesis, Brno University of Technology, Brno
133. Neuber H (1958) Kerbspannungslehre. Springer Verlag, Berlin
134. Rice J R, Thomson R (1974) Phil Mag 29:73–79
135. Rice J R, Beltz G E, Sun Y (1992) Peierls Framework for Dislocation Nucleation from a Crack Tip. In: Argon A S (ed) Topics in Fracture and Fatigue. Springer, Berlin
136. Pokluda J, Šandera P (1995) Metall Mater 33:375–383 (in Czech)
137. Ewalds H L, Wanhill R J H (1989) Fracture Mechanics. Edward Arnold Publ, Delft
138. Lin I H (1983) Mater Sci Lett 2:295
139. Rice J R (1992) J Mech Phys Solids 40:239–271
140. Gorgas I (1986) Scr Metall 20:113–117
141. Hora P, Pelikán V, Machová A, Spielmannová A, Prahl J, Landa M, Červená O (2008) Engng Fract Mech 75:3612–3623
142. Krenn C R, Roundy D, Cohen M L, Chrzan D C, Morris J W Jr (2002) Phys Rev B 65:134111
143. Göken M, Kempf M, Nix W D (2001) Acta Mater 49:903–911

144. Göken M, Kempf M (2001) Z Metallkd 92:1061–1067
145. Horníková J, Černý M, Šandera P, Pokluda J (2007) Acta Metall Slov 13:65–69
146. Horníková J, Černý M, Šandera P, Pokluda J (2007) Key Engng Mater 348/349:8001–8004
147. Bei H, Lu Z P, George E P (2004) Phys Rev Lett 93:125504
148. Horníková J, Šandera P, Černý M, Pokluda J (2008) Engng Fract Mech 75:3755–3762
149. Pokluda J, Kroupa F, Obdržálek L (1994) Mechanical properties and structure of solids. PCdir, Brno (in Czech)
150. Dekker A J (1957) Solid State Physics. Prentice-Hall, New Jersey USA
151. Jokl M L, Vitek V, McMahon C J Jr, Burgers P (1989) Acta Metall 37:87–97
152. Pokluda J, Vitek V (1999) Metall Mater 37:67–70
153. Wnuk M P (1974) J Appl Mech 41:234–242
154. Neimitz A, Aifantis E C (1987) Engng Fract Mech 26:505–518
155. Unger D J, Aifantis E C (1983) Acta Mech 47:117–151
156. Cottrell A H (1953) Dislocations and Plastic Flow in Crystals. Oxford University Press, Oxford
157. Hahn G T, Rosenfield A R (1975) Metall Trans A6:653–668
158. Kroupa F, Vitek V (1967) Can J Phys 45:945–971
159. Felkins K, Leighly H P Jr, Jankovic A (1998) J Metals 50:12–18
160. Hondros E D, Seah M P, Hofmann S, Lejček P (1996) Interfacial and surface microchemistry. In: Cahn R W, Haasen P (eds) Physical Metallurgy, 4th edition, Chapter 13. North-Holland, Amsterdam
161. Kalderon D (1972) Proc Inst Mech Eng 186:341–375
162. Broek D (1982) Elementary engineering fracture mechanics. Martinus Nijhooff Publ, Dordrecht
163. Anderson T L (1995) Fracture Mechanics. Fundamentals and Applications. CRC Press, Texas
164. Magill M A, Zwerneman F J (1997) Engng Fract Mech 56:9–24
165. Ritchie R O (1988) Mat Sci Eng A103:15–28
166. ASTM E 399-72 Standard method of test for plane-strain fracture toughness of metallic materials
167. Pokluda J (2001) Mechanika 67:277–296
168. Pokluda J, Šandera P, Horníková J (2004) Fat Fract Engng Mater Struct 27:141–157
169. Faber K T, Evans A G (1983) Acta Metall 31:565–576
170. Murakami Y (ed) (1987) Stress Intensity Factors Handbook: Vol 1. Pergamon Books
171. Pokluda J, Podrábský T, Slámečka K, Pospíšilová S, Chai G C (2008) Mater Sci Forum 567/568:101–104
172. Ingrafea T *et al.* (1996) FRANC3D, 3D FRacture ANalysis Code. The Cornell University Fracture Group, Cornell University, Ithaca NY
173. Pokluda J, Šandera P, Horníková J (2005) J ASTM International 2:1–15
174. Chawla K K (1993) Ceramic Matrix Composites. Chapman and Hall, London
175. Boccaccini A R (2001) J Ceram Soc Jap 109:99–109
176. Boccaccini A R, Trusty P A (1996) J Mat Sci Lett 15:60–62
177. Pokluda J, Švejcar J (1999) Structural analysis of fatigue crack growth in ferritic ductile iron. In: Miannay D, Costa P, Francois D, Pineau A (eds) Fatigue 99. EMAS, Beijing
178. Horníková J, Šandera P, Pokluda J (2005) Mater Sci Forum 482:311–314
179. Todd R I, Boccaccini A R, Sinclair R, Yalle R B, Young R B (1999) Acta Mater 47:3233–3240
180. Boccaccini A R, Winkler V (2002) Composites Part A33:125–131
181. Lange F F (1971) J Am Ceram Soc 54:614–20

182. Chaim R, Talanker V (1995) J Am Ceram Soc 78:166–72
183. Horníková J, Šandera P, Pokluda J (2006) Modelling Crack-Tip Shielding Effects in Particle Reinforced Composites. In: Gdoutos E (ed) 16th European Conference of Fracture. Springer
184. Kotoul M, Pokluda J, Šandera P, Dlouhý I, Chlup Z, Boccacini A R (2008) Acta Mater 56:2908–2918
185. Winn A J, Boccaccini A R, Imam N, Trusty P A (1997) J Microscopy 186:35–40
186. Knott J F(1983) Mechanics of Fracture. In: Latanision R M, Pickens J R (eds) Atomistics of Fracture, Plenum Press, New York
187. Kerlins V (1994) Modes of Fracture. In: Mills K *et al.* (eds) ASM Handbook: Vol 12 Fractography. ASM International
188. Šandera P, Poníž il P, Pokluda J (1997) Three dimensional model of intergranular crack development in the process zone. In: Parilák L' (ed) Fractography '97. ÚMV SAV, Stará Lesná
189. Pokluda J, Saxl I, Šandera P, Ponížil P, Matoušek M, Podrábský T, Horníková J (2000) Real-like polycrystal with intergranular crack. In: Miannay D, Costa P, Francois D, Pineau A (eds) Advances in Mechanical Behaviour, Plasticity and Damage, Euromat 2000. Elsevier
190. Walin K (2008) The Use of the Master Curve in Structural Integrity Assessment. In: Pokluda J, Lukáš P, Šandera P, Dlouhý I (eds) Multilevel Approach to Fracture of Materials, Components and Structures (ECF17). VUTIUM, Brno
191. Kim Y J, Lin G, Cornec A, Schwalbe K H (1996) Int J Fract 78:21–34
192. Suresh S (1998) Fatigue of materials. Cambridge University Press, Cambridge, UK
193. Zeman J, Rolc S, Buchar J, Pokluda J (1990) Microstructure and fracture toughness of cast and forged ultra-high-strength, low-alloy (UHSLA) steels. In: Gudas J P, Joyce J A, Hackett E M (eds) Fracture Mechanics: Twenty-First Symposium, ASTM STP 1074. ASTM, Philadelphia, PA
194. Petit J (1998) Metall Mater 3:220–232
195. Cook J R, Irwing P E, Booth G S, Beewers C J (1975) Engng Fract Mech 7:67–77
196. Pokluda J, Siegl J (1990) Fat Fract Engng Mater Struct 13:375–385
197. Kunz L, Lukáš P (1982) Fatigue crack growth in 15CH2NFA steel. In: Karel V (ed) Fractography '82. DT, Žilina, Czechoslovakia (in Czech)
198. Kolednik O (2000) Int J Sol Struct 37:781–808
199. Sun Z, de los Rios E R, Miler K (1991) Fat Fract Engng Mater Struct 14:277–292
200. Pokluda J, Šandera P, Horníková J (2000) Statistical model of roughness-induced crack closure. In: Fuentes M, Elices M, Martín-Meizoso A, Martínez-Esnaola J M (eds) Fracture mechanics: applications and challenges, ECF 13. Elsevier
201. Saltykov C A (1970) Stereometric metallography. Metalurgija, Moscow (in Russian)
202. Saxl I, Ponížil P (1998) Acta Stereol 17:247–252
203. Avery I, Hall F R, Sturgess C E N (1998) J Mater Proc Techn 80/81:565–571
204. Pokluda J, Šandera P, Pippan R (2006) Analysis of crack closure level in terms of crack-wake plasticity. In: Johnson W S (ed) Fatigue 06, Elsevier, Atlanta, Georgia
205. Lee S, Maino L, Asaro R J (1985) Metal Trans 16A:1633–1648
206. Tomita Y (1988) Metal Trans 19A:2313–2321
207. Datta K P, Wood W E (1981) J Test Evaluation 9:111–117
208. Guimaraes R, Savedra A (1985) Metall Trans 16A:329–336
209. Server W L (1992) Charpy impact testing. In: Newby J R *et al.* (eds) ASM Handbook: Vol. 8 – Mechanical Testing

210. Pokluda J, Šandera P, Zeman J (1992) Effect of intergranular crack branching on fracture toughness evaluation. In: Ernst H A, Saxena A, McDowell D L (eds) Fracture Mechanics: Twenty-Second Symposium, ASTM STP 1131. ASTM, Philadelphia PA
211. Šandera P, Pokluda J (1992) Computer simulation of intergranular crack branching. In: Aliabadi H M, Cartwright D J, Nisitani H (eds) Localized Damage 92. CMP Elsevier, Southampton-Boston-London-New York
212. Šandera P, Pokluda J (1996) Metall Mater 34:230–240 (in Czech)
213. Pokluda J, Vlach B, Šandera P, Jurášek L (2001) Antagonism in Fracture and Notch Toughness of UHSLA Steels. In: Francois D, Pineau A (eds) Charpy Centenary Conference SF2M. Les Fontelles, France
214. Underwood E E, Banerjee K (1992) Quantitative fractography. In: Mills K *et al.* (eds) ASM Handbook: Vol. 12 – Fractography. ASM International, Metals Park, Ohio, USA
215. Janovec J, Jenko M, Lejček P, Pokluda J (2007) Mater Sci Eng A462:441–445
216. Lejček P (1994) Anal Chim Acta 297:165–178
217. Lejček P, Pokluda J, Šandera P, Horníková J, Vlach B, Jenko M (2008) Fracture Behavior of Phosphorus-Doped Polycrystalline Fe-2.3%V Alloy. In: Pokluda J, Lukáš P, Šandera P, Dlouhý I (eds) Multilevel Approach to Fracture of Materials, Components and Structures (ECF17). VUTIUM, Brno, Czech Republic
218. Lejček P, Hofmann S, Janovec J (2007) Mater Sci Eng A 462:76–85
219. Wu S J, Ding R G, Knott J F (2007) Mater Sci Technol 23:1262–1268
220. Janovec J, Pokluda J, Jenko M, Lejček P, Vlach B, Horníková J (2006) Surf Interf Anal 38:401–405
221. McClintock F A (1968) J Appl Mech 35:362–371
222. Rice J R, Tracey D M (1969) J Mech Phys Solids 17:201–2017
223. Gurson A L (1977) J Engng Mater Techn 99:2–15
224. Tvergaard V (1981) Int J Fract 17:389–407
225. Staněk P, Pokluda J (1980) Metall Mater 18:709–720 (in Czech)
226. Staněk P, Pokluda J (1983) Metall Mater 21:383–395 (in Czech)
227. McClintock F A (1969) Ductile Fracture. In: Argon A S (ed) Physics of Strength and Plasticity. MIT Press, Cambridge, MA
228. Brown L M, Emburry J D (1973) Inst Metals 1:164–169
229. Rousselier G (1987) Nucl Engng Design 105:97–111
230. Tvergaard V, Needleman A (1984) J Mech Phys Solids 32:461–490
231. Argon A S, Im J, Safoglu R (1975) Metall Trans A6:825–834
232. Lui M W, LeMay I (1976) Eng Mat Techn 4:179
233. Jonas J J, Baudelet B (1977) Acta Metall 25:43–50
234. Staněk P, Pokluda J (1979) On the mechanism of ductile fracture in metalic materials. In: Staněk P, Pokluda J (eds) Prediction of Mechanical Properties of Metallic Materials from Structural Characteristics (Prediction79). DT, Brno (in Czech)
235. Staněk P, Pokluda J (1980) Metall Mater 18:471–481 (in Czech)
236. Gilman J J (1968) Micromechanics of Flow in Solids. Mc Craw-Hill, New York
237. Argon A S, Im J, Needleman A (1975) Metall Trans A6:815–824
238. Bridgman P W (1975) Studies in Large Plastic Flow and Fracture. Mc Craw-Hill, New York
239. De Gennes P Q (1978) Czech J Phys A28:549 (in Czech)
240. Smirnov-Aljajev G A (1961) Resistance of Materials to Plastic Deformation. Metallurgia, Moscow (in Russian)
241. Staněk P, Pokluda J (1984) Metall Mater 22:710–719 (in Czech)
242. Kraft J M, Irwin G R (1965) ASTM STP 381:84–113
243. Rice J R, Johnson M A (1970) The Role of Large Crack Tip Geometry Changes in Plane Strain Fracture. In: Kanninen M F *et al.* (eds) Inelastic Behavior of Solids, McGraw-Hill, New York

244. Hancock J W, Cowling M J (1980) Metal Sci 14:293–304
245. Pokluda J, Šandera P (1990) A Simple method for Calculation of K_{Ic} values of Ductile Materials. In: Koutsky *et al.* (eds) Brittle Fracture of Materials and Structures. Skoda, Pilsen, Czechoslovakia (in Czech)
246. McEvily A (2002) Metal Failures: Mechanisms, Analysis, Prevention. John Viley, New York
247. Miller K J (1993) Mat Sci Technology 9:453–462
248. Zapfe C A, Clogg M Jr (1945) Trans ASM 34:71–107
249. Underwood E E, Banerjee K (1992) Fractal analysis of fracture surfaces. In: Mills K *et al.* (eds) ASM Handbook: Vol 12 – Fractography. ASM International, Metals Park, Ohio, USA
250. Dong W P, Sullivan P J, Stout K J (1994) Wear 178:29–43
251. Gadelmawla E S, Koura M M, Maksoud T M A, Elewa I M, Soliman H H (2002) J Mater Proc Tech 123:133–45
252. Suh A Y, Polycarpou A A, Conry T E (2003) Wear 255:556–568
253. Bakolas V (2003) Wear 254:546–554
254. Sherrington I, Howarth G W (1998) Int J Mach Tool Manufact 38:599–606
255. Okabe A, Boots B, Sugihara K (1992) Spatial Tessellations: Concepts and Applications of Voronoi Diagrams. Wiley, New York
256. Petropoulos G P, Vaxevanidis N M, Pandazaras C N, Antoniadis A T (2004) Wear 257:1270–1274
257. Geometrical Demands on Production (GPS) – Surface Structure: Profile Method – Terms, Definitions and Parameters. CSN EN ISO 4287, 1999 (in Czech)
258. Wang S H, Müller C (1998) Mater Sci Eng A255:7–15
259. Mandelbrot B B (1984) The Fractal Geometry of Nature. Freeman, New York
260. Balankin A S (1997) Engng Fract Mech 57:135–203
261. Camastra F (2003) Pattern Recogn 36:2945–2954
262. Zaiser M, Bay K, Hahner P (1999) Acta Mater 47:2463–2476
263. Xie H, Wang J (1999) Int J Sol Struct 36:3073–3084
264. Yuan C Q, Li J, Yan X P, Peng Z (2003) Wear 255:315–326
265. Charkaluk E, Bigerelle M, Iost A (1998) Engng Fract Mech 61:119–139
266. Campos I, Balankin A S, Bautista O, Ramirez G (2005) Theor Appl Fract Mech 44:187–191
267. Guerrero C, Reyes E, Gonzalez (2002) Polymer 43:6683–6693
268. Weiss J (2001) Engng Fract Mech 68:1975–2012
269. Pokluda J, Staněk P (1980) Strojirenstvi 30:485–496 (in Czech)
270. Antolovich S D, Saxena A (1992) Fatigue Failures. In: Powel G W, Mahmoud S E (eds) ASM Handbook: Vol 11 Failure Analysis and Prevention. ASM International
271. Beachem C D (1967) Trans ASM 60:324–343
272. McEvily A (2007) Mater Sci Forum 567/568:397–400
273. Paris P C, Erdogan J (1963) J Basic Eng 85:528–537
274. Rice J R (1967) Mechanics of Crack Tip Deformation and Extension by Fatigue. In: Fatigue Crack Propagation, ASTM STP 415, Philadelphia
275. Petit J, Hanaff G, Sarrazin-Badoux C (2000) Fatigue Crack Threshold, Endurance Limits and Design, In: Newman J C, Piascik R S (eds) ASTM STP 1372, ASTM
276. Lipetzky P, Knésl Z (1995) Int J Fract 73:81–92
277. Simha N K, Fischer F D, Kolednik O, Chen C R (2003) J Mech Phys Solids 51:209–240
278. Pokluda J, Staněk P (1985) Scr Metall 19:435–439
279. Pokluda J (1994) Mater Engng 1:21–29 (in Czech)
280. Menčík J. (1994) Strength and Fracture of Mass and Ceramics. SNTL, Prague (in Czech)

281. Polák J, Klesnil M (1982) Fat Fract Engng Mater Struct 5:19–32
282. Kenedy A J (1953) Nature 71:927–928
283. Benham P P (1961) J Inst Mater 89:328–338
284. Elfmark J, Pokluda J, Staněk P (1980) Jader Energ 10:361–374 (in Czech)
285. Bezděk L *et al.* (1973) Fatigue Fractures of Military Steels and Components. Res Rep IA17-73, Military Academy, Brno (in Czech)
286. Pokluda J, Staněk P (1978) Kovove Mater 16:583–598 (in Czech)
287. Pokluda J (1984) On the cyclic creep of metalic materials. PhD Thesis, Masaryk University, Brno (in Czech)
288. Bartík L, Pokluda J (1998) Metall Mater S36:53–56
289. Klesnil M, Lukáš P (1975) Fatigue of metallic materials under cyclic loading. Academia, Prague (in Czech)
290. Pokluda J, Staněk P (1980) Probl Prochn 9:16–29 (in Russian)
291. Lorenzo F, Laird C (1984) Acta Metall 32:671–680
292. Oldroyd P W J, Radon J C (1979) Fatigue Fract Engng Mater Struct 1:297–306
293. Mròz Z (1969) Acta Mechanica 7:199–212
294. Dafalias Y F, Popov E P (1975) Acta Mechanica 21:1973–1992
295. Pokluda J, Staněk P (1982) Kovove Mater 20:521–523 (in Czech)
296. Bartík L (1999) Mechanical properties of ultra high-strength steels of the Czech provenience. PhD Thesis, Brno University of Technology, Brno (in Czech)
297. Mughrabi H (1985) (ed) Dislocations and Properties of Real Materials, The Institute of Metals, London
298. Stroh, A N (1957) Adv Phys 6:418–465
299. Tanaka K, Mura T (1981) J Appl Mech 48:97–103
300. Liu W, Bayerlein M, Mughrabi H, Day A, Quested P N (1992) Acta Metall Mater 40:1763–1771
301. Van der Ven A, Ceder G (2004) Acta Mater 52:1223–1235
302. Wejdemann C, Pedersen O B (2004) Mater Sci Eng A 387:556–559
303. Gerold V (1982) Scr Metall 16:405
304. Lukas P, Klesnil M, Krejci J (1968) Phys Stat Sol 27:545–558
305. Finney J M, Laird C (1975) Phil Mag 31:339–366
306. Laird C, Finney J M, Kuhlmann-Wilsdorf D (1981) Mater Sci Eng 50:127–136
307. Kuhlmann-Wilsdorf D, Laird C (1980) Mater Sci Eng A46:209–219
308. Rasmussen K V, Pedersen O B (1980) Acta Metall 14:1467–1478
309. Eshelby J D (1957) Proc Roy Soc London A 241:376–396
310. Sauzay M, Gilormini P (2002) Theor Appl Fract Mech 38:53–62
311. Sauzay M (2009) Influence of Slip Localization on Crack Initiation at the Grain Scale. In: Elboujdaini M (ed) 12th International Conference on Fracture (ICF12). NRC, Ottawa, Canada
312. Polák J (1991) Cyclic Plasticity and Low Cycle Fatigue Life of Metals. Academia, Prague
313. Polák J (1987) Mater Sci Eng A92:71
314. Polák J, Lepistö T, Keitunnen P (1985) Mater Sci Eng A74:85–91
315. Essmann U, Gösselle U, Mughrabi (1981) Phil Mag 44:405–426
316. Antonopoulos J G, Brown L M, Winter A T (1976) Phil Mag 34:549–563
317. Ford F P (1996) Corrosion 52:375–395
318. Miller K J, de los Rios E R (eds) (1986) The behaviour of short fatigue cracks. Mechanical Engineering Publication, London, UK
319. Kitagawa H, Takahashi S (1976) Applicability of fracture mechanics to very small cracks or the cracks in the early stage. In: Promisel N E, Weiss V (eds) Mechanical Behavior of Materials. ASM, Metals Park, OH
320. Riemelmoser F O, Pippan R, Stüwe H P (1998) Acta Mater 46:1793–1799
321. Dai S A, Li J C M (1982) Scr Metall 16:183–188
322. Riemelmoser F O, Pippan R, Kolednik O (1997) Comput Mech 20:139–144
323. Pelloux R M N (1969) Trans Am Soc Met 62:281–89

324. Wanhill R J H (1975) Metall Trans 6A:1587–95
325. Neumann P (1969) Acta Metall 17:1219–25
326. Elber W (1970) Engng Fract Mech 2:37–45
327. Toyosada M, Gotoh K (2005) Mater Sci Forum 482:95–102
328. Suresh S, Ritchie R O (1982) Metall Trans 13A:1627–1631
329. Wang S H, Müller C (1999) Analytical evaluation and experimental study of roughness-induced crack closure. In: Wu X R, Wang Z G (eds) Fatigue 99: Proceedings of the seventh international fatigue congress. Emas, Beijing
330. Noda N, Oda K (2001) Zig-zag Edge Crack in Semi-infinite Plain under Uniaxial Tension. In: Murakami Y (ed) Stress Intensity Factors Handbook: Vol IV. TSMS Elsevier
331. Hanlon T, Suresh S (2005) Fatigue of Nanostructured Metals and Alloys. In: Carpinteri A (ed) International Congress on Fracture (ICF11). Politechnico di Torino, Torino Italy
332. Kondo Y (2005) Mater Sci Forum 482:89–94
333. Pippan R (1993) Engng Fract Mech 44:821–829
334. Slámečka K, Ponížil P, Pokluda J (2007) Mater Sci Eng A462:359–362
335. Horníková J, Šandera P, Pokluda J (2006) Key Engng Mater 324–325:803–806
336. Carter R D, Lee E W, Starke E A Jr, Beevers C J (1984) Metall Trans 15A:555–563
337. Pippan R (1998) Phil Mag 77A:861–873
338. Robinson J L, Beevers C J (1973) Metal Sci J 7:153–159
339. Gregory J K (1996) Fatigue Crack Growth of Titanium Alloys. In: Lampman S R *et al.* (eds) ASM Handbook Fatigue and Fracture: Vol 19. ASM, Metals Park, OH
340. Boyce B L, Ritchie R O (2001) Engng Fract Mechanics 68:129–147
341. Pippan R (1991) Mater Sci Eng A138:1–13
342. Pippan R, Berger M, Stüwe H P (1987) Metall Trans 18A:429–435
343. Shang J K, Tzou J L, Ritchie R O (1987) Metall Trans 18A:1613–1627
344. Taylor D (1989) A Compendium of Fatigue Thresholds and Growth Rates. EMAS, Warley
345. Wang Z G, Al S H (1999) ISIJ Inter 39:747–759
346. Pokluda J, Kondo Y, Slámečka K, Šandera P, Horníková J (2008) Key Engng Mater 385/387:49–52
347. Pook L P (2002) Crack Paths. Wit Press, UK
348. Pokluda J, Pippan R (2005) Fat Fract Engng Mater Struct 28:179–186
349. Nayeb-Hashemi H, McClintock F A, Ritchie R O (1983) Int J Fract 23:163–185
350. Tschegg E K (1982) Mater Sci Eng A 54:127–136
351. Tschegg E K (1983) J Mater Sci 18:1604–1614
352. Tschegg E K (1983) Acta Metall 31:1323–1330
353. Tanaka K, Akinawa Y and Yu H (1999) The propagation of a circumferential fatigue crack in medium-carbon steel bars under combined torsional and axial loadings. In: Miller K J and D L McDowel D L (eds) Mixed-mode Crack Behaviour. ASTM 1359, West Conshohocken, PA
354. Murakami Y, Kusumoto R, Takahashi K (2002) Growth mechanism and threshold of mode II and mode III fatigue crack. In: Neimitz A *et al.* (eds) Fracture Mechanics Beyond 2000 (ECF14). EMAS, UK
355. Murakami Y (2002) Private communications
356. Nayeb-Hashemi H, McClintock F A (1982) Metall Trans A 13:2197–2204
357. Ritchie R O, McClintock F A, Nayeb-Hashemi H, Rittler M A (1982) Metall Trans A 13:101–110
358. Tanaka K (2009) Crack propagation in lead-free solder under cyclic loading of mode I and II. In: Atzori B, Carpinteri A, Lazzarin P, Pook L P (eds) Crack Paths 2009. University of Padua, Vicenza

359. Vaziri A, Nayeb-Hashemi H (2005) Engng Fract Mech 72:617–629
360. Makabe C, Anggit M, Sueyoshi T, Yafuso T (2006) J Test Eval 34:423–429
361. Pokluda J, Slámečka K, Šandera P (2009) On the mechanism of factory-roof formation. In: Atzori B, Carpinteri A, Lazzarin P, Pook L P (eds) Crack Paths 2009. University of Padua, Vicenza
362. Miller K J *et al.* (1985) Multiaxial Fatigue. In: Miller K J, Brown, M W (eds) ASTM STP 853, ASTM, Philadelphia
363. Matake T (1977) Bulet JSME 20:257
364. Vaziri A, Nayeb-Hashemi H (2006) Int J Sol Struct 43:4063–4081
365. Holáň L, Pippan R, Pokluda J, Horníková J, Hohenwarter T, Slámečka K (2009) Near-threshold propagation of Mode II and Mode III cracks. In: Atzori B, Carpinteri A, Lazzarin P, Pook L P (eds) Crack Paths 2009. University of Padua, Vicenza
366. Hellier A K, Corderoy D L H, Mc Girr M B (1987) Int J Fatigue 9:95–101
367. Hellier A K, McGirr M B, Corderoy D J H, Kutajczyk L A (1990) Int J Fract 42:R19–R23
368. Pokluda J, Trattnig G, Martinschitz C, Pippan R.(2008) Int J Fatigue 30:1498–1506
369. Horníková J, Šandera P, Pokluda J (2010) Key Engng Mater 417/418:321–324
370. Slámečka K, Pokluda J (2003) 3D Analysis of Fatigue Fracture Morphology Generated by Combined Bending – Torsion. In: Nilsson F *et al.* (eds) Advanced Fracture Mechanics for Life and Safety Assesments (ECF15). KTH, Stockholm
371. Benthem J P, Koiter W T (1973) Asymptotic approximations to crack problems. In: Sih G C (ed) Method of Analysis and Solutions of Crack Problems. Noordhoff International Publishing, Leyden
372. Lampman S (1996) Fatigue and Fracture Properties of Stainless Steels. In: Lampman S R *et al.* (eds) ASM Handbook Fatigue and Fracture: Vol 19. ASM, Metals Park, OH
373. Sander M, Richard H A (2002) Effects of Block Loading and Mixed Mode Loading on the Fatigue Crack Growth. In: Blom A F (ed) Fatigue 2002. Stockholm, Sweden
374. James M, Herman D J, Scott F (2003) Fatigue crack growth rate and stress intensity factor corrections for out-of-plane crack growth. In: Daniewitz S R, Newman J C, Schwalbe K H (eds) Fatigue and Fracture Mechanics. ASTM, West Conshohocken
375. Socie D F, Marquis G B (2000) Multiaxial fatigue. SAE International, Warrendale
376. Gough H J, Pollard H V (1935) Proc Inst Mech Eng 131:3–103
377. Gough H J (1950) J Appl Mech 50:113–125
378. Sines G (1959) Behavior of Metals Under Complex Statis and Alternating Stresses. In: Sines G, Waisman J L (eds) Metal Fatigue. MsGraw-Hill, New York
379. Crossland B (1956) Effect of large hydrostatic pressure on the torsional fatigue strength of an alloy steel. In: Fatigue of Metals. Institution of Mechanical Engeneers, London
380. Findley W N(1959) J Engng Ind 7:301–306
381. Papadopoulos I V, Davoli P, Gorla C, Filippini M, Bernasconi A (1997) Int J Fatigue 19:219–235
382. McDiarmid D L (1991) Fat Frac Eng Mat Struct 14:429–453
383. Carpinteri A, Spagnoli A (2001) Int J Fatigue 23:135–145
384. Dang Van K (1999) Introduction to Fatigue Analysis in Mechanical Design by the Multiscale Approach. In: Dang Van K, Papadopoulos I V (eds) High-Cycle Metal Fatigue in the Context of Mechanical Design. Springer, Vienna
385. Papadopoulos I V (1998) Fat Frac Eng Mat Struct 21:269–285

386. Gonçalves C A, Araujo J A, Mamiya E N (2005) Int J Fatigue 27:177–187
387. Major Š, Papuga J, Horníková J, Pokluda J (2008) Strength Mater 40:64–67
388. De la Cruz P, Oden M, Ericsson T (1998) Mat Sci Eng A242:181–194
389. Genel K, Demirkol M, Çapa M E (2000) Mat Sci Eng A279:207–216
390. Pokluda J, Dvořák I, Horáková H, Major Š (2006) Influence of Plasma-Nitriding Surface Layer on Fatigue Life of Steel Specimens under Push-Pull and Bending-Torsion. In: Johnson W S (ed) Fatigue 2006. Elsevier, Atlanta, Georgia, USA
391. Carpinteri A, Spagnoli A, Vantadori S (2002) Fat Fract Engng Mater Struct 25:619–627
392. Carpinteri A, Spagnoli A (2004) Int J Fatigue 26:125–133
393. Slámečka K, Pokluda J, Ponížil P, Major Š, Šandera P (2008) Engng Fract Mech 75:760–767
394. Limodin N, Verreman Y (2006) Mat Sci Eng A435/436:460–467
395. Sirin S Y, Sirin K, Kaluc E (2008) Mat Charact 59:351–358
396. Murakami Y (2001) Mechanism of fatigue silure in ultralong life regime. In: Stanzl-Tschegg S, Mayer H (eds) Fatigue in the very high cycle regime. Institute of Metrology and Physics, Vienna, Austria
397. Terentev V F (2004) Metal Sci Heat Treat 46:244–249
398. Kuroshima Y, Harada S (2001) Fatigue crack growth mechanism of high strength steel in gigacycle fatigue region. In: Stanzl-Tschegg S, Mayer H (eds) Fatigue in the very high cycle regime. Institute of Metrology and Physics, Vienna, Austria
399. Murakami Y (2002) Metal Fatigue: Effects of Small Defects and Nonmetallic Inclusions. Elsevier, Oxford, UK
400. Shiozawa K, Lu L T (2001) Superlong fatigue behaviour of shot-peened high-strength steel, In: Stanzl-Tschegg S, Mayer H (eds) Fatigue in the very high cycle regime. Institute of Metrology and Physics, Vienna, Austria
401. Gong Y, Norton M P (1996) J Test Eval 24:263–267
402. Slámečka K, Pokluda J, Kianicová M, Major Š, Dvořák I (2010) Int J Fatigue 32:921–928
403. Onuki A, Yanagi N, Satoh N, Takase F (1992) Fatigue strength of a ion nitrided steel. In: Sedmak (ed) 9th European Conference on Fracture (ECF9). Varna, Bulgaria
404. Holemář A, Hrubý V (1989) Ion nitriding in the practice. SNTL, Prague (in Czech)
405. Forsyth P J E, Ryder D A (1960) Aircraft Engng 32:96–99
406. Jacoby G (1965) Exper Mech 5:65–82
407. Peel C J (1972) An Analysis of a Test Fatigue Failure by Fractography and Fracture Mechanics. Royal Aircraft Establishment, TR 72034
408. Ryder D A, Bunk W (1974) Zeit Flugwiss 22:223
409. Uchimoto T, Sakamoto A, Nagata S (1977) Trans ISIJ 17:1
410. Pokluda J, Staněk P (1981) Probl Prochn 4:13–20 (in Russian)
411. Pokluda J, Staněk P (1981) Acta Techn 4:415–428
412. Nedbal I, Siegl J, Kunz J, Lauschmann H (2008) Fat Fract Engng Mater Struct 31:164–176
413. Nedbal I, Lauschmann H, Siegl J, Kunz J (2008) Fat Fract Engng Mater Struct 31:177–183
414. Staněk P, Pokluda J, Kopeček K (1980) Strojírenství 30:681–686
415. Pokluda J, Staněk P, Unčovský A (1984) Failure of Compressor Rotor of Aircraft Engine M-301. Res Rep VU23-84, Military Research Institute 070, Brno (in Czech)
416. Pokluda J, Šandera P (2006) Simple Method for Estimation of Fatigue Loading and Useful Life from Fracture Morphology. In: Carpinteri A (ed) 11th International Conference on Fracture (ICF11). Torino, Italy

417. Forman R G, Kearney V E, Engle R M (1967) J Basic Engng 89:459–470
418. Klesnil M, Lukáš P (1972) Engng Fract Mech 4:77–92
419. Murakami Y *et al.* (1987) Stress Intensity Factors Handbook. Pergamon Press, Oxford-New York
420. Prokopenko A V (1981) Probl prochn 4:105 (in Russian)
421. Thomas L H (1927) Proc Cambr Phil Soc 23:542–548
422. Slater J C (1951) Phys Rev 81:385–390
423. Hartree D R (1928) Proc Cambr Phil Soc 89:426–437
424. Ceperley D M, Alder B J (1980) Phys Rev Lett 45:566–569
425. Perdew J P, Wang Y (1986) Phys Rev B33:8800–8802
426. Perdew J P, Burke K, Wang Y (1996) Phys Rev B54:16533–16539
427. Ziesche P, Kurth S, Perdew J P (1998) Comp Mater Sci 11:122–127
428. Andersen O K (1975) Phys Rev B12:3060–3083
429. Andersen O K (1981) Europhys News 12:4–8
430. Skriver H L (1983) The LMTO Method: Muffin-Tin orbitals and Electronic Structure. Springer Verlag, Berlin
431. Krier G, Jepsen O, Burkhardt A, Andersen O K (1994) The LMTO-ASA program. Max-Planck Institut für Festkörperforschung, Stuttgart
432. Andersen O K, Jepsen O (1984) Phys Rev Lett 53:2571–2574
433. Blaha P, Schwarz K, Madsen G, Kvasnicka D, Luitz J (2001) WIEN2k Users Guide. Vienna University of Technology, Vienna, Austria
434. Blaha P, Schwarz K, Luitz J (1997) A Full Potential Linearized Augmented Plane Wave Package for Calculating Crystal Properties. Vienna University of Technology, Vienna
435. Singh D J (1994) Planewaves Pseudopotentials and the LAPW Method. Kluwer Academic Publishers, Boston-Dordrecht-London
436. Kresse G, Hafner J (1994) J Phys Cond Matter 6:8245–8257
437. Vanderbildt D (2003) Phys Rev B67:104105
438. Blöchl P E (1994) Phys Rev B50:17953–17979
439. Sih G C J (1973) Engng Fract Mech 5:365–377
440. Richard H A, Fulland M, Schöllmann M, Sander M (2002) Simulation of Fatigue Crack Growth using ADAPTCRACK3D. In: Blom A F (ed) Fatigue 2002. EMAS, Stockholm
441. Gurland J, Plateau J (1963) Trans ASM 56:442–454
442. Cox T B, Low J R (1974) Metall Trans 6A:1457–1470
443. Bergh S (1962) Jemkont Allnaler 146:748–762
444. Kleverbring B I, Bogren E, Mahrs R (1975) Metall Trans 6A:319–327
445. Staněk P, Pokluda J (1978) On the theory of ductile fracture in metals. In: Staněk P, Pokluda J (eds) Prediction of Mechanical Properties of Metallic Materials from Structural Characteristics (Prediction78). DT, Brno (in Czech)

Index

C

N

O

P

Q

R

MIX
Papier aus verantwortungsvollen Quellen
Paper from responsible sources
FSC® C105338

If you have any concerns about our products,
you can contact us on
ProductSafety@springernature.com

In case Publisher is established outside the EU,
the EU authorized representative is:
Springer Nature Customer Service Center GmbH
Europaplatz 3, 69115 Heidelberg, Germany

Printed by Libri Plureos GmbH
in Hamburg, Germany